Lectures

on

GEOMORPHOLOGY

Lectures

on

GEOMORPHOLOGY

V.K. Verma, Ph.D., F.N.A.Sc.
Professor of Geology, University of Delhi

PILGRIMS BOOK PVT. LTD.
DELHI - 33

LECTURES ON GEOMORPHOLOGY

Pilgrims first edition - June, 1999

Published by
PILGRIMS BOOK PVT. LTD.
416, Express Tower, Azadpur Commercial
Complex, Delhi-110 033
Phone: 7132459 Fax: 91-11-7249674
E-mail: pilgrim@ndb.vsnl.net.in

Distributed by
PILGRIMS BOOK HOUSE
P.O. Box 3872, Kathmandu, Nepal
Tel.: 424942, 425919 Fax: 977-1-424943
E-mail: pilgrims@wlink.com.np
Web Site: http://gfas.com/pilgrims

Varanasi Branch
PILGRIMS BOOK HOUSE
P.O. Box 38, Durga Kund,
Nababganj, Varanasi
Phones : 314059, 314060

Distributor for West Bengal and North-East
NP Sales Private Limited
206 Bidhan Sarani
Calcutta 700 006

ISBN 81-7624-073-7
Price: Rs. 350/- (PB)
Rs. 425/- (HB)

Produced by Graphic Prints House, Delhi
Printed at J.K. Offest

PREFACE

I have been inspired by suggestions received from large number of teachers, students and professionals to write a book on Geomorphology, an inter-disciplinary exciting field-science. Packed with enormous goodwill and good wishes from fellow scientists, I have ventured to bring out this little volume out of more than three decades of classroom teaching through preparing, modifying and redrafting the lecture-notes to catch up with the emerging newer concepts concerning landforms and/or related process dynamics to make handy the available published literature as the base and recasted, where necessary, to present in the form of a text book for M.Sc/M.A. students of Geology/ Geography of Indian Universities as also for advanced data-base seekers at the B.Sc/B.A. (Honours) levels.

There are quite a number of very good books on the subject by authors both, from India and abroad. The systematic treatment of this book is structured and assigned to cover the major topics of landform-study related processes and products, into twelve well-woven chapters with illustrative figures and data-studded tables (their source often referred just below the title within brackets). While no originality is claimed, the book can be regarded as distinctively different from other similar inputs on the subject in as much as

(a) the generally descriptive aspect is splitted as far as feasible into points and subpoints to facilitate improved grasping and

(b) by citing numerous Indian examples with which one may be more familiar and may arouse more interest

to promote better appreciation and assimilation by the readers and yet another singularity provided by two appendices aimed to document focus on the varied nature of research work being carried out in the Indian context with a view to

(c) acquaint the unfamiliar user with the nature of Indian output, and

(d) identify the obvious gaps for further probe in this vast field.

I would like to place on record my sincere thanks to Professor S. S. Merh and his co-authors of M. S. University of Baroda, Vadodara for contributing the paper at Appendix I and to Professor R. A. Vaidyanadhan for granting permission to reproduce his paper at Appendix 11. Permission sought from Professor S. N. Pandey of the University of Saugor for using some tables given in his book *(Principles of Applications of Photogeoloty.* Wiley Eastern Limited, 1987) is gratefully acknowledged. I record my sense of appreciation to Shri Amar Jeet Roop Rai, Draftsman, Department of Geology, University of Delhi for illustrations and diagrams. It is my previledge to record my gratitude to Professor V. K. Srivastava of Department of Geology. Aligarh

Muslim University for kindly going through the manuscript, offering helpful criticism and several useful suggestions. I thank Professor B. Thakur, Head, Dr. R. B. Singh, Research Scientist and Miss Monika Rana, Research Scholar of the Department of Geography, University of Delhi and Dr. Rameshwar Thakur and Shri Ramashray Prasad, Department of Geography, B. R. Ambedkar College, New Delhi, Professor Vinaya Jhingran, Shri Shashank Shekhar and the Museum Curator Shri K. N Kandwal of Deparment of Geology, University of Delhi, who all have helped me in care fully reading the proofs.

I cannot claim that the book is free from mistakes, despite every care. Response from the users-teachers and students in this regard will be thankfully acknowledged and suitably incorporated in the subsequent edition. I shall also be glad to receive suggestions for improvement as well.

Delhi : June 1, 1999 V. K. VERMA

CONTENTS

CONTENTS

CHAPTER I

INTRODUCTION

Geomorphology is a composite term christened after three Greek root words, *viz*

(i) *ge* meaning the earth,

(ii) *morphe/morphos* meaning the "form", "face" or the "shape", and

(iii) *logos* meaning the "reason" or "discourse" or the "science".

Thus, Geomorphology is that branch of Geology or Earth sciences which deals with the study of the earth's surface configuration. Geomorphology "is the interpretative description of the relief features "(Worcester), "studies relief features of lithosphere" (Holmes), "is generally a science of landforms, but it also includes submarine terms (Thornbury), or "the study of landforms is a branch of geology" (Lobeck). According to Strahler, "the science of geomorphology treats the origin and systematic development of all types of landforms and is a major part of physical geography". It is often used as synonym for physiography, which embodies in its study the landform description. In contrast, Geomorphology is much wider in its scope and embraces the

(a) distinct morphologic expression including spatial variability of landforms,

(b) material that constitutes the landorm-the lithology,

(c) process(es) of which the landform is a product and/or which tends to modify the landform, besides

(d) interpretation and analysis, and

(e) evolution of the landscape.

The first three falls within the common domain of Geology and Geography. The last two were the areas of study in France under 'Geologie Dyanamique '. As such, geomorphology is more comprehensive in nature than either of the above two approaches and includes the study of its various landforms-development of their geometry, attendant processes and their evolution through time.

The forces that bring about changes to give shape to the earth's surfaces may be collectively grouped into two broad categories:

a) endogenetic (Gr. *endo* means inside) processes which originate from within the earth such as tectonism that leads to folding, faulting, fracturing, or like volcanism, seismicity *etc.* and

b) exogenetic (Gr. *exo* means outside) processes which operate only upon the earth's surface. These include the work done by the atmospheric agencies like, wind, ice and water.

In response to these processes the earth's surface is modelled uneven, with some highlands (+ve anomaly) and some lowlands (-ive anomaly) over its general surface. The natural agencies are, thus, called upon to level up (*grading*) the surface by removing the materials that constitute highland by *degradation,* i.e.

$$G \propto cl,\ b,\ \text{and}\ g$$

where, G = geomorphology
cl = climate
b = biota
and g = geology(lithology and structure)

and transport down the wastes to lowlands as supplies (*aggradation*) to fill up the depressions. This calls for conjunctive studies of denudational processes with the studies of sedimentary sequences which provides a time scale, much in contrast to erosional landforms alone where the time scale derivation is not so obvious and, thus, not easily tied up. Accordingly, the process-response investigations in geomorphology (Thornes, 1979) is concerned with

a) interpretation of evolutionary sequence of forms,
b) studies of spatio-temporal variations in the magnitude and frequency of geomorphic processes for establishing the pattern of synergic forms, and
c) evaluation of causal relationship between contemporary processes and landforms.

In order to aim the objective of the study of landforms-its geometry and evolution, both

a) structure and
b) related process

as well as

c) systematic analysis of its evolution through time, constitute the investigation.

Relief and Order :

Depending upon the magnitude, Salisbury (1919) introduced the concept of hierarchy of relief features of the earth in which 'each successive level includes features of successively smaller size, greater unity of origin and more transient durability'. Bloom (1978) divided the modulations of the earth's surface into three orders :

a) *First order relief forms :* Two very broad features, continents and ocean basins, constitute the earth's surface. The former is referrable as area of positive features and the latter, the basin, as negative features. The continental margin is at the junction of the two, which though appear fairly stable, but indeed have a long geological history at their back reflected by the changes in

   i) forms,

   ii) extensions and

   iii) geographic location.

b) *Second order relief forms* : These are provincial features developed on the continents and ocean basins which are originated generally in response to endogenetic forces especially diastrophic movements and volcanic activity, both on the terrestrial surface as also on ocean floors. The main features include

	Terrestrial surface	*Ocean floor*
Modulations :	Mountains	Mide-oceanic ridges
	Plateaus	Guyots
	Plains	Oceanic plains and trenches

c) *Third order relief forms* : Superimposed on the second order relief form, they are still smaller features, owing their origin essentially to destructive or erosive activities of exogenetic processes, such as carved mountain hills, alluvial fans, valleys, waterfalls, beaches, sand-dunes, glacial eskers and drumlins *etc.* as well as some constructional forms - volcanic mountains and the like ones. A major bulk of the study in geomorphology is related to third order relief forms. An appreciation of these features facilitates better understanding of landforms of higher orders. Wider and intensive studies are directed towards terrestrial landforms, though some information is now forthcoming on marine landforms as well. From the status of knowledge acrued so far, it is obvious that much need to be known about the landforms buried below the ice or glacier, on the one hand, as also those below the marine waters on the ocean floor, on the other hand.

In fact, there is long drawn history through which the basic concepts of geomorphology developed. In this respect, much is credited to the ancient Great Roman thinkers and philosophers. However, the dawn of geomorphic ideas in themselves root about five and a half century back.

History of Development of Geomorphic Ideas

A. Views of the Ancients

1. **Herodotus** (485-425 BC) - Father of History made geological observations as well . He

   i) realized the importance of yearly increments of silt and clay deposited by the Nile,

   ii) deduced that earthquakes were responsible for rending asunder of mountains, and

   iii) noted shells in the hills of Egypt and concluded from their presence the prevalence of sea over the area - a concept of changing sea level.

2. **Aristotle** (384 - 322 BC) - He identified source of water in the springs as

   a) rain water which has percolated downward, and

   b) water formed within the earth on condensation

      i) of air that entered the earth

      ii) vapour of uncertain origin

   and also noted that

   c) sea-covered tracts were earlier dry land and new land would emerge where sea exists now, and

   d) materials removed by streams from land are laid down by them as alluvium.

3. **Strabo** (54 BC - AD 25)

   i) observed local sinking and rise of land,

   ii) considered Vale of Tempe as a result of earthquakes,

   iii) rightly inferred the nature of the summit of Mt. Vesuvius as of volcanic origin, although it was never active in his life time, and

   iv) recognized the importance of river alluvium which constitute the delta of a river that varied in size according to the nature of the region drained.

4. **Seneca** (20 BC - AD 85)

   i) recognized the local nature of earthquakes and attributed it as an effect of internal struggle of subterranean activities, and

   ii) thought that rainfall alone could not feed the rivers.

B. Dawn of Geomorphic Ideas

Though Geographers made a big contribution to geomorphology, but, this as an explanatory science had its beginning in Geology.

1. Leonards de Vinci (1452 -1519) observed that valleys were cut by streams and they carried materials from one part of the earth and deposited elsewhere.

2. Buffon (1707-1788) recognized the stream ability to destroy the land which eventually would be reduced to sea level.

3. Nicolaus Steno (1638- 1687) believed that running water was probably the chief agent of sculpturing the earth's surface.

4. Tarqioni-Tozethi (1715 -1786) discerned that the erosion by storms and their irregular courses are related to the principle of differential erosion and established such response to varying geologic materials and structures.

5. Guethard (1715 -1786)

i) discussed degradation of mountains by streams and postulated that all removed material does not reach sea but some form flood plains,

ii) believed sea to be more powerful destroyer than streams, and

iii) recognized volcanic origin of numerous hills.

6. Desmarest (1725 -1815) propounded that the valleys of central France were *products of stream occupying them* and conceived the development of landscape through successive stages of its evolution.

7. De Saussure (1740 -1799) a Swiss was the first to use the terms Geology and Geologists in their present sense. He

i) believed streams sculpture mountains and produce valleys, and

ii) also visualized the ability of glaciers to carry on erosional work.

His observational power, not necessarily the interpretational values, was great as reflected through the great stock of information he had amassed, upon which Hutton later drew heavily in developing the doctrine of uniformitarianism.

8. James Hutton (1726 -1797) was a trained physician, but had interest in Chemistry and Geology, too.

He argued in his *'Theory of Earth, with Proofs and Illustrations'* (1795) for the constancy of natural forces through long geologic time, as against his contemporaries who generally held that the changes in the earth are due to catastrophic events. This led to the development of what is called *uniformitarian doctrine.*

9. John Playfair (1748- 1819) a young friend of Hutton and Professor of Mathematics at Edinburgh, wrote a livcly dcfcncc of the Huttonian Theory in 1802. Illustrations of the Huttonian Theory of the Earth stated many principles, still acceptable like -

i) the origin of the valleys as the work of the streams that flow in them,
ii) the adequacy of the ocean waves to shape coasts,
iii) the constant renewal of the soil by weathering of rocks, and
iv) the transportation of huge boulders by Alpine glaciers.

In fact, Playfair's work lends valuable support to Huttonian's Theory which otherwise would have been lost.

10. Sir Charles Lyell (1748-1819) The above ideas of Hutton and Playfair became the chief concept of his book *'Principles of Geology* (1830-32)' being an attempt to explain the former changes of the earth's surface by reference to the causes now in operation and this doctrine has since been called *uniformitarianism.* He carried his advocacy of uniformitarianism to such extremes that he became to be known as its high 'priest'.

11. A. C. Ramsay (1814-1891) championed marine erosion as an extremely potent agency in levelling down of lands and introduced the concept of glaciers as sculptures of lake basins.

12. In 1869, **Peschal** attempted to bring together the principles of landform development into systematic form and attempted generally a genetic discussion.

13. In 1886, **Richthofeu** succeeded to even a greater degree.

14. In 1864-**A. Penck,** father of **W. Penck** (both great Geomorphologists) published *"Morphologie der Erdoberfldche"* wherein the genetic treatment of landforms was largely stressed upon.

15. Major Powell (1834-1902) one armed civil war veteran laid the foundation of the American school of thought. He proposed the classification of stream valleys:

i) based upon the relationships between valleys of the strata which they cross, and
ii) according to their origin.

In his classification, he recognized the terms antecedent, consequent, superimposed valleys which are still widely used. He believed the relief of any region, however, great its original volume of altitude will be reduced by the subaerial degradation to a low lying plain for which the sea will be the ultimate base level.

16. G.K. Gilbert, (1843 -1918) merits the designation of first true geomorphologist of the United States. He

i) confirmed Powell's concepts and elucidated the processes, and
ii) worked out histories of land sculpture for many areas.

17.W.M.Davis (1850 -1934) an acknowledged master geomorphologist developed the techniques 'explanatory description' of landscapes to high state of perfection. He is regarded as father of the American school, often referred to as Davisian school. He recognized cycle of erosion and geomorphic cycle in his studies. His principle was that one could describe the evolutionary history of a landscape in terms of three factors (i) structure (ii) process and (iii) time.

No experimental work was done by him and only it was his intuition by which he propounded the hypothesis, some of which were lately realized as practically true. Developed in 20th century, Davis book *"Geographical Essays"* embodied his various thoughts.

18. In post-war years, beach studies led trend to quantification and experimental verification. Soil erosion and river flood became theme of focal attention now.

19. The post-war trend toward quantification grew so rapidly that in 1950's, geomorphology threatened to become a branch of topological mathematics. Landforms, sediment grain size, the hydraulic properties of streams and flood frequencies and duration were all reduced to statistical probabilities. Authors described rivers in terms of entropy increase, steady-state systems, and random-walks *etc.*

20. More recent developments started during World War II with the advent of interpretation of aerial photographs.

Major landmarks in geomorphology in the present century are chronologically arranged in Table I.1

TABLE I.1 THE MAJOR LANDMARKS IN GEOMORPHOLOGY IN THE PRESENT CENTURY.

Year	Landmark
1909	Davis's Geographical Essays; Penck and Bruckner model : four glacials
1914	Gilbert's transportation of debris by running water
1924	Walter Penck's morphological analysis of landforms (in German)
1928	Baulig : the origin of denudation chronology
1941	Bagnold's Physics of blown sand and desert dunes
1945	Horton : drainage basin morphometry and the origin of quantification (?)
1950	Strahler : early application of Statistics L. C. King's pediplanation theory
1953	Penck's Morphological analysis of landforms (English translation)
1955	Wooldridge and Linton : denudation chronology reaches its peak
1960-65	A large number of statistical concepts introduced in geomorphology; Chorley : geomorphology and systems theory
1969	Chorley's Water, Earth and Man
1970	Applied studies begin to multiply noticeably Craik's Environmental Psychology on perception
1971	Chorley and Kennedy's Physical Geography: A Systems Approach, advanced statistical techniques introduced
1974	Cooke and Doornkamp's synthesis of applied studies in geomorphology
1976	Whalley's properties of materials and geomorphological explanation
1980	Coates and Vitek's Thresholds in Geomorphology
1980	Man-environment studies come to the fore

We are now at the threshold of a new-golden era of Geomorphology. Recent trends include :

A. tendency for geomorphology to become more strictly geological than geographical as a result of increasing application of other phases of geology to these studies, *i.e.* mineralogy, in the study of weathering, stratigraphic principles in palaeogeomorphology, palaeontologic techniques in glacial deposit studies *etc.*

B. development of regional geomorphology which divides areas of similar geomorphologic features and history.

C. increasing recognitions of practical applications of geomorphic principles to such fields as groundwater geology, soil sciences, engineering and military geology.

D. dawn of quantitative and experimental stage with attempts to apply the laws of hydrodynamics to a better understanding of the geomorphic processess. Experimental laboratories have been set up, and others are coming up to determine more accurately, the application of hydraulic laws to waves, current and streams action.

The following tables (1.2-6) attempt to present a synoptic overview of the various landforms and their associated features developed in response to varying lithology, structural framework and by different geomorphic agencies.

SOME FUNDAMENTAL CONCEPTS

The studies so far have led to some very basic generalizations referred to as fundamental geomorphic concepts as below :

1. *"The same physical processes and laws that operate today operated throughout geological times, although not always with same intensity" - James Hutton* (1785).

It is considered to be fundamental in the modern understanding of geology and related events and is well contained in Hutton (1785)'s theory of uniformitarianism which states *the present is the key to the past.* He further added that "no vestage of a beginning ; no prospect of an end". John Playfair was a strong supporter of Hutton and advocated that continental drifting is still going on though the causative intensity and effects thereof may not be as it was. Seismicity in the Himalaya is an illustrative example in this connection. Plates slowly move (an ongoing process), but it is realized only when under the cumulative stress (intensity becomes more) the earth tremors or shakes.

2. *"Geologic structure is a dominant control-factor in the evolution of landforms and is reflected in them".*

TABLE I.2 : SHOWING COMMON LANDFORMS DEVELOPED IN PLUTONIC, VOLCANIC, TECTONIC AND EROSIONAL TERRAINS

(after S. N. Pandey 1987)

Terrain type Relief features	Plutonic	Volcanic	Tectonic	Erosional
Mountain	Batholith	Volcanic cone	Anticline	Monadnock, inselberg or bornhardt
	Dome	Volcanic dome	Syncline	Exfoliation dome
	Laccolith			Cirque
	Stock			Horn
				Mesa
Hill	Batholith	Volcanic cone	Structural dome	Monadnock, inselberg or bornhardt
	Boss	Volcanic dome	Salt dome	Exfoliation dome
	Dome	Volcanic island	Island	Mesa
	Island	Volcanic plug	Horst	Butte
	Laccolith			Flatiron
	Steptoe			Hums (pepins)
	Stock			Inlier
				Outlier
				Volcanic neck
				Pinnacles
Ridge	Dyke	Coulee	Anticline	Anticlinal ridge
		Dyke	Syncline	Synclinal ridge
		Lava flow	Homocline	Cuesta scarp
		Lava tongue	Dip slope	Hogback
			Fault scarp	Dip slope
				Ridge crest
				Fault line scarp
				Homoclinal ridge
				Sill scarp
				Rock Drumlin
				Clastic dyke

Plateau	Dome	Volcanic plateau	Shield plateau	Topographic bench
				Karst plateau
				Plateau
Plain	Island shield	Lava flow	Island	Topographic bench
		Island plain	Karst plain	Karst plain
			Plain	Pediment
Valley	Dyke valley	Fracture or joint valley	Fault line	Anticlinal valley
	Fracture or joint valley		Joint line	Synclinal valley
			Fracture line	Homoclinal valley
			Rift valley	Glacial trough
				Canyon
				Gully
Basin		Caldera	Graben	Water gap
		Crater	Cirque	Sinkhole

TABLE I. 3 : SHOWING TOPOGRAPHY AND LANDFORMS DEVELOPMENT UPON TECTONIC STRUCTURES

(after S. N. Pandey 1987)

Structures		Common Erosional landforms developed
Dip		
Horizontal and sub-horizontal rocks	:	Plateau, plain, escarpment, step-like topography, mesa and butte
Low and gentle dipping beds	:	Cuesta, asymmetrical ridge and valley topo-elongated dip facets.
Medium dipping beds		Hogbacks, symmetrical ridge and valley topo-elevated dip facets
Steep and vertical beds	:	Straight ridge and valley topography
Folded		
Anticlines	:	Parallel, diverging or converging; Symmetrical or asymmetrical disposition of cuesta or hogbacks (with obsequent slope facing core of the fold) and ridge and valley topography. Eroded fold closures, drainage adapted to the underlying structure, development of anticlinal valley or anticlinal ridge
Structrual domes and salt domes	:	Cuestas and hogbacks (with obsequent slope facing the core) developed in annular fashion. Annular and radial drainage topographic inversion.
Synclines	:	Parallel diverging or converging symmetrical or asymmetrical disposition of cuestas and hogbacks (with obsequent slope facing opposite to the core). Development of ridge and valley. synclinal ridge and inversion of topography.
Structural basins	:	Annular series of cuestas and hogbacks (with obsequent slopes facing opposite to the core) and ridges separated by valleys with annular and radial drainage.
Homoclines	:	Homoclinal ridges and valleys corresponding to cuestas and hogbacks (with obsequent slopes facing in the direction opposite to dip direction) development of ridges and valleys
Faulted		
Faults	:	Fault scarps, fault blocks or block fault topography, fault line scarps, resequent fault line scarps, obsequent fault line scarps, tectonic scarps, scarplets, resurrected fault scarp, resurrected fault line scarp.
Other associated forms	:	Tilted or monoclinal blocks, block mountain, tilt block basins, resequent tilt block mountain, resequent tilt block valleys.
Horst and graben	:	Horst and graben rift blocks, rift block mountains, rift block basins.

TABLE I. 4 : SHOWING LANDFORMS RESULTING FROM VOLCANISM
(after S. N. Pandey 1987)

Main Features		Associated landform features
Volcanic cones	:	Volcanic cone, volcanic dome, shield volcano, cinder cone, ash cone, composite or strato-volcanoes, lateral or parasitic cone, rebuilt volcano, volcanic vents.
Volcanic craters	:	Explosive crater, crater with rim, collapse crater, volcanic sinks, nested crater.
Calderas	:	Explosive caldera, subsidence caldera, composite caldera, volcano-tectonic depression.
Lava fields	:	Volcanic dust, volcanic ash, lapilli, scoria, volcanic bombs, volcanic blocks.
Lava surface features	:	Lava blisters, spatter cones or scoria mounds, flow lines
Erosional and other associated features	:	Dykes, ring dykes, radial dykes, intersecting dykes, coulees, sills, volcanic necks, spines, dyke ridges, ash shower, volcanic mudflow, plug dome.
Volcanic lava flows	:	Volcanic plateau and plains

TABLE I. 5 : SHOWING DEPOSITIONAL LANDFORM TYPES PRODUCED BY DIFFERENT GEOMORPHIC AGENCIES

(after S.N. Pandey 1987)

Agencies Landforms	Gravitational	Fluvial	Glacial	Aeolian	Lacustrine	Marine	Littoral
Hill	Talus cone Talus shape	Alluvial fan Alluvial cone Pingo	Drumlin Kame	Sand dune Complex dune Barchan Clay dune Loess hill			
Ridge	Debris slide	Escarpment	Lateral moraine	Longitudinal dune	Beach ridge	Coral reef	Bar
	Rock slide Rock glacier	Natural levee River bar	Terminal moraine	Transverse dune	Strand	Barrier reef Fringing reef	Baymouth bar Offshore bar
		Point bar island	End moraine	Complex dune Loess ridge Whaleback		Atoll Beach ridge	Beach-ridge Spit, Submarine bar, Tombolo, Wave-built delta, Strand terrace
Plain	Debris flow	Apron	Ground moraine	Desert pavement		Estuarine plain	Coastal bench
	Earth flow	Arcuate delta	Outwash plain	Gassi plain		Lacustrine terrace	Coastal plain
	Mud flow Slump terrace	Digital delta	Esker delta Kame terraces	Loess plain Erg		Lacustrine plain	Coastal terrace Lagoonal plain Marine bench [Marine terrace]
	Solifluction	Estu-marine delta		Hammada		Lagoonal plain	Wave-built delta
	Apron	Flood plain Outwash plain Fluvial plain Terrace plain					

Valley	Avalanche	Canyon Gully Rill Valley	Esker trough Kettle chain		Inlet	Inlet	Inlet

TABLE I. 6 : SHOWING LANDFORMS ASSOCIATED WITH DIFFERENT AGENCIES
(modified after S. N. Pandey 1987)

Erosional landforms	Depositional landforms
A. AEOLIAN FEATURES	
Desert peneplain and pediplain	Major plains : river flood plain, structural plain, playas, bajadas, pediment
Inselberg (or bornhardt)	Bolson, alkali flats, salinas
Yardang, pedestal rocks, blowouts	Large scale forms, sand deposits, sand shadows and sand drifts
Desert pavement or desert armour	Barchan or crescentic dunes
Hammadas	Seif or longitudinal dunes
	Transverse dunes
	Whaleback or sand levees
	Parabolic dunes, sand undulations and sand sheets; small scale forms : small ripples and ridges on sand surface
	Vast deposition of loess or blown silt
B. GROUNDWATER IN LIMESTONE TERRAIN (Karst topography)	
Lapies	
Sink holes and associated forms	Cave deposition -
Dolines and solution pan, collapse sink holes,	Cave travertine
Swallow holes, sink hole ponds or karst lakes,	Dripstone
Compound sink-holes, uvala, polje	Stalactites and Stalagmites
	Columns, pillars and flowstone
Karst plain, sinking creek, dry bed, blind valley	
Solution valley or karst valley, rise and resurgence	
Subterranean cutoff, natural tunnel, natural bridges or karst bridges.	

Erosional remnants : Hums, caves, and caverns and associated features	

C. SEAS AND LAKES FEATURES

Sea shore, shore line, sea coast, coast line, sea cliff Wave cut bench, abrasion platform, high water platform	Beaches : Bayhead beach, pocket beach or crescent beach, storm beach Bars : Spit, recurved or hook, compound recurved spit or compound hook, winged headland
Marine cut terraces or marine built terraces	Bayhead, mid bay or baymouth bars, cuspate bars, cuspate foreland, tomblos, offshore bars
Continental terraces or continental shelf Retrograding shore line Features of sea coast, caves, bays or bights, headlands, sea arches and caves, stacks, chimneys, skerries or islands	

D. GLACIAL FEATURES

Formation of cirque, glacial trough,glacial stairway aretes or serrate ridges and horns Truncated fault spurs	Deposition of recessional moraine Formation of knob and basin topography, deposition of ground moraines, medial and terminal moraines.
Typical U-shaped valleys and hanging valleys Development of ice scoured plain with mammiliated surface Fjords, piedmont lakes and trough lakes developed	Sub- glacial moraine and englacial moraine End moraine formed as lodge dumped and push moraines. Interlobate moraines, till plain and glacial drift, drumlins

E. FLUVIO-GLACIAL FEATURES

Development of typical pitted outwash plain Formation of kettles, drainage lines and sluice ways	Fluvio-glacial deposit Valley trains, eskers, kames, kame, terraces and outwash fans and deltas Kame moraines Glacial lacustrine features and lacustrine plains Eskers

The rock structures like folds, faults, unconformities, joints and even their nature in terms of relative hardness play considerably significant role, in as much as in certain given conditions a soft, say clay rock like shale would be eroded faster and form a valley than a hard, say sandy rock, like sandstone which may resist erosion to develop into risen land like mounds of hills. Likewise fold-axis, being generally weaker would yield faster to erosion, resulting in a valley over a pre-existing anticline.

3. *"Geomorphic processes leave their distinctive imprint upon landforms and each geomorphic process develops its own characteristic assemblage of landforms".*

Broadly, two geomorphic processes- endogenous and exogenous operate on, during which they sculpture the rocks of the earth's surface which is a function of its (i) physical state-solid, liquid and/or gas (ii) environmental conditions-temperature, pressure and rainfall *etc.*

4. *"As the different erosional agents act upon the earth's surface, there is produced an orderly sequence of landforms having distinctive characteristics at the successive stage of their development".*

This concept recognizes the development of landforms in stages - one following the other. For instance, alluvial fans in river system are developed at the young stage, and delta at the last stage of its journey, although, both maintain a lot of geometrical similarity. It never occurs in the reverse way. Similarly, hanging valley is observable only in the early stage of glacial movements. However, on account of rejuvenation in the source area, the earlier stage may be lengthened and the older stage progressively reduced, but it cannot reverse its direction. That is to say, the sequence of landforms development is in a way irreversible. This leads to the development of Davis's *geomorphic cycle* and identification of young, mature and old stages of landforms development. As opposite to the concept of Davis, Penck opined that the development of landform is a function of rise and levelling rates of low earth's surface.

5. *"Complexity of geomorphic denudation is more common than simplicity".*

The landform seen to-day, often reflects many cycles of erosion, each cycle is represented by its characteristic show. Reminiscences of older cycle which may reveal the conditions of its development is exposed through newer sets. This brings in complexity when the areal landforms are considered in totality. Five types of landforms can be recognized in field -

(a) Simple landforms developed due to single geomorphic process.

(b) Compound landforms are those which are the result of two or more geomorphic agencies, such as seen in the Himalaya. In such landforms combined effects of glacier, rivers *etc.* can be easily noted.

(c) Monocyclic landforms represent those which owe their origin to a single geomorphic cycle like newly emerged area along a coast. River profiles without a break are examples of monocycle type.

(d) Multicyclic landforms are due to more than one geomorphic cycle. River/marine terraces illustrate this type of landforms. The Ranchi plateau is regarded as of multicyclic type.

(e) Exhumed or resurrected landform exposes the landscape concealed below the newer sets, which are exposed only after the partial removal of the younger geomorphic features. For instance, in Sagar and Bhopal in Central India at some places horizontal basalts are found in the valleys of horizontal Vindhyan rocks without any tectonic disturbance. The Deccan Trap basalts once must be occupying the entire terrain including the Vindhyan valleys. On removal of top basalts, the original Vindhyan landscape is again exposed with the difference that the valleys still have some basalts. The Pachmarhi hills in Central India are good examples of exhumation.

6. *"Little of the earth's topography is older than Tertiary and most of it is no older than Pleistocene".*

Most (say around 90%) of the landscapes are estimated to be of recent origin, dating not much beyond Pleistocene. At the end of Cretaceous and beginning of Tertiary the earth witnessed tremendous change when more or less present distribution of land and sea took place, deep oceans (like Tethys), shallowed and in due course its sediments blown out to form high mountains like the Himalaya. Such a vast global event obliterated more or less completely the pre-existing topography. The Pleistocene time imprinted its signature through development of new landscape over those formed in the Tertiary times. The development of the Thar desert on the fluvial land drifted the course of Jamuna, once a tributary of the Indus river system, to join the Ganga river system. It is rather of recent origin and well-borne by the history as well.

7. *"Proper interpretation of present day landscapes is impossible without a full appreciation of the manifold influences of the geologic and climatic changes during the Pleistocene".*

The present topographic expressions are the products of agencies that prevailed at the time of their formation together with those which were brought into interplay subsequent to the redistribution of land and ocean which drastically affected the climatic conditions. As indicated earlier , there is indisputable evidence that many regions which are arid to-day enjoyed humid climates during the glacial ages.

8. *An appreciation of world climate is necessary to a proper understanding of varying importance of different geomorphic process".*

Processes that operate in humid climates are often the same as in arid climates, but it is their intensities and conditions that produce entirely dissimilar topographic expressions in arid and humid terrains.

9. " *Geomorphology, although concerned primarily with present day landscapes, attains its maximum usefulness by historical extension".*

For proper appreciation of geomorphology of a terrain in its best form is to study its landforms with the background of its geological history.

APPLICATIONS OF GEOMORPHOLOGY

The significance of geomorphology in present day context has enormously increased. Its analysis and applications thereof have found ways in understanding and solving problems in the fields of

(a) Planning and development-besides often fixing up political limits of a state, district or town, geomorphic features like hills or river courses are often very handy and convenient; water divide is considered these days best to demarcate limits of any region. Such natural divides have their own characteristics, which can be developed best to the advantage of the concerned region and people who live, generating employment, reducing goods transportation and consequently cost thereof.

(b) Mineral prospecting and exploration-many a times geomorphic features and processes play a vital role, by way of mineral concentrations of otherwise their sporadic distribution through secondary sulphide enrichment process which would be most effective in gentler (rather than steep) slopes to horizontal surface; placers-beach or inland are geomorphologically controlled, both in respect of their occurrence as also for their concentration.

(c) Hydro-geology - more often the water table follows surface topography of the region and when the water table is intersected on excessive erosion, the water oozes out in the form of springs on the surface. The nature of the landforms (calcareous, glacial or flood-area) determines the type of profitable resource exploitation.

(d) Engineering projects - a study of geomorphic features in civil construction works (like dams, tunnels, roads, large buildings) offers vital clues in the selection of sites, construction and even later in maintenance, from economic considerations; for instance, dams are preferred where the valley is rather narrow. But the nature of river, its flow, water content and rate of sedimentation are also important considerations for site selection and stability of the constructions. Aspects like availability of broad wide planar surface free from flood-waters, relatively deeper groundwater availability, clear visibility even in winter, places the geomorphic investigations at high priority for airport construction.

(e) Military science - ports, islands, bays assume their significance in wartime to play safe with arms and ammunition. Terrain appraisal and evaluation determines the nature of army movement. Thus, geomorphology has its own role in military science. In predicting terrain conditions of virtually the "unknown" regions, geomorphic tools - topographic and geologic maps and some air photos have been used successfully.

CHAPTER II

GEOMORPHIC SUBDIVISIONS OF INDIA

Vast country, as India is, occupying an area of about 32.9 lakh (or 3.29 million) sq km between latitudes 7° 15' N to 35° 45' N, and longitudes 68° 15' E to 97° 15' E, northward convex crescent majestically capping the northern border and defending the Great Himalaya, and slender gradually tapering southwards, the southern part emerging out of the Indian ocean, along 5,700 km of coastline washed by the Arabian sea in its west and Bay of Bengal in the east.

Three strikingly contrasting second order relief units are recognizable at a first glance on the physical map of India, *viz.*

(a) The young fragile mountainous Himalaya with broad shoulders at the western and eastern extremity and rising head in between consisting of Proterozoic and Phanerøzoic sediments of not very great antiquity, which have been folded, faulted, thrusted and parts of which sometimes have moved over hundreds of kilometers apart, caused by diastrophism, physically attaining enormous heights. It separates India from the rest of the Asian mainland.

(b) The triangular Peninsula and old stable plateau essentially of Pre-Cambrian mass which has been virtually peneplaned through long span of time. Some Phanerozoic events in this region include Gondwana sedimentation and the unprecedented volcanism in outpouring of the Deccan lava. Mostly, the relief carved on its surface are circum - denudation mountains.

(c) The intervening Sindhu (Indo)-Ganga plains of Quaternary origin, the surface of which is enveloped by Holocene or Recent age, brought out by the Sindhu (Indus) - system in the west, the Brahmaputra system in the east, the Ganga system joining the two without any intervening surfacial gap.

The following notable broad characteristics attest the salient distinctive signatures between the Peninsular and Extra-Peniusular regions :

A. PHYSIOGRAPHIC CONTRASTS

The Peninsula, broadly, exhibits the physiographic characters of ancient terrestrial surface, such as

(i) gently undulatory scenery

(ii) mountain formation and hill development due to its differential denudation of extensive table mass instead due to orogenic movements, but for the Aravalli range

(iii) the flat-topped topography over successive basaltic lava flows covering large tracts of land in the central and western part are practically undisturbed since they erupted in late Cretaceous to early Tertiary times

(iv) gentle average gradient of the easterly drainages, tilt being an expression of mild flexuring influence upon the watershed system, with a couple of exceptions (rivers Narmada and Tapti), having open generally wide valleys.

In contrast, the extra-Peninsular terrain in the Himalaya exhibits

(i) highly dissected mountain-chain including the loftiest peaks of the world. This chain of compressional force acting on the constituent sedimentaries at right angles, developed during contrasting orogenic movements, setting in them the tectonic framework of folding, faulting and thrusting involved in the transporation of uprooted sheets wide apart. The forces also brought in some parallelism between the ranges and sub-ranges and in the separating intervening valleys

(ii) the rivers in the region are swift torrents engaged actively in deepening further the narrow, steepsided valleys, producing unstable slopes, ready to be pulled down to match up with the angle of repose

(iii) the rocks in the fragile belt are always ready to slide down in destructive landslip, when additional moisture or water becomes available to lubricate the surface, as it does during the rainy season

(iv) during their readjustment to the stress and strain, they are subjected to release the locked up rock-energy through seismicity causing tremors of appreciable magnitude.

B. STRATIGRAPHICAL CONTRASTS

By and large, since the culmination of the Vindhyan sedimentation referable to the early Palaeozoic time or even earlier, the Peninsula remained a terrestrial landmass, an object of continued attack by the atmospheric processes (agencies) that have largely flattened the surface. The Indian Peninsula is an ancient Pre-Cambrian shield area, the like of which is present in Africa, Madagaskar, Arabia, Brazil, Uruguay, Guiana, western and central Australia. This warped peneplain is the outcome of strong continued attack of erosion and is so dominant that little depositional sediments could be laid on it since the commencement of Palaeozoic era; but for some isolated transgression in northern India during the Permo-Carboniferous period, a few Cretaceous occurrences in South India and in the middle Narmada valley and coastal Mesozoic and Tertiary deposition resting over older rocks, the latter thinning landward proving the point that those shorelines were not very distant from their present position. Extensive, more or less horizontal lava sheets of Deccan Trap concealed much of the post-Vindhyan and post-Gondwana topography.

Contrastingly, the extra-Peninsular northern or Tibetan zone, cited in literature as Himalayan geosynclines, has a good representative succession of marine sequences of Palaeozoic, Mesozoic and Tertiary times, which records their widening or narrowing, deepening or shallowing and extending or withdrawing, limited to small extent in north-south direction. The mountain core is made up of granitic intrusions of Tertiary (?) age. However, in recent years doubts have been expressed in its persistence itself. On the southern flank of the chain in Kashmir, too, some of these fossiliferous rocks are recognized and zonal mixing observed in some sections.

C. CONTRAST IN STRUCTURE

The tectonic history of the Peninsula is rather much simpler when contrasted with that of the extra-Peninsula. The immunity of the Peninsula dates back to the commencement of the Purana era, after which there has been practically no to little contortions, a fact well borne and documented by the beds referrable to the Cuddapahs and the Vindhyans which are generally horizontal to sub-horizontal in nature throughout their greater part of exposure and tilting as also some folding, whatever little they show, locally do not stand anywhere in contrast to what is seen in more recent Pliocene-Pleistocene Siwaliks at the foot of the Himalaya. However, after Palaeozoic era, most noticeable tectonic expression seen in the Peninsular part is the linearly arranged tensional faulting, some of them making troughs, which assumed great economic significance as much as they act as the store-house basin for preservation of most of the Indian coals. Compared to the size of the Peninsula, these basins occupy small areas, essentially limited in its eastern and north-eastern part. Even Deccan Trap volcanism in upper Cretaceous and Tertiary times produced no appreciable tectonic effect on the rocks on which it was poured. The eruption was indeed only a tensional phenomenon. Although subsequent scission and submergence of large western part of the, then, landmass drowned into the Arabian Sea, brought about profound land and ocean distribution changes, the rest of the remaining landmass practically was unaffected. Numerous evidences are forthcoming to demonstrate that the present day relatively straight western peninsular coast is of large-scale normal faulting with seaward downthrow :

(a) comparatively simple and straight western margin is strongly suggestive of faulted nature developed in not very distant past

(b) absence of large creeks or inlets of sea at this margin

(c) presence of lineraly arranged series of thermal springs along the western coast reflective of recent sub-surface tectonism or earth movement

(d) lack of any large delta formation, often met near the river mouth at ancient sea coasts, implying thereby that the streams had not much time to develop them

(e) strong presumptions are provided by the peculiar drainage system developed on the peninsula, wherein the watershed for all larger easterly flowing rivers are so close to the western coast and practically none flowing westward (the Narmada and the Tapti, two westward flowing rivers occupy only faulted course); obviously westerly flowing rivers must have been drowned into the sea on faulting of which only the traces are now seen which have yet to make their delta afresh

(f) existence of Deccan Trap down to a depth of over 600 meters in Arabian sea appear to be continuity of terrestrial Deccan Trap and their present position may be due to strongly faulted downthrown block

(g) cutting across different geological formations from the Deccan Trap in the north to the granites and gneisses in the south, could be achieved by an effective faulting extending from Mumbai right upto the peninsular tip in the south.

The most vital and fundamental contrast between the Peninsular and extra-Peninsular regions is portrayed by the fact that the Indian Peninsula has behaved practically like an unyielding rigid "horst" or shield like mass against the massive compressional stresses, initiated with the close of Cretaceous times. It is these forces which are reasoned for the piled sediments buckling into folds, faults and nappes to which the resistant promontories of the Peninsular shield offered as an yielding block along which the limbs of Himalaya arc-shaped from north to south near Sidya in the east and north to south near Gilgit, then southwestward and ultimately westward orientation in the western sector. These bends are referred to as *syntaxial bends.* Examining the trends of hills and mountains more carefully, their main alignments in the Peninsular regions is virtually normal to that in the extra-Peninsular region, again pointing at dissimilar tectonic history of the two parts.

The extra-Peninsular region comprises a series of more or less east-west parallel mountain ranges or variant thereof, including Hindukush and Karakoram, Kailash Range, Ladakh Range of the extensive Himalaya trending east-west in Darjeeling-Nepal section, following ENE to NE orientation eastward, and westward takes a turn WNW to NW directions. Four parallel longitudinal zones are identified in the Himalaya proper :

(i) Siwalik zone - consisting mainly of Tertiary sediments, the zone is 8 to 50 km wide and rises up from the adjoining plains in the south to terrain elevation of 600 to 1800 meters elevation. Within this zone are located tectonic flat valleys referred to as *duns.*

(ii) Lesser Himalaya or sub-Himalayan zone - further northward 65 to 80 km wide at an elevation upto 3,750 to 4,500 meters including Pir Panjal and Dhauladhar ranges; this zone is made up of ancient sediments severely disturbed tectonically as shown by the magnitudes of overthrusts and nappes.

(iii) Great Himalaya or Central Himalaya - included in it are the high snow-clad peaks, the zone generally occurs at an elevation of even more than 6,000 meters, spread over 130 to 150 km in width, comprises the same type of sediments which got metamorphosed on either sides of the granitic intrusions.

(iv) Trans-Himalayan or Tibetan zone - acting as a major water-divide between the southern and central Asia from which many important rivers of the Indian sub-continent originates. It also houses several important glaciers. The zone is situated over 6,000 meters high terrain within which are Zanskar, Ladakh, Kailash and Karakoram ranges *etc.* that contain fossiliferous marine sediments laid down in the Tibetan sedimentary zones. These sediments are of various ages. 12 out of 14 peaks of the world at an elevation of more than 8,000 meters are located in this Himalaya—

	meters		*meters*
Annapurna	8,078	Hidden Peak	8,068
Broad Peak	8,047	K_2	8,611
Cho-Aon	8,153	Kanchenjunga	8,598
Dhaulagiri	8,172	Makalua	8,481
Gasherbrum II in Karakoram	8,035	Manasul	8,156
Gosainath	8,013	Mount Everest	8,848

and it is for this reason, this zone is referred to as *roof of the earth.*

ORIGIN OF THE HIMALAYA

This lofty chain of mountains owes its origin to some four or five major, perhaps still ongoing, upheavals commencing from the Upper Cretaceous onwards:

(i) The first movement dated in Upper Cretaceous seems to have determined the trend of the Himalaya

(ii) the second episode is regarded to be of great magnitude which completely destroyed the Tethyan floor in the upper Eocene that marks the general marine sedimentation except limited to some smaller basins. The Murree and Nari-Gaj sediments were laid down during the succeeding quiescent interval

(iii) the third, presumably, the mightiest upheaval occured in the Middle Miocene times, which caused a conspicuous linear or a series of connected trough(s) or furrow (s) that received the fresh water essentially fluviatile Siwalik sediments

(iv) the fourth episode dates back in the Upper Pliocene to Lower Pleistocene. Magnitude of this upheaval is noticeably lesser than the third one which caused the folding, faulting and overthrusting (as *nappe sheets*) of rocks, as also inversion in the successive beds.

(v) still feebler movements are recorded during the late Pleistocene times.

PENINSULAR INDIA REGION

The Western and Eastern Ghats, Vindhya, Satpura, Aravalli and Assam ranges constitute the topographic highs in the Peninsular region.

(a) Western Ghats - extending over 1,600 km in north-south, these ranges are parallel to the western Peninsular coast along the Arabian sea, separated by a narrow coastal strip generally not more than 50km wide. Their general elevation is 1,000 to 1,300 meters but for some exceptions which rise up even more than 2,500 meters above the sea-level. The Deccan trap prevails in its northern part, while in the southern part from Cape Comorin to Dharwar, the constituent rocks are ancient crystallines and metamorphics. In different segments they are called differently as the Anamalais, Cardamom hills, Nilgiris and Sahyadris.

(b) Eastern Ghats - meeting the Western Ghats near Nilgiris, the Eastern Ghats comprises a series of disconnected ranges upto more than 1,500 meters at several places north of the river Kistna (Krishna) made up of charnockites and khondalites and to the south of the drainage by the charnockites and gneisses rising above, but, to relatively lesser extent. Paralleling the Bay of Bengal, it extends from Orissa, through Andhra Pradesh and Tamil Nadu to the Nilgiris.

(c) Assam ranges - composed mostly of ancient gneisses and schists, the Garo, Khasi, Jaintia and Mikir hills together constitute this range that shapes as a wedge- like mass in the north-eastern corner.

(d) Aravalli Mountains- rising from 210 to 1,500 meters in the north towards south to about 1,722 meters at Mount Abu, this NE to SW trending mountain extend from near Delhi to Gujarat, spreading southwards into two wings

(i) one leading towards the Western Ghat, and

(ii) the other towards the Satpuras in the Madhya Pradesh. Essentially, the crystallines and metamorphics and to some extent older sedimentaries constitute the Aravalli mountains.

(e) Vindhya Mountains - separating the north and south India this ENE - WSW mountain range is made up of many plateau units extending from Jobat in Gujarat in the west to Sasaram in Bihar in the east and is located to the north of the Narmada river. It's general elevation is from 450 to 600 meters and rises upto 900 meters in some parts. These are covered by the Deccan lava in the west and by the limestone and sandstone with some shales in the eastern part.

(f) Satpura Mountains - also ENE-WSW trending mountains of 1,220 meters height is confined between the Narmada and Tapti, west flowing rivers, made up of Gondwana sandstone and Deccan basalts in higher reaches and granites and metamorphics rises above horst-like from Rajpipla on the west through Pachmarahi, Mekal, Sarguja, Ranchi and Hazaribagh is distributed in the north-east, east-west and south-west directions.

An interesting lineament Narmada-Son-Damodar graben extends right from its western extremity and suddenly terminates eastward under the Ganga-alluvium in West Bengal. In its eastern sector it is met by the sea coast Mahanadi graben and further down south by the parallel Godavari graben which is indeed surficially separated by the Deccan synclise. These grabens are sunken blocks in the Archaean basement. The upper Palaeozoic and Mesozoic coal-bearing Gondwana rocks are deposited into these three grabens in the Peninsular India.

Narmada-Tapti Rifts - the youthful Narmada and Tapti rivers unlike peninsular eastward flowing rivers like Mahanadi, Krishna and Godavari, particularly take an opposite route towards the west and discharge into the Arabian sea, following comparatively straight path in deep narrow valleys distinctly different from the broad, shallow, meandering and open valleys of other peninsular drainage. To explain this peculiar situations of these two valleys, a theory has been advanced that these drainage have not carved their own valleys and, instead, they flow through two parallel rifts developed in response to the say-faulting in the rocks caused by the upheaval of the Himalaya in the north. Spectacular gorge near Jabalpur where the Narmada cuts across the marble rocks in otherwise Deccan country has become a great tourist interest locale.

The Table II. 7. sets the sequence of the geomorphic history of the Peninsular India.

SINDHU (INDO)—GANGA PLAIN

The third segment, the Sindhu (Indo) - Ganga plain, is a large alluvial aggradational tract, altogether differently constituted and nothing in common with the other two, intervenes between the Peninsula and Extra-peninsula. It is a composite alluvial deposits of the three great drainage systems *viz.* the Sindhu (Indus), Ganga and Brahmaputra which lay down their detritus brought from the Himalaya, concealing below the northern edge of the peninsular geology along its southern margin and the southern extension of geological formations of the Extra-Peninsula along its northern margin. Some alluvium is also contributed by the northerly flowing rivers from the peninsular region to its south. Near Delhi, where it transposes the remains of the Aravalli mountains, the alluvial tract is the narrowest, as it is relatively shallow between the Rajmahal and Garo hills where it overlies a concealed ridge connecting the two hills. Although, the basinal floor configuration is not yet fully known, investigations indicate maximum depth of the alluvium even upto 2,000 meters or more (in upper Assam, the basement rocks are estimated at a depth of 7,000 to 8,000 meters and in the Bengal basin at about 7,000 meters) It is believed to be asymmetrical in nature generally thicker towards its northern fringe near the foredeep folded belt of Siwalik hills and the southern boundary may be relatively thinner under gently northward dipping and sloping Indian Peninsula.

The bottom of the Sindu-Ganga plain is rippled by three transverse "highs" (burried hills) (i) Delhi- Haridwar ridge (ii) Faizabad ridge and (iii) the Munger-Saharsa ridge from west to east separating four shelf areas.

(a) Lying to the west of Delhi - Haridwar ridge which is believed to be the extension of NW-SE trending Delhi orogen, the Punjab shelf comprises Pre- Cambrian basement over which lies the Palaeogene and Neogene sediment succession.

(b) The Faizabad ridge separates east and west U.P. shelfs of which the northern part is deeper into which the Mesozoic and Palaeogene, deposits are buried under the Quaternary sediments.

Table II. 7 : SEQUENCE OF EVENTS IN THE GEOMORPHIC HISTORY OF PENINSULAR INDIA *

(* source ?)

AGE	WEST COAST	WESTERN GHATS	EASTERN GHATS AND PIEDMONT BELT	EAST COAST
	Backwaters, Lagoons (Cochin, Quilon)			Beach sands, Coastal dunes (Tiruchendur) Pediments (Bellary), Teris-red blown sand
	Barrier beaches		Badlands	
RECENT	Drowning of river valleys (Karwar)			
	Submergence of coast			Coral reefs (Pamban), Growth of deltas Lagoons (Pulicat), Cave deposits
	Detrital laterite		Moderate dissection	Detrital laterite(Nellore)
PLEISTO-CENE	Oscillations in sea level			Oscillations in sea levels
	Lateritization-erosion surfaces now between 80 and 30 meters above M. S. L.		Accumulation of bedload in rivers	Lateritization-erosion surfaces now between 50 to 17 meters above M. S. l.
		Intensive downcutting		
		Rejuvenation of rivers (gorges, wind gaps, water gaps)		
UPPER PLIOCENE		EROSION SURFACAE- now around 400 meters above M. S. L. (long belts close to and within the Eastern Ghats range)		
		Minor cymatogeny on the following axes :		
	(i) along 13° 30', N. Latitude across southern Karnataka State and Andhra Pradesh (W-E)			
	(ii) from Sandur to the northern part of Nallamalai hills in the northern part of Cuddapah Basin (WSW-ENE)			
MIDDLE PLIOCENE	(iii) from Belgaum towards Sholapur in Maharashtra State (SW-NE)			
	(iv) from Vijayawada towards the eastern part of Orissa State (SW-NE)			

LOWER PLIOCENE	(part of Karnataka State, Recession of scarps, piracies valley excavátion, Palghat Gap (Palghat)	EROSION SURFACE-now around 600 meters above M.S.L. (with laterite cover) part of Cuddapah Basin and hills close to the east coast of Andhra Pradesh) Pediments, Inselbergs (Arni, Gingee) Gravels (Conjeevaram)		
	Rejuvenation of certain rivers (Cauvery R. ?)			
UPPER MIOCENE	Warkala beds and similar ones with lignite		Uplift along the Eastern Ghats	Rajahmundry sandstone Cuddalore sandstone
		EROSION SURFACE-now around 1,000 meters above M.S.L. (with laterite cover) (Karnataka Plateau, Pachaimala, Shevaroy, Javadi, Nallamalai hills of Cuddapah Basin, Khondalite range of Andhra Pradesh)		
MIDDLE MIOCENE		NW-SE 'thrust' zone across southern part of the Western Ghats		
LOWER MIOCENE	Estuarine and marine deposits (Quilon, Ratnagiri)			Sedimentation beneath the present deltas of Cauvery, Krishna and Godavari R.
	Swift flowing short coastal rivers	(Godavari, Krishna, Cauvery and Vagai rivers) Monsoon conditions followed by superposition of an easterly drainage		(Marine deposits Pondicherry, Vriddhachalam, Triuchcherappalli)
		Uplift of charnockites (?);	Disturbances resulting in block faulting in granitic area (Bukkapatnam, Pakala)	
UPPER EOCENE	Uplift and faulting(?) along the axis of the arch NNW-SSE and foundering of the western segment.			
		Consequent drainage on the gentle regional slope		
LOWER	EROSION SURFACE - now around 2,000 meters above M.S.L. (with laterite cover)			

EOCENE	(Agastyamalai, Biligirirangan and Bababudan hills) Fissure eruptions in the northwestern part in different centres and extensive lava flows (plateau type)	Trap flows in Rajahmundry with inter-trappeans, infra-trappeans Cenomanian transgression along the coast Initiation of coastal flexure
MIDDLE CRETA-CEOUS	Upwarp and major cymatogeny along NNW-SSE, the axis being a little west of the present coast.	
UPPER JURASSIC	Minor uplift in the Cuddapah Basin. EROSION SURFACE- now around 2,500 meters above M. S. L. (Nilgiri, Anamalai and Palni hills)	Estuarine and marine deposits in a few places (Gondwana) Coast fairly defined
MIDDLE JURASSIC	The entire land was a pediplain/peneplain with a few inselbergs/ monadnocks here and there. Dismemberment of Gondwanaland.	Coast slowly emerges

Elsewhere on either side of this ridge is the Pre-Cambrian basement under the cover of Vindhyan rocks and Neogene sediments.

(c) Succession of Gondwana rocks and elements of Rajmahal Traps lie to the east of Munger -Saharsa ridge. Rao (1973) classified the Sindhu-Ganga depression from east to west into five divisions.

(i) The Brahmaputra basin in Assam - debouching from the Mishmi Hills, its basement too dips steeply northward, believed to be 7,300 meters NE of Tezpur, gets shallower towards Mikir Hills. These alluvium overlie the Tipam series (Miocene), Barail (Oligocene) and Dihing formation (Pleistocene) without any significant unconformity with the latter, though minor unconformity is recorded between the Dihing and the Tipam at some places.

(ii) The Bengal (Ganga- Brahmaputra) Basin of West Bengal and Bangla desh - bounded by Naga - Lushai orogenic belt to the east and the peninsular shield to the west, the southern limit seems to be opening into Bay of Bengal for considerable stretch. It includes the Bengal Delta and conceals below a thick succession of Tertiary and Cretaceous sediments over Rajmahal (?) Trap basement. The basin is 900 meters deep near the Archaean shield in the peninsula and deepens to 11,000 meters in the eastern and southeasten part.

(iii) The Ganga basin in Uttar Pradesh and Bihar - is rippled by a series of three transverse "highs" (buried hills) from west to east (i) Delhi- Haridwar ridge (ii) the Faizabad ridge and (iii) the Munger- Saharsa ridge

(iv) The Punjab basin NW - SE, ESE- WNW and E-W trending this basin parallels the Siwalik hills, and is limited by the Punjab ridge comprising surface or subsurface Archaean. Its maximum thickness is about 4,800 meters towards the hills.Below this alluvium two tongues of ridges presumably extend :

   (i) trending NW towards Lahore and

   (ii) the other northwards towards Dehradun; the former Delhi - Lahore basement presumably separates the Rajasthan platform of the Sindhu basin from the Punjab platform of the Himalayan basin.

(v) The Sindhu basin - the above NW trending Delhi - Lahore ridge basement presumably separates the Rajasthan platform of the Sindhu basin from the Punjab platform of the Himalayan basin.

The discerning of the precise nature of this furrowed depression, an absorbing problem for geoscientists to probe, is interesting both from academic and economic significance viewpoint.

(a) The earliest view considered it to be a filling of an ancient sea by the load of the numerous rivers that flow into and through it.

(b) Middlemiss (1900) opined that it was sinking plain belt which kept subsiding simultaneously with the rise of the Himalaya.

(c) Sir Burrard (1917) postulated that a depression is developed as a *rift valley* and is a sunken block about 30 km in depth and 3,200 km in length between two normal faults, produced as a consequence of the rise of the Himalaya. It subsequently got filled up with alluvium. He attributed this rifting as induced by the Himalayan compression to the north.

(d) However, Oldham (1917) disagreed and on the basis of available geodetic and geophysical data concluded in favour of northward sloping floor to maximum depth of 5,000-6,000 meters about 15-50 km from the mountains and, therefore, it was not a rift or crack of the type Burrard visualized.

(e) Suess, a Swiss geologist regards this extensive linear depression as a great geosyncline of a foredeep sag in the crust at the junction of the ancient peninsular mass and the soft younger sediments of the Himalayan region produced on account of the advancement of tangential thrusts from the northern or Tibetan side southward which was arrested by the rigid and stable peninsular mass. Some seismic foci are located in the alluvium, filled up gradually by stream deposits during the Quaternary-Recent times offer clues to indicate that the basinal bottom contains weakness zones which are indeed the sites of dislocations and responsible for some slow but ongoing changes still taking place.

Sands, silts and clays are the essential components of the alluvial deposits within which lenticular bodies of the gravels, pebbles and peat beds occur. These alluvium are divisible into two groups.

(a) The Older Alluvium (Bhangar of the Ganga valley). It

(i) is generally dark in colour

(ii) has zones of *kankar* - nodules and concretions of calcareous matter

(iii) often forms elevated mounds and terraces

(iv) is assigned Middle to Upper Pleistocene age

(v) encloses Pleistocene mammal fossils - *Palaeoloxodon, Elephas Equus, Rhinoceras* etc.

(b) The Newer Alluvium (called Khadar in Punjab). It is the youngest unit and

(i) is light coloured

(ii) contains lenses of sand and gravels which act as a good reservoir and source of underground water

(iii) rarely, if at all, contains the calcareous kankar

(iv) is referrable to Upper Pleistocene and imperceptibly grades into Recent alluvium

(v) animal remains met in it have close resemblance with the living species today, and may be identical as well.

GLACIERS

The extra-Peninsular region comprising the Himalaya appears to be the only home of glaciers on this vast subcontinent, the high ranges of which are always perpetually blanketed below the great magnitude of snowfields, from which tongues of the small (as most glaciers generally are) and large (Sicahen glacier as much as 70 km long and Fedechenks glaciers of Pamirs, about 65 km long) glaciers emerged. With the exception of polar glaciers, by far the Himalayan and Tibetan glaciers are the largest in the world even though much shrunken as they are to-day and limited in extent in relation to their Pleistocene dimensions. Occurrences of terminal moraines at 2,700 meters elevation against glacier termination today at 4,300 meters in Sikkim and are 1,800 meters during the Pleistocene times bear testimony to the glacier recession since then. In fact, in Kangra it descended down even to 1,000 meters.

Some important glaciers are

Sikkim Himalaya		*Kashmir Himalaya*		*Kumaon Himalaya*	
Zemm	25 km	Rimo	40 km	Gangotri	25 km
Kanchenjunga	16 km	Punmali	26 km	Milam	18 km
		Rundun	18 km	Kedarnath	15 km

Ordinarily, the levels to which glaciers descend at a given time in the Eastern and Western Himalaya is expected to be of the same order , but for

(a) latitudinal variations - Kanchenjunga (Eastern Himalaya) is located at around 28^0 N while as Karakoram (Western Himalaya) is at 36^0 N; 8^0 latitude-difference is rather considerable for temperature variation, and

(b) the stronger summer monsoon considerably raises the temperature in the Eastern Himalaya causing greater ice melting. It becomes practically ineffective by the time it reaches the Western Himalaya.

There is hardly any glacier movement in the terrain and whatever it is, it is not at a uniform rate even in the same glacier mass. It moves faster in the middle, about 20-30 cms per day, but closure to the resistance site, that is towards the side and near the bottom, it may barely be 7-12 cms.

RIVERS

Extra - Peninsular River

The Sindhu (Indus) flow west-south westwards, whereas the Ganga and Brahmaputra river systems constitute the drainage of the Himalaya from west to east with the water shed in the trans-Himalayan range between Tibet and India and not in the great Himalayan range with some of the highest peaks. Several of these while traversing through the mountainous course have cut often hundreds of meters of deep and steep gorges, a feature

(i) considered to be due to the *antecedent* (implying that the mountain are younger than the flowing river (s)) nature or

(ii) as in some cases the drainage may be following the radially disposed, with reference to mountain arcs now rendered weak, fault zones

The Himalayan drainage are torrential in the hilly terrain, at times cascading, but on reaching plains the flow virtually becomes laminar. Some of these as the river Arun, a tributary of Kosi, excellently illustrate the "headward erosion" and "river capturing", a characteristic of youthful terrain, as the Himalaya is.

Delimiting the political boundary between India and Pakistan, the Sindhu drainage system to which the rivers Kabul and Kurrum meet from the northwest includes five famous Punjab rivers, *viz,* Jhelum, Chenab, Ravi, Beas and the Sutlej. Rising from near Mount Kailash, the river Sindhu takes a southerly turn near Nanga Parbat to flow down to the Arabian sea after a long westward journey. The celebrated river Saraswati of Vedic lore originates from Sirmur hills and drains down to Rann via Adhbadri, Karnal , Rasula, Sirsa, Patiala, Bikaner and Bahawalpur as also through the present day bed of the Eastern-Nara. Near Sirsa it was joined by the Sutlej earlier through Sirhind and Naiwals channels, which sometime in the 12th century migrated to flow into Beas, rendering the entire tract till then drained by the river Saraswati, to be over-run by the desert.

The rivers Yamuna, Ganga, Kali, Ramganga, Karnali, Gandak and Kosi are the important rivers of the vast east-west trending Ganga drainage system. The sacred name Ganga, the river takes from Rudraprayag at the confluence of Alaknanda and Bhagirathi following a necklace-like route from the Gangotri glacier- their source.

Flowing over 1,600 kilometers from its source close to the Mansarover where it is known as Tsang-po, it takes sharp southward flow near Namcha-Barwa peak, until emerging into plains, when it takes the name Dihang, the Brahmaputra drainage system comprises, the Tista, Subansiri, Brahmaputra, Dibang and Luhit.

Peninsular Rivers

Important rivers of this region include

(a) the eastward flowing rivers, the Subarnarekha, Damodar, Brahmani, Mahanadi, Godavari, Krishna and the Cauvery,

(b) the westward flowing, the Narmada and Tapti rivers and

(c) the northward flowing the Chambal, Betwa and the Son.

Some rivers originating with their watersheds in the Western Ghats are almost within the sight of the Arabian sea, have their sizeable course well graded and tend to reach the base level of erosion. The Narmada and Tapti flow westward, their courses determined by the faults developed during the Tertiary times, when presumably the peninsula also was uplifted, a fact supported by some Tertiary deposits in the regional narrow coastal strip. In the Peninsular region, the Aravalli acts as a water divide separating the courses of westerly and easterly drainage. Likewise, the Vindhya and Satpura separate the northerly drainage from southerly drainage. To the north of it flows the Chambal , Betwa and the Son.

The remarkably eastern flowing rivers of this region, as also of the Ganga system undoubtedly points to the eastward slope, a course they must be flowing over a long geological times, a fact borne out by their U-shaped valleys, gentle profile and developments of delta indicating detrital supply in excess over the tidal removal. Perhaps the rise of western flats during the Tertiary may have provided the eastward tilt.

LAKES

EXTRA-PENINSULAR REGION

Located in higher regions of the Himalaya and Tibet, some of these numerous lakes are indeed of large dimensions, *e.g.* Koko Nar (which encompasses an area of 4,075 sequare kilometers) or the salt - laden Tsai-dam large depression are considered either

(a) barrier lakes, which develop on stream damming by the alluvial or glacial debris

(b) as rock basins, the hollows of which are formed by glacier-scooping during the Pleistocene period

(c) as a result of differential earth movements like faulting *etc.* as are Kumaon lakes.

The Kashmir fresh water lakes, may be large depressions in Jhelum alluvium, are of ox-bow type. The sacred Mansarover lake, the source of Sutlej, Sindhu and Brahmaputra in the Sikkim region, the Wular and Dal lakes of Kashmir Himalaya and Tso-morari and Pong-Kong lakes of Ladakh [saline in nature, developed (i) due to increasing desiccation initiated by the rising Himalaya, which obstructs the rain-bearing winds reaching this side in Kashmir and (ii) also due to ice recession from the area], are important ones of the region.

PENINSULAR REGION

The lakes in the peninsular parts are generally salt bearing or saline in nature. The Lonar lake, about 100 meters deep and 1.5 km in diameter (explosive caldera or crater). in the Buldana district contains sodium carbonate and chloride derived from the decomposition of volcanic rocks. Besides a group of backwaters (kayals) of Kerala coast, the Chilka and Pulicat owe their genesis to obstructions of seawater by bars and/or spits.

The Sambhar inland lake in Rajasthan is known for its salinity caused on partial evaporation of the water concentrating behind the salt and saline substances brought in during the occasional showers over the terrain. Small fresh water Puskhar lake is also in Rajasthan.

VOLCANOES

India was theatre of volcanicity, when it witnessed outpouring of Deccan lava on grand scale. Relatively, small scale activity is recorded even earlier, in Mesozoic times (Rajmahal Traps), Palaeozoic times (Panjal vocanics) and even to Proterozoic times in the Jammu region between the rivers Tawi and Ravi. Some similar activities did occur even during the Dharwar period. The dormant volcanoes of the Barren Island close to the Andaman Islands in Bay of Bengal is regarded to be of Vesuvian type blowing off large portion of the crater formed earlier, enlarging its amphitheatrical diameter to about 3 km within which is situated a smaller symmetrical cratered cone. It is known to have been active in 1979, besides earlier eruptions in 1789, 1795 and 1803.

CHAPTER III

SLOPES AND MASS MOVEMENTS

Slope is the lowest common denominator of which all the landscapes are made up of. By definition, it is the inclination of a surface or plane with respect to horizontal. A slope may or may not be parallel to the constituent bed. The landscape of a region may comprise of several segments of sloping surfaces which may be horizontal plane, inclined or even curved. Any face of the landscape is composed of slope(s) of variable

(i) inclination

(ii) orientation

(iii) dimensions and

(iv) shape

A permutation and combination of these parameters offer variety of landscapes sculptured either by

(a) endogenetic (or hypogenetic) activities such as orogeny, epeirogeny, tephrogeny or volcanism

or

(b) exogenetic (or epigenetic) processes operating at or near the surface such as weathering, mass-wasting, erosion, transportation and deposition.

Slopes are rather extremely dynamic and, therefore, transitional in nature irrespective of their genesis-endogenetic or exogenetic, rendering their study rather complicated. Even, large-scale topographic maps manifest only broad framework of a feature with approximations, within which are included infinite slope elements. They are therefore not expressible in terms of simple mathematical equations. The interaction of erosional processes with tectonic deformation produces infinite variety of landscape slopes. Without interaction by the erosional processes, the landform would be unmodified endogenetic, composite structural surfaces and volcanic slopes and those developed without endogenic processes would assume a monotonous level surface. Slopes, therefore, are regarded as geomorphic unit in terms of which any natural surface can be expressed. As such, in general, slope refers to a geomorphic element of the earth's landscape of any origin, but in a limited restricted sense, it also connotes the angle earth's surface makes with a horizontal datum which is better expressed as inclination, slope angle or slope gradient.

At a given time any landform/landscape in response to weathering, mass-wasting and erosion is regarded as an aggregate of small but related slope elements in equilibrium and slightest disequilibrium in any part, affects the neighbouring slope segments, developed due to rain wash or even apparently insignificant animal burrowing may destablize the slope in its own way and thus a dynamically stable to graded slope is an example of an open physical system through which both energy and matter move, a system that tends to maintain itself in the most efficient possible configuration. Gradation of slopes includes both lowering of some parts of the surface by degradation (or erosion) and raising its level by aggradation (or deposition) and the slopes as such dynamically integrate the landscape by closely knitting together the entire profile-degradational and /or aggradational. However, rates of weather may not always balance rates of erosion. Soil and regolith develop on the hillslopes when the rate of weathering is faster and when the erosion is faster than the rate of weathering, more materials developed at hill slopes are removed than generated, exposing bare surface hillslopes (Figure 1).

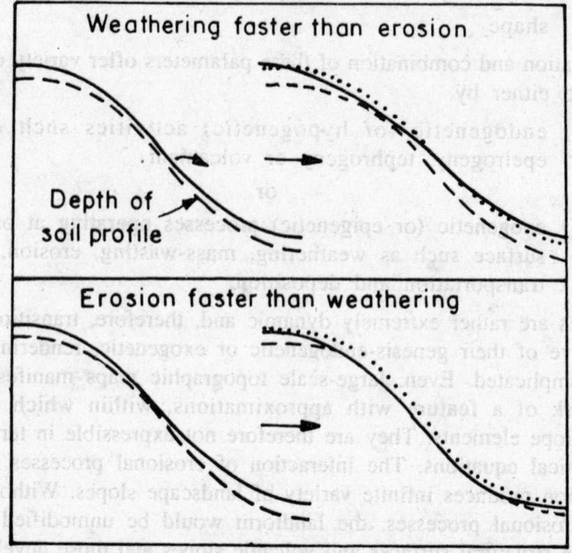

Figure 1 : Slope Development related to Rates of Weathering and Erosion

Some basic considerations in slope study are related to

(i) the real form of slopes in profile

(ii) variation of slope profiles with climate, lithology, structure, time, tectonism, relief and other factors (Figure 2).

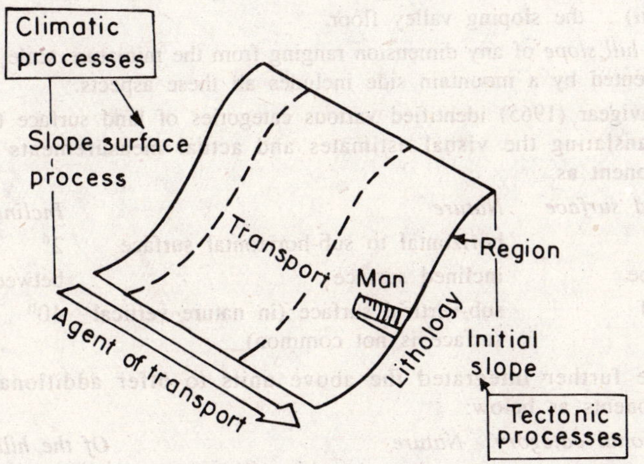

Figure 2 : Forces operating on Slope

(iii) parallel retreat slopes (pediments) versus slope decline (peneplains)

(iv) retreat of structurally controlled escarpments

(v) significance of various processes upon slope form and slope development.

(vi) rates of operations of processes and change of slope and

(vii) configuration and role of slopes as part of open system

Slopes reflect the interaction of

(a) the rate of uplift of subsidence

(b) the resistance to weathering of the underlying rock type

(c) the weathering and transport processes operating on the slope and

(d) the effect of running water; glaciers or the sea, wearing away or depositing at the slope foot.

The multivariate nature of the factors introduces complexity in evaluating the inputs (weathered sediment, rainwater, atmospheric conditions) : outputs (transported sediments and water) ratio of the slope system as also in determining definitive relationships between slope form, process, rock-type and time, of which the landform is a product.

In practical geomorphic terms, the slope study would extend from a valley floor up onto the watershed interfluves and would include

(i) the upland interfluve

(ii) the valley sides and

(iii) the sloping valley floor.

A *hill slope* of any dimension ranging from the miniature scale or a scale represented by a mountain side includes all these aspects.

Savigear (1965) identified various categories of land surface (Figure 3) by translating the visual estimates and actual measurements of select component as

Land surface	*Nature*	*Inclination*
Flat	horizontal to sub-horizontal surface	2^0
Slope	inclined surface	between 2^0 -40^0
Cliff	sub-vertical surface (in nature vertical surface is not common)	40^0

He further integrated the above units to offer additional several components as below:

Additional category	*Nature*	*Of the hill slope*
Crest slope	merging into upland flat	upper convex part
Mid slope	central part separated by cliffs	gently (smoothly) curving or straight (rectilinear)
Foot slope	merging into the valley bottom	lower concave part at the toe

Slope merges at either extremity at the crest or foot slope may be either abrupt and *angular* or it may be *smooth* (Figure 4) depending upon the nature of lithology and slope developing process(es).

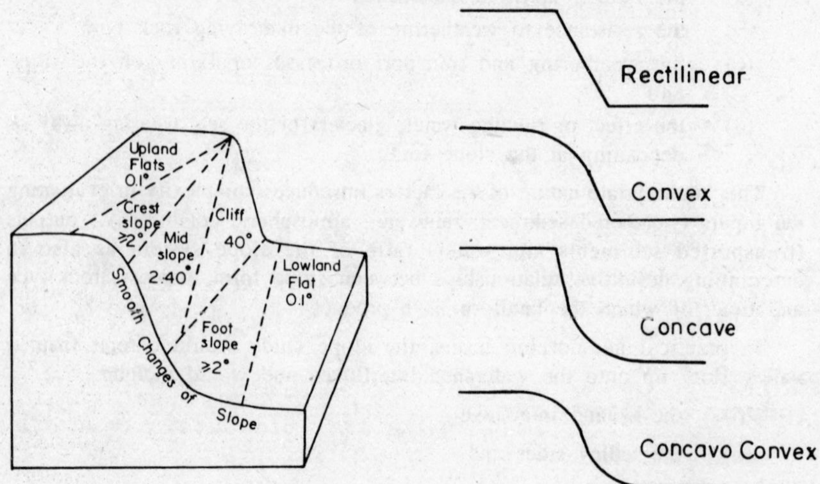

Figure 3 : Elements of Slope

Figure 4 : Slope Forms : Categories

Straight or rectilinear slopes are characterized by angular crest-and foot slopes both, but if the crest slope is smooth curvilinear outward, they are referred to as *summittal convexity*. The slope is termed as *convex* or upper wash slope.

Just below the summittal convexity a steep wall- like slope is called the *scarp* or *free face slope* which consists of bare rock. Because of the steepness, it does not support any rock-debris or regolith.

When the crest-slope is angular, but the foot slope is smooth curvilinear downwards, it forms a *concave* or lower wash slope or valley floor basement slope. Both curvilinear upward crest - and curvilinear downward foot- slopes constitute a *concavo-convex* slope.

In practice, a general slope of any type may be interrupted by minor, step like *facets* resulting in angular deviation or segments when the departure is smoothly curved. The profile or vertical section depicts the shape of the slope. A slope profile constitutes a series of slope units (Table III. 8)

Slope evolution is best studied in the field by the measurement or process and the related landform (Table III. 9).

In a dipping sedimentary sequence of hard and soft beds, the soft rocks are eroded away rather faster than the hard rocks that are exposed, which reflects in its slope the attitude of the beds. Such a type of land shape in which the ground slope and the bed inclination coincide is known as a *dip-slope*. Three types of slopes are developed along a ridge formed by resistant dipping rocks (Table III. 10)

(a) one sloping in the dip direction is called a *resequent slope*

(b) while the slope opposite to the dip is known as *obsequent slope*

(c) the ridge upheld by a single uniformly resistant rock is characterized by the smooth slopes developed on its both sides and differs from the one developed due to differential erosion of certain interbedded hard and soft rocks, where several beds of differing resistances are involved, the secondary resistant beds below and above the main resistant beds stand to break the slope continuity. A *compound slope* is one constituted by the primary (main) as well as secondary resequent and obsequent slopes, and is steeper in relation to the secondary resequent slope on both the primary slopes. This criterion helps in determining the nature of dip of the beds.

HILLSLOPE ANALYSIS : TECHNIQUES AND THEORIES (APPROACHES)

Two different philosophical approaches are at work in the study of slope and their analysis

TABLE III. 8 : SLOPE UNITS*

(*source ?)

Slope Unit	Angle approximately Segment constant	Maximum segment	Steeper than above or below
		Minimum segment	Gentler than above or below
		Crest segment	
		Basal segment	
	Segment approximately Element constant	Convex element	Angle increases downslope (positive curvature)
		Concave element	Angle decreases downslope (negative curvature)
		Frequent changes of angle or curvature	

TABLE III. 9 : SLOPE FORMS-METHODS AND TYPES*

(*source ?)

Problem	---------------------- in study --------------- Method	Type
The field evidence of a former process which demands explanation	The means of analysis ; the factors and variables analyzed and the methods used in explanation	The general, common form for the approach type of study
Forms of slopes, *e.g.* convexo-concave	Observed slope forms, and processes in the field, the use of the inference, the link between process and morphology, process response systems	*Inductive :* the process-form approach.
Decline and retreat of slopes	Evolutionary or sequential approach. Use of mathematical and geometric analysis of forms	*Geometric :* the slope evolution approach

TABLE III. 10 : SHOWING CONTRASTING CHARACTERISTICS OF RESEQUENT AND OBSEQUENT SLOPES DEVELOPED ON LOW TO MEDIUM DIPPING BEDS (CUESTAS)
[after S. N. Pandey 1987]

Characteristics		← Slopes → Resequent	Obsequent
1	Ground slope	Corresponds to the dip of beds and is gentler	Slope is much steeper than the dip slope
2.	Lithological aspects	The slope is held normally by only one litho-unit which is more resistant to erosion and generally it is sand-stone. In certain cases more than one bed is present, if the resistant beds are thin and interbedded with soft rocks like shales or clays.	The slope clearly displays several resistant litho-units and the bedding planes and rock contacts are intersected by sloping surface. If the beds are thick and uniform in nature only one or two rocks may be distinguishable.
3.	Drainage	More widely spaced and uniform in distribution. The streams are longer.	More closely spaced and shows remarkable parallelism. The streams are shorter.
4.	Contrast in slopes	Secondary small scale and local obsequent slopes may be noticed corresponding to interbedding with soft rocks.	Secondary small scale and local resequent slopes may be present projecting through the obsequent slope. This is corresponding to interbedding of hard rocks.
5.	Dip facets	Triangular dip facets are very prominently developed where inter bedding of hard and soft rock exists.	Dip facets are not developed
6.	Lithological contacts	Lithological contacts are more irregular and serrated due to low dips, gentle slope, presence of dip facets and long stream valleys.	Lithological contacts are more straight and prominent due to steep slope, shallow and small stream valleys.
7.	Vegetation	Is more often uniform in distribution as only one rock constitutes the slope	Distribution is not uniform and corresponds to varying lithology.

(a) The deductive school attempts to understand systematic slope form variations in response to long continued sub-aerial weathering and erosion. The effect of such process(es) though is, too slow to be witnessed and, therefore, necessitates certain assumptions that, of course, could be tested. This leads to a generalization stating that the slopes particularly in humid terrain tend to be lowered and become more broadly rounded with time.

However, some insist that the slopes are practically stable with angles being function of

(i) rock type and

(ii) operating processes of weathering

and so evolved stable process persists through the vigour, time and slowly retreating parallel to its own face until obliterated by the intersection of other slopes.

(b) The other school of thought, without involving much with the theoretical considerations, proposed empirical slope descriptions including slope formation process and slope geometry.

However, with both deductive and empirical studies, some understanding in the slope characteristics is evolved, and it is generally realized that terrestrial slope profiles often have upper segment convex and lower segment concave skywards and may at times be intercepted by an intervening straight segment. A variety of schemes have been evolved to classify slopes. King and Wood (in Young, 1972) gave a classification based on slope facets (Figure 5 inset A) whereas Troch (1965 in Bloom, 1968) brought slope elements in landscape classification. Dalrymple recently (1968) identified nine slope segments (Figure 5 inset B) as below :

A segment, termed as *free fall*, composed of weathered debris is introduced above the straight segment, called *talus* (it is a slope landform and the material, designated as *slide rock or scree*) in the profile when a cliff interrupts the slope. Talus is a constructional landform, but usually it reflects bed rock surface below a thin surfacial cover of a layer of sliderock. The creepings under the action of gravity, armours its surface and fresh debris falls onto the talus aims at balancing the mass by replacing the down- moving fragments. The talus angle is a function of

(i) fragment size

(ii) angularity of the slide rock

(iii) climate

(iv) vegetation and

(v) the slide rock dynamics - supply and removal

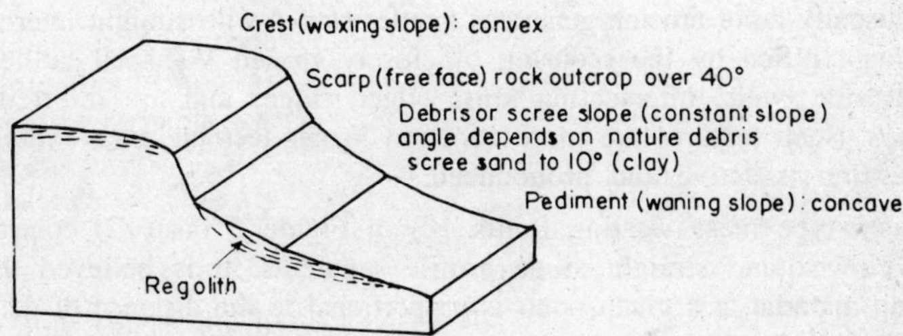

Figure 5A : Slope Facets

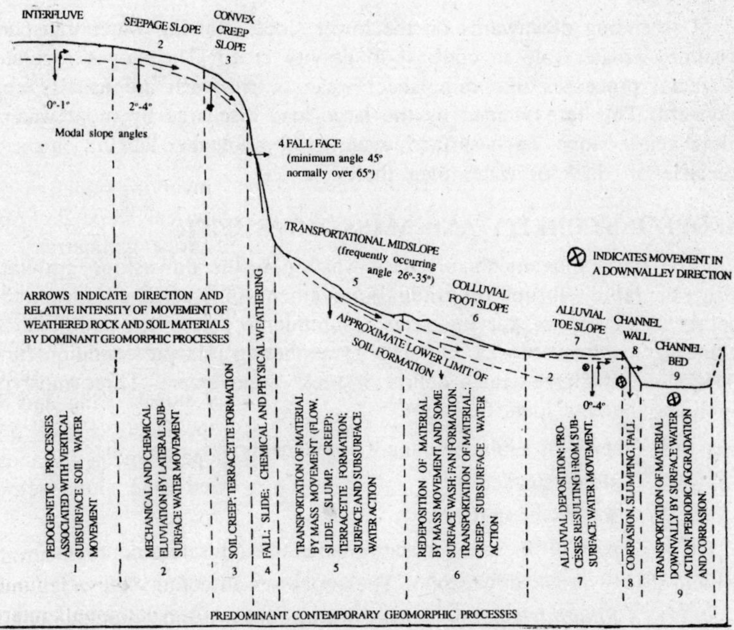

Figure 5B : Dalrymple *et* al's Hypothetical Nine Unit Landform Model (ref. Young

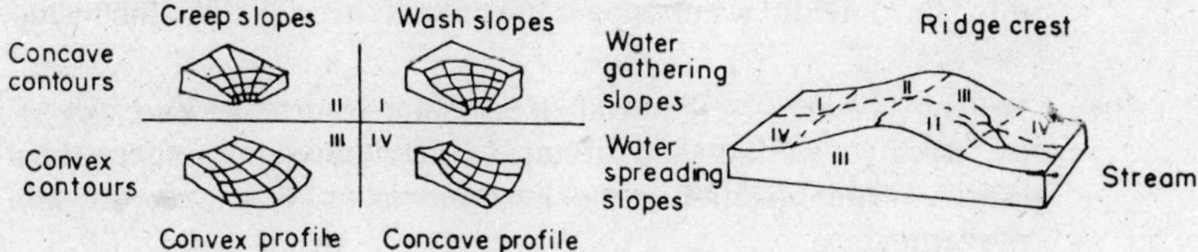

Figure 5C : Slope Elements in Landscape Classification

Unusually rapid erosion generates gentler slopes with straight intermediate segments typified by the scouring of closely spaced V-shaped gullies with straight side walls, intersecting knife edged ridges and are referred to as *badlands.* Such features are more common in neo-tectonic areas where river downcutting is active and pronounced.

Creep type mass wasting (units 1-3 in Figure 5 inset C) controls the upper convex and straight slope profile segments. It is believed that the creeping material at a given point is proportional to the distance of that point from the hill summit to enable progressively longer amount of debris to move.

Descending downwards on the lower slopes flowing water transportation assumes greater role in contrast to gravity creep. The slopes regulated by surfacial processes like rain, sheet wash or rill-wash are usually concave skyward. This is explained by the large load discharge by more water on a more gentle slope. Such surfaces seem to be altogether lacking on the moon because of lack of water over there.

SLOPE INSTABILITY AND MASS MOVEMENTS

The mass movements or mass wasting is the downslope gravitational, slow or rapid, abrupt or gradual movement of weathered loosened rock debris or masses of varying sizes : sub-microscopic silica colloid to hudge giant sized jointed blocks. Obviously, weathering is a pre-condition for slope instability which generates debris or rock movement. Three most typical such movements identified are:

A. Soil creep - It embraces small movements involving the

(i) slope surface

(ii) soil layer and

(iii) part of the subsurface weathered and saturated rocks

which slowly move downslope. The processes affecting soil creep include-

(a) *Raindrop spatter* - Mass transfer downslope though on a small scale takes place when the raindrop that falls on the surface produce mudspatter, because of which smaller particulate materials are thrown out, not far off, may be hardly a few centimeters. In this, the muddy water moves down hill over the inclined slope surface.

(b) *Moisture changes* - Removal of moisture from otherwise wet soil by repeated wetting and drying or dehydration developes small cracks, which facilitates the loosened material for net downhill movements.

(c) *Frost heaving* - Thin layer of water in rock slabs or in pore-spaces tend to expand on freezing and, consequently

(i) rocks are somewhat lifted from their position

(ii) and with, thawing and contracting

the loosend individual particles, singularly or in group on ice melting move downslope.

(d) *Temperature changes* - In this process, too, a net downslope movement on minute expansion or contraction of soil materials take place. Interestingly, the expansion and contraction amounts along the slope may be different. Maximum expansion is observed toward the downhill side and great retardation from the uphill side.

(e) *Weathering* - Particles detached from the parent body on weathering on an inclined slope has tendency to roll downslope under the influence of the gravity, after overcoming the forces preventing such movement.

(f) *Biogenetic process* - Plant roots on their growth, as also some worms and burrowing animals disturb the soil around and the particles move downhill under the influence of gravity.

B. *Landslides* - these are larger movements of surface materials comprising the slope-rock. This movement of rocks extends to a considerable depth.

C. *Rock avalanches* - It involves sudden slumping of enormous rock masses, particularly in young mountain region, that tumbles down and fragments. Such motions are usually engineered by the action of water on the rock by

(i) reducing their strength and

(ii) lubricating the surface to facilitate the movement of rocks. Evidently, mass - wasting process(es) combine(s) elements of weathering (*i.e.* dislodging the material from the parent body) on the one hand and erosion (in-as-much as in removing the dislodged material from the site of dismemberment) on the other hand and yet is distinctly different from those two processes which constitute a different category, termed *'transitional phenomena'*.

Two sets of stresses operate on a sliding mass over a surface/ slope of failure:

(i) One which prevents or resists the motion, due to inertia corresponding to its own weight, called *'sticky component'* perpendicular to the sloping surface;

(ii) the other which generates or initiates the motion due to the gravity. The component of gravity acting along the sloping surface is proportional to the sine of the slope angle and is referred to as *slide component* (Figure 6). The co-efficient of sliding friction, μ_s may be expressed as

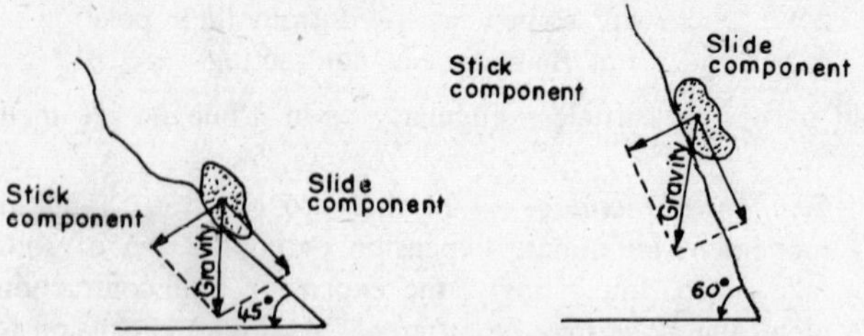

Figure 6 : Forces acting on a rock mass on a sloping surface

$$\mu_s = \frac{\text{downslope component of gravity}}{\text{gravity component perpendicular to the slope}}$$

or the tangent of the slope angle, when the particle or mass is moving

It is defined as stability factor, F_s where

$$F_s = \frac{\text{forces resisting slope failure}}{\text{disturbing forces}}$$

in which case the motion initiates soon when the numerator exceeds the denominator causing mass movement. It is theoretically considered that the slopes greater than 45^0 may not be able to hold weathered rocks. There are not many materials in nature having co-efficient of friction greater than one. Depending upon

(i) the slope and

(ii) roughness of the mass,

debris-covered surfaces may have maximum slope angles ranging between 25^0 to 40^0. Water of moisture accentuates the mass mobility over the sloping forces. Sites of unstable slopes may be identified and measures of slope stability may be taken up (Prasad and Verma, 1979)

LANDSLIDE MORPHOLOGY

Often mass-wasting processes or large landslides cause considerable changes in the surface configuration of a terrain, which is more visible along the slopes. The plane or surface along which the rock mass is detached and rolls down under the action of gravity is termed *slip-plane*. Such surfaces generally show effects of scouring or polishing due to the friction offered by the downward sliding of the rock mass and resemble slickenside developed due to faulting. The rolling down material forms landslide deposits or *slump body*. Above the slump, the upper part of the slope is referred to as *landslide scarp*. The rolled rock fragments accumulate at the bottom of the slope with irregular hummocks, or more or less smooth surfaced, parallel to the crest above the slope with a backward tilt towards the remaining undisturbed standing part of the slope. These accumulations are called *landslip-terraces.* At the slope-foot, where the slip plane emerges is referred to as the *sole of the landslides* and the junction between the slumped body and the scarp is termed *rear joint of the landslide* (Figure 7)

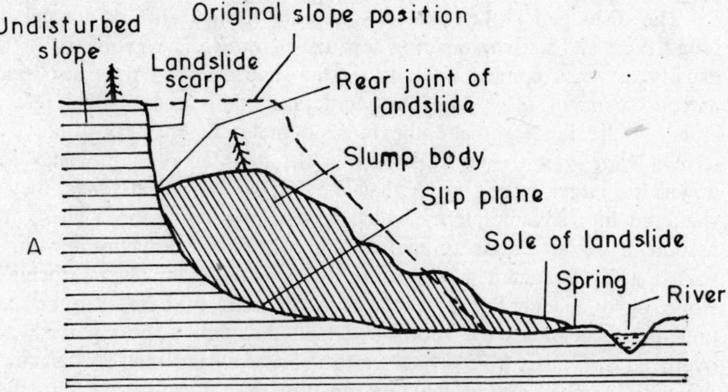

Figure 7 : Elements of landslides

Slumping of rocks or landslides, generally comprise a series of blocks that slip down under their own weight with undisturbed bedding but with backward tilted bed-surface. This portion developed due to gravity action constitutes *delapsive part of the landslide.* The slide debris are heavily crushed by pressures due to the pushing action of the falling overlying rocks and at times when the overlying pressure is great, the rocks at the slope-foot tend to bulge accompanied by the development of friction breccia along the slip plane. This crushed lower part is called *detrusive part of the landslide* (Figure 8).

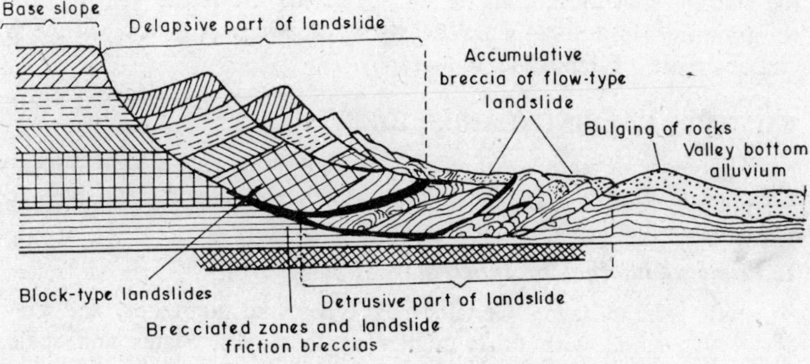

Figure 8 : Landslide morphology

Trees assuming odd positions deviating from their vertical disposition, and sometimes splitting are good indicators of landslides, besides the landslide relief form or morphology.

The dislodged broken materials usually form a cliff or conical hill with broad base and narrow tapering top, called *talus*. The geometry of the talus usually assumes conical or at times fan-shaped, when it is also referred to as *talus cone* or *talus fan*. The word talus, talus cone or fan refers to the slope or the landform and the loosend materials that constitute a talus is termed *slide rock or scree*. Broken rocks that fall onto the talus top, slide down, the larger falling faster than the finer ones, but at times, they may be shattered into smaller pieces which gets strewn downslope. These fragments assume a certain angle to maintain a continuous downhill movements. As such a talus is regarded as a slope of transportation. Oversteepening of the slope by a chance fall of rock from the cliff or basal sapling or stream undercut, accelerates the slide movement downhill, exposing fresh and more overhead cliff face from which more debris fall and nourish the talus. When the talus reaches almost the hill top, the slide rock supply virtually ceases and the talus in due course, on surficial weathering becomes covered with soil and vegetation, a characteristics that distinguishes them from newly generated or younger talus surface. Rockslides and avalanches are threat to human life, particularly in areas of high relief, steep slopes and local special conditions, where their recurrences are more, but weathered rock and soil, though less conspicuous, due to subtle evidences that are often overlooked. Certainly, creep is the dominant mass wasting process.

The surface from which the slide rocks are drawn, becomes rather barren spots of varying sizes on the cliff depending upon the nature of dislodgement. They are termed, *landslide scar.* Irregular accumulations of debris at the base indicate lanslides unfailingly. Of these, many do not move far off from the site and that too not all (i) at once or)ii) with same velocity, giving rise to gigantic rippled appearance by a mountain side or valley wall, so characteristic of landslide topography.

FACTORS CAUSING LANDSLIDES

There are a variety of attributes that may cause landslides or mass wastings. These may be (i) primary which are inherent and passive or (ii) secondary which are essentially activating.

I. *Primary, Internal or Inherent*

a) *Lithologic* - Certain rock types like quartzites, and granites do stand with fairly higher angle slopes. Slates and shales slope seldom exceed 37^0 for considerable distance. Limestones and dolomites do constitute slopes of over 45^0. Thus, different lithologies support different optimum slopes. The slopes exceeding this limit dislodge the material which falls down under the action of gravity, tending to restore the sustaining slope. Further, the state of preservation of rocks is also an important consideration. Obviously, crushed and brecciated mass of a rock is more susceptible to landslides than its uncrushed counterpart.

(b) *Stratigraphic* - A sequence of rocks may comprise relatively strong competent and weak incompetent beds. If a competent rock underlies an incompetent rock it gives a strong basal support on which the latter rests. Contrastingly, on removal of the underlying incompetent rock by weathering or erosion, the overlying competent mass looses its support and topples down. For instance, a sandstone overlying a shale would break on removal of even, a part of the basal support on the free side of the slope.

(c) *Structural* - Bed-dip is a very important factor in landslide. If the dip of the bed is more than or equal to the hill slope, it facilitates landslide when the two inclinations are in the same direction. Even if the dip of the bed is against the free sloping surface, into the hill, but the rock is jointed, the joint set parallel to the slope direction too aids the landslides. Likewise crust and shear zones as also foliation planes parallel to the free sloping surface, facilitates sliding of rock mass.

(d) *Topographic* - There is an empirical relationship of rock stability with lithology and the surface inclination it can sustain and the slopes become steeper or surface inclination increases beyond the rock surface destablizes. Rock surface is a topographic feature. Steep slopes and vertical cliffs generally are prone to break off from the parent body causing landslides.

(e) *Climatic*-Subaerial - temperature variation causes surface rock expansion in summers on heating and contraction in winter. The rates of expansion and contraction

(i) dislodge the affected part from the parent body. The loosend materials move down the hillside along the bedding plane in the direction of least resistances.

(ii) also the differential rates of expansion and contractions produce cracks and openings which facilitate further loosening of the materials, particularly along wall rock, such as by the frost action *etc.* for further downslope debris movement.

(f) *Organic*-Vegetation

(i) holds the surface cover by binding the material with it,

(ii) tends to reduce direct impact of raindrops on ground, as also

(iii) delays the surface runoff.

Removal or uprooting the vegetation necessarily follows freeing the bonded surface cover materials permitting their downslope movement under the action of gravity.

II. *Secondary or Passive*

These attributes are further regrouped as internal or external depending upon the responses of which they are the derivatives.

A. Internal -

(a) *Water content* : water tends

(i) to reduce the shearing strength and

(ii) consequently, the binding capacity of rocks.

Further, water

(iii) smears the rock surface and, thus, act as a lubricating agent.

These are conditions conducive to downhill mass movement. Heavy and prolonged rainfall activates the landslides.

(b) *Composition of mass* : certain minerals like salt (halite), gypsum *etc*. are susceptible to

(i) dissolution under surface conditions and, thus, develop open spaces for further water action on the one hand and

(ii) dislodging the associated grain on the other hand. Yet some minerals provide

(iii) slippery base when in contact with water. Montmorillonite, gypsum, rock salt *etc*. are accordingly injurious minerals as they facilitate debris movement.

(c) *Rock texture* : Water bearing and transmitting in rocks depends upon the porosity and permeability of mass which, in turn, depend on rock texture.

Thus, rock texture is related to water, and an attribute which contributes to landslide occurrences.

B. *External* -

(a) *Surface water* : depending upon the lithology and topography of the terrain and its own velocity, running water is capable of

(i) deepening the rivers by erosion or

(ii) undercutting/ basal sapling the river valley walls or sea-shore.

It produces unsupported overhanging rocks or steep slope beyond the angle of repose, developing in rock instability and, therefore, causing mass movements.

(b) *Suffosion* : (L. suffosion, meaning undermining) : is caused by the subsurface water movement in an aquifer. This water movement carries with it small particles and dissolved substances of aquifer rocks and weakens it which after quite sometime is unable to substain the weight of the deposits that constitute the higher parts of the slope and causes their slumping. "*The mechanical and chemical leaching of the small particles and dissolved substances which loosens the water bearing layer*" (Gorshkov and Yakushova (1972), is called *suffosion.*

(c) *Consistency of clayey rock* : exposed surface clays are subjected to various external and internal agents, such as surface and subsurface waters and weathering *etc.* Differential responses to these agents, their

(i) chemical composition

(ii) swelling on being wet and

(iii) differential contracting on their drying

are some such causes that brings in inconsistency in clayey rock and, thus, in loosening the sediments which on suitable slopes slides downward as landslides.

(d) *Support removal*

(i) mining of underground rocks often leaves unsupported roof above, as also

(ii) overloading the bed, too, disturbs the rock equilibrium and favours landslides.

(e) *Seismicity :* also disturbs the rocks causing instability and, thus, sliding of rocks.

(f) *Artificial vibrations :* vibrations caused by heavy traffic and explosion *etc.*, too, generate movements on the slopes.

Various factors stimulating mass-movements and the mechanism of landslides are listed in table III.11.

TABLE III. 11 : AGENTS INFLUENCING LANDSLIDES OCCURRENCE
(after Petak and Atkisson, 1982)

Agent	Event or process that brings the agent into action	Mode of action of agent
Transporting agent	Construction operations or erosion	1. Increase of height or rise of slope
Tectonic stresses	Tectonic movements	2. Large scale deformation of the earths crust
Tectonic stresses or explosives	Earthquake or blasting	3. High frequency vibrations
Weight of slope forming materials	Process that created the slope	4. Creep on slope 5. Creep in weak stratum below foot of slope
Water	Rain or melting snow	6. Displacement of air in voids
		7. Displacement of air in open joints
		8. Reduction of capillary pressure associated with swelling
		9. Chemical weathering
	Frost	10. Expansion of water due to freezing
		11. Formation and subsequent melting of ice layers
	Dry spell	12. Shrinkage
	Rapid drawdown	13. Seepage towards foot
	Rapid change of elevation of water table	14. Rearrangement of grains
	Rise of water table in distant aquifer	15. A rise of piezometric surface in slope forming
	Seepage from artificial source of water (reservoir or canal)	16. Seepage towards slope
		17. Displacement of air in the voids
		18. Removal of soluble binder
		19. Subsurface erosion

TYPES OF LANDSLIDES/MASS WASTINGS

Landslides are identified on the basis of the mass motion into following broad types:

A. CREEP -The slowest downgrade mass-wasting is referred to as *creep* and depending upon the major constituent of moving mass, it is called *rock creep* or *soil creep*. Relatively, the creep movement is more at the surface and slowly diminishes to practically zero with depth, with the consequence that creep

(i) do not shear across immobile rock at depth or

(ii) are incapable of wearing off a burried surface.

Volumetric mass (soil or rock, as the case may be) expansion by

(a) freezing and thawing or

(b) wetting and drying, causes displacement of detached particles, toward

(i) the free face of the mass or

(ii) at right angles to the sloping surface.

On contraction, however, the particle, instead pulled back, settles down under the influence of gravitational component of the motion. However, the cohesive forces of soil and water are only rarely so strong as to pull the particles back to the surface during contraction without some net downslope motion. The downslope creep movement is measured in centimeters per year, and, thus, practically imperceptible motion is often documented by the uphill bend of tree trunks.

B. FLOWAGE - It is irregular downgrade mass-movement of the incoherent rock-debris that are sufficiently mobilized to flow like a viscous fluid due to internal turbulence along *no* definite surface failure. Flows generally are saturated with water, though dry flows are also known, move as lobes or tongues with very variable rates-may be easily noticeable as in the case of earth and mudflows (rapid flowage) or imperceptibly slow (slow flowage) as in the case of solifluction *etc. Debris flow, earth flow* and *mudflow* are three terms that form a series of progressively higher water content in which mud flow is the most liquid end member, 20-80 percent of the particles are coarser than sand-sizes in debris flow, whereas 80 percent or more mud and sand constitutes the other two. The flows generally

(i) form planar hillsides or

(ii) at the toe of large slump, but by and large

(iii) their development at small gully heads are most common.

Several case histories reveal that a combination of

(i) steep slopes, weak or easily weathered rock or surface cover, perhaps along seepage lines at the head of the incipient drainage net work and

(ii) abnormal precipitation

may be the reason for triggering for the high frequency of flow initiation on small gully heads. Water content of these flows are enough to move the solids in turbulent flow and any further water addition may end up with their transporation by flowing water instead by mass-wastings.

Solifluction (literally meaning, *soil flow*) is a process in which the saturated soil or regolith, that is the soggy mass, moves downhill, at a rate varying from a few millimeters or centimeters per hour to per day. This process of mass-wasting is common in terrain characterized by saturated impermeable bed rock, surface layer, as clay hardpan would not allow water percolation. In tropical rainforests, steep slopes are scarred by large debris slides in deeply weathered rocks and lower slopes with masses of solifluction lobes.

Solifluction can take place anywhere and it has no connotation of climate. However, in contrast, the term *gelifluction* refers to flowage of saturated debris in periglacial regions, related to frozen ground. During the short summer in higher altitude, the meltwater may penetrate below and saturate the frost shattered rocks and other weathered debris, over the frozen ground which does not allow further downward movement of water.

So generated active layer cannot flow over the frozen subsoil by itself. Flowage is induced by the capillary uprise of water towards the active layer from the frozen basal soil. In this water migration process, the melt water layer is reduced to a low plastic consistency, and behaves like *'gel'*. Gelifluction can take place on gentle slopes, provided the rate of upward movement of water exceeds the rate of meltwater production in the active layer.

Debris flow or gelifluction takes place at practically negligible downslope gradient when the rate of water uprise is more than meltwater production, with a kind of rolling motion like the endless tread of tracked vehicle. The toe or lower part of the mass is marked by arcuate ridges and troughs.

C. SLIDING- When the movement of a rock mass is liberated off or broken loose from its bed, move downhill, sliding over the substratum always along a *definite plane* or *surface* of failure, the failure or motion is referred to as *sliding*. This movement is characterized by

(i) the surface of failure beneath a slump block, a spoonshaped concave upward or outward and

(ii) backward tilted upper surface of the slump block

Such a motion is engineered by streams or waves undercutting the foot of a slope. *Debris slides, rock slides* and *rock falls* are examples of sliding. When the involved rock is compact, it is referred to as rock slides and when loose material is involved it is called debris slides. A fall from exceedingly steep slope similarly is termed, *rock fall* or *debris fall,* depending upon the nature of the material. Thus, steepness of slopes determines the nomenclature of "slides", 'or falls'.

D. SUBSIDENCE - In this case, the movement of surface materials on failure is practically in a vertically down direction, extending over to a few to several centimeters. Sinking or setting of the ground could be due to

(a) natural circumstances, such as
  (i) dissolution of subsurface rocks or
  (ii) geological constitution of the rocks or

(b) artificial reasons, such as
  (i) removal of the material
  (ii) overloading the surface
  (iii) high trafficability, beyond the rock capacity to hold itself.

E. ROCK AVALANCHE - Rock avalanche, junction of continuous and discontinuous joint-patterns in the rock body, is most rapid form of mass movement along a deep-seated slip-surface. They move so rapidly that the entire collapsed mass is practically shattered and pulverized. Accordingly, they are also designated as 'rock fragment flow'. The possible mechanism of the movement of failed mass may be grouped into

(i) slide
(ii) flow and
(iii) momentum transfer

to explain the streaming of debris in avalanches (sturzstorms).

(a) Shreve (1968) proposed *air lubrication hypothesis* which envisaged that rockfalls rapidly "slide" over the bed. They travel long distances due to compressed air held briefly in the mass of rock rubble. The rushing effect of the pressure-heated compressed air is enough to slide the debris at high speed over the surface without escape of the trapped air and .much frictional contact with ground.

(b) Heim (1882) and later revived by Hsü in 1975, suggested that the dispersion of interstitial fine debris and rock dust reduces the frictional resistance of colloiding blocks. The interstitial material reduces the effective normal pressure of the cohesionless entrained grains and, thus, friction on the bed.

(c) Kent (1966) subscribes to the *flow hypothesis* and explains the flow mechanism by postulating fluidisation of rock grains after the collapse of the dry material. Fluidisation is a process in which interstitial air in debris reduces the internal friction between grains and keeps the mass in suspension. The process gives high mobility to the failed mass.

(d) Eisbacher (1979) contends that streaming of a disintegrated rock mass may be explained by a *momentum transfer mechanism.* Two-thirds of the failed debris mass lies above the ramp, and the momentum transfer of this huge mass probably causes streaming of debris downslope of the ramp. The argument is based on (i) the nature of ramp that marks the transition between the coherent but broken rock slabs at the proximal, and (ii) more fragmental debris at the distal, slope. Downslope of ramp, the debris stream out laterally and sideways under the influence of *momentum transfer mechanism.* The final debris on the distal slope is possibly 'kinetically sieved" off particles through an open pore space of coarse debris upslope of the ramp.

CLASSIFICATION OF MASS WASTINGS

An ideal classification of mass wastings would provide at a glance appreciation of

(i) the spectrum of its various types

(ii) processes involved

(iii) elements - nature of movements and

(iv) the types of materials.

of the constituting slide mass.

Sharpe (1938) proposed a classification keeping in view the following factors for sub-dividing mass wasting :

(a) *extent of lubricant*, which includes the amount of ice or water in the system

(b) *nature of movement*, such as sliding or falling, as a coherent mass or falling as small pieces due to internal deformation, and

(c) *speed of movement* of mass downgrade

He gave an useful simplified classification (Table III. 12) which serves as a guiding reference in landslide studies :

TABLE III. 12 : CLASSIFICATION OF MASS WASTING

(simplifed from C. F. S. Sharpe, 1938)

	Nature and rate of movement	With increasing ice content	Rock or soil	With increasing water content
	Imperceptible	Solifluction Gelifluction	Creep (Rock creep Soil creep)	Solifluction
Flow	Slow to rapid	Debris Avalanche		Earth flow, Mud flow, Debris avalanches
Slide	Slow to rapid		Slump, Debris slide, Debris-fall, Rock slide, Rock fall	

Varnes (1979) presented different groupings (Table III. 13.). His proposal included in addition

(i) types of movements,
(ii) materials and
(iii) textural sub-divisions

as well

TABLE III. 13 : TYPES OF SLOPE MOVEMENTS, ABBREVIATED VERSION

(after Vernes, 1978)

Types of movement			Type of material		
				Engineering soils	
			Bed rock	Predominantly coarse	Predominantly fine
Falls			Rock fall	Debris fall	Earth fall
Topples			Rock topple	Debris topple	Earth topple
Slides	Rotational	Few units	Rock slump	Debris slump	Earth slump
	Transitional	Many units	Rock block slide	Debris block slide	Earth block slide
			Rock slide	Debris slide	Earth slide
Lateral spreads			Rock spread	Debris spread	Earth spread
Flows			Rock flow (heap creep)	Debris flow	Earth flow (soil creep)
Complex			Combination of two or more principal types of movements		

Strahler in 1952 constructed yet another classification (Table III. 14.), a more generalized and widely applicable as below, based on strain types involved that may form the dynamic basis of geomorphology including mass-wastings.

Table III. 14 : THE DYNAMIC BASIS OF GEOMORPHOLOGY
(after Strahler, 1952)

A. GRAVITATIONAL STRESSES

Material involved in motion	Properties of material	Types of failure (strain)	Geomorphic processes and forms
1.Crystalline rocks; arenaceous rocks, limestone; dry soil	Rigid, elastic solid or "elastic continuum" obeys Hooke's law strain α stress	Sudden rupture along shear surface or tension fractures	Landslides:slump slide (compressional stress); rock fall (tensional stress)
2.Glacial ice near surface	Plastic solid in region of elastic behaviour	Sudden rupture along shear surfaces or tension fractures	Crevassing, overthrusting, calving of glaciers
3. Argillaceous rocks and soils	Plastic solid in region of slow creep	Continuous, slow laminar flow (distributed shear) Plastic	Large scale creep phenomena, Superficial deformation clays
4.Unconsolidated rock, soil and water	Plastic solid in region of flowage. Obeys Binghams law-rate of shear proportional to stress, above a yield limit	flow (shear between grains) when stress exceeds yield limit. Movement ceases below yield value.Flow laminar or turbulent	Solifluction, earthflow, highly turbid streams, turbidity current
5.Glacial ice under heavy load	Plastic or quasi-plastic solid. Nonlinear increase of shear rate with increase of stress	Plastic or quasi-viscous flow above 'yield' limit	Continental and alpine glacier. Erosion form due to ice abrasion. Depositional forms molded by ice flow.
6. Water film on sloping surface	Newtonian fluid. No yield value. Linear increase of shear rate with stress, subject to capillary influences	Laminar flow, ceasing when water thins below capillary control limit	Sheet run off-on slope reduction by removal of ions, colloidal clays. Fluting grooving of limestone
7. Water impermeable rock, soil (no surface slope)	Newtonian fluid, subject to capillary influences	a. Silts: laminar flow following Darcy's law.	Infiltration of precipitation, carrying down of ions, colloid
		b. Sands: mixed laminar	Clays, silts (illuviation). or turbulent flow General slope reduction.

			c. Gravels : turbulent flow Karstic form in highly soluble rocks.
8. Water on sloping surface	Newtonian fluid	Sheet run off in turbulent or mixed turbulent laminar flow	Slope erosion transportation slope forms of fluvial drainage basin
9. Water in sloping linear channel	Newtonian fluid	Stream flow, turbulent except in bed layer	Stream erosion, transportation, deposition. Drinage systems. All fluvial landforms.

Indirect Responses to Gravitational Stresses

10. Standing water bodies; Oceans and lakes	Newtonian fluid	Turbulent flow as a. Pulsating or oscillating currents caused by waves. b. Tide-induced current	Shoreline processes of erosion transportations and deposition. Shoreline landforms : Cliffs, benches, beaches, bars, spits.
11. Air	A gas : Compressible fluid of extremely low viscosity	Turbulent flow induced by pressure gradients (gravitational stress on air masses)	Wind erosion, transportation, and deposition. Depositional and abrasional landforms, Dunes, loess

B. MOLECULAR STRESSES

Materials involved	*Properties of material*	*Stress and Cause*	*Kinds of failure*	*Weathering process*
1. Rock : strong hard crystalline glassy or crystal aggregate	a. Elastic solid non-homogeneous	Shear stress due to non-uniform expansion-contraction in cyclic temperature changes	Rupture by shear or tension fractures between grains, along cleavages, joints, bedding planes	Granular or blocky distingration of rocks, *esp*. coarse grained crystalline rocks.
	b. Elastic solid, homogeneous	Shear stress set up by thermal gradient from surface heating	Rupture between layers paralleling surface	Exfoliation of rock by fire lightening, solar or atmospheric heating-cooling
2. a. Permeable rock+water	a. Elastic solid	a. Shear stress set up by interstitial ice crystal growth	a. Rupture between grains, cleavage pieces, joint blocks, beds	a. Frost disintegration of rock Felsenmeer.

b. Clay soils	b. Plastic solid	b. Stress from growth of ice lenses, wedges	b. Plastic deformation of clays adjacent to ice	b. Heaving of clay soils, frost mounds, polygons.
3. Permeable rock or soil+water and salts	Elastic Solid or elastic continum	Shear stress set up by interstitial growth of salt crystals	Rupture between grains cleavage, pieces, joint, blocks or beds	Efforescence, granular dis-integration in dry climates, calicheheaving
4. Rock or soil+ colloids and water	Elastic or plastic solid	Shear stress set up by dilitations accompanying water adsorption and drying	Rupture between grains, plastic deformation of clays during swelling	Exfoliation of basaltic, granitic rocks upon alteration of silicates. Slaking of shales and argillaceous sandstone/siltstone
5. Rock or soil+ capillary water	Elastic or plastic solid	Shear stress set up by dilitations accompanying changes in capillary film tension	Rupture between grains or masses of soil	Disintegration of granular permeable rocks. Heaving or subsidence or clays, sills
6. Rock or soil+ plant root	Elastic or plastic solid	Shear stress set up by swelling of rootlets under osmotic pressure	Rupture between grains cleavage pieces, joint blocks or beds.	Disintegration of rock by prying of roots. Deformation of soil
7. Strong, hard monolithic bed rock	Elastic solid	Shear stress of tectonic origin stored as elastic strain at depth pressure	Rupture of rock on planes paralleling surfaces after release of confining	Exfoliation of domes, slabs, shells. Quarry rupture, rock burst

C. CHEMICAL PROCESSES

Material	Processes	Forms Produced
1. Soil, rock+ acid water	Reaction between acid ions and mineral surfaces. Removal products in solution	Lowering of rock and soil surface, pitting, cavitation of rocks carbonates, caverns and karst forms. Weakening of bonds between mineral grains
2. Soil, rock+ water	Simple solution (ionization) of susceptible minerals	Cavitation of soluble salt formations. Slow attrition of exposed mineral surfaces.

IDENTIFICATION OF LANDSLIDE ZONES : HAZARD ZONING MAPS

A comprehensive investigation (Table III. 15) embracing the following help in identification of zones prone to mass- movements.

(a) structural and lithological settings of the investigation terrain

(b) such geomorphic features apprehended to contribute to slope instability

(c) regional seismicity pattern, epicentral distributions and ground acceleration due to quakes

(d) instability signatures like

  (i) tremors cracks on the ground and walls

  (ii) distortion of structural frames

  (iii) tilting of trees and poles

  (iv) leakage of underground pipes and

  (v) *tautness* in pipes and wires, *etc.*

(e) measurements of residual horizontal stresses and electrical potential in the soil, and

(f) determination of cohesion and strength of soil.

and preparation of a landslide hazard zone map. Such maps are useful in evaluating real threats and, thus, elucidates measures of investigation to meet the challenge.

Photogeological Investigation in the Study of Landslides : In addition, the following factors may be suggested to study on aerial photographs (Pandey 1987; Valdiya 1987)

(a) "Seepage from lakes, muskages, swamps, canals, ditches *etc.*

(b) Active slope downcutting and toe unloading by streams or by wave erosion, or caused by man-made excavations

(c) Natural or artificial loading of the upper portion of a slope that has failed in the past.

(d) Change in land use practices that affect the ratio or runoff to deep percolation of surface waters.

(e) Relationship between the trend of steep slope and the orientation of prominent near surface joints and figures.

(f) Tensional effects caused by notable rebound or heavily preconsolidated clays and shales following unloading ensuring development of undesirable strength."

TABLE III. 15 : INFORMATION NEEDED FOR LANDSLIDE-HAZARD MAPPING AND RISK EVALUATION

(after Petak and Atkisson, 1982)

Information Needed	Uses
Geological Maps	For understanding causes of landslides, for predicting density, severity and frequency of landslides in large areas. Aerial photographs enable recognition and mapping of landslides in larger areas.
Topographic Maps	Provide a base for plotting landslide information. Geomorphic (topographic) forms are used to identify landslides. Topo-maps provide basic data for slope
Slope Maps	Slope-stability maps combine steepness, land-sliding and physical characteristics of different materials
Climatic Data	Climatic data based on specific events and annual climate are needed for assessments of influence for case histories and probabilistic predictions
Subsurface Water	Buildup of pore-water pressure in subsurface materials and changes in water level needed for initiation of specific landslides and predicting probability of failure
Seismic Data	Enable preparation of maps and susceptibility to landsliding by earthquakes.
Vegetation maps	Vegetation tending to support and stabilize slopes needed to be depicted in terms of types, density and canopy cover.
Weathering of Rocks	Information of thickness and kinds of weathered rock products is used for estimating hillslope strength and landuse susceptibility.
Landuse Maps and Socio-economic Data	Learning present and planned development for assessment of rocks.
Chronology Data	Determining past intensities of landsliding for comparison to climatic and human activity data through use of historical records.
Computerization	Data collected are stored and manipulated to check for sensitivity of different map parameters and to superimpose large sets in different ways.

Future of likely landslides can be estimated by the presence of ponds, diverted streams of drainage ways, faint dark lines showing sub-surface channels and dark patches of vegetation.

REMEDIAL MEASURES

One or more of the following measures that may prevent their occurrences are adopted as remedial measures to control the recurrence of mass-wastings in an area depending upon the terrain under consideration :

(a) *Drainage* - *'Water is the cause and drainage the cure of landslide'.* Water present between the open or pore spaces tend to lubricate the surface and/or loosend rock debris facilitate unlocking the mutual grain-bonds or the mass and that present on the surface lubricates to reduce the frictional forces. Therefore,

(i) subterranean underground water accumulations are removed through adits driven parallel to the direction of dip of the beds and

(ii) surface water is removed with the help of diverting waterways. Obstructions like trenches and ditches *etc.* are constructed to help in reducing the water flow regime by reduction in its velocity and, thus, depositing the transporting load which acts as erosion-tool into the depression. Also

(iii) grouting of joints and shear planes prevents percolation of water to depth.

(b) *Retaining structures* - Construction of buttresses retaining walls and revetments *etc.* also help in holding the rocks to their position and, thus, prevent mass-movements. Retaining wall is preferred and is successful where

(i) ground is neither too fine nor too plastic

(ii) sliding mass is likely to remain dry and

(iii) movement is of shallow nature.

(c) *Chemical treatment* - Chemical treatment of rock mass is also attempted in recent time. Oiling the surface is one such practice to prevent the water

(i) to percolate down the depth or

(ii) even stay at the surface to interact with the mass.

It is, however, an expensive method.

(d) *Slope treatment* - It is a common practice to dress the rocks that may prevent downgrade movement. It includes

(i) flattening the slope

(ii) decreasing external load

(iii) increasing the shearing strength of the mass by decreasing the water content

(iv) obtaining additional shearing strength by use of piles and

(v) grouting the joints and shears on the free surface.

(e) *Vegetation* - With its firmly penetrating roots plantation tend to hold the rock or soil and the root grip is often so strong that it resists downgrade movement of mass. It is often most powerful remedial measure, but it takes sometime to become affective after plantation.

LANDSLIDES IN HIMALAYAN REGION

Mass wasting and landslides are serious high altitude environmental hazards and the Himalaya is one of the most sensitive areas of the world, dotted with ugly looking repulsive scenes, a constant threat to trafficability and life.

Krishnaswamy (1980) discussed at length the landslide occurrences of Indian Himalaya. He classified them into two types : those effecting

(i) the bed rocks and

(ii) the overburden.

Important landslide occurrences are given in Table III. 16

A detailed geological/structural mapping in the area would help in unravelling the presence and extent of the fracture zones and thrusts often affected by reactivation and neotectonism of the terrain that can easily wash down the crushed rocks on the slopes, during rains triggering off landslides (Mittal and Chansarkar, 1970) and eventually causing devastating floods including flash floods. Such zones are lying concealed below the Quaternary sediments (Prasad and Verma, 1982). The pulverized rock material when soaked in water turns to slippery clays and further aid in sliding along the fracture zone (Prasad and Mathur, 1976). These zones in view of rock instability are prone to landslides causing natural disaster in contrast to those caused due to unwise human intervention leading to the toppling of mass. Devastating continued massive landslides from August 18-20, 1998 near Malpa village, 50km from Pithoragarh, Kumann Himalaya and near Ukhimath, not far off from Rudraprayag, Grahwal Himalaya in Uttar Pradesh, recorded

(i) the total to go upto 225 lives

(ii) swept off several villages

(iii) thousands rendred homeless

besides further endangering on bursting of a lake formed by toppling mass in Madhya Maheshwar river

Landslides are very common during wet season in the Himalayan terrain. The maximum rainfall is recorded at around 2,000 meters, whereas there is hardly any precipitation above the altitude of 3,000 meters (Purohit, 1978) At times, landslides are caused due to cloud bursts. Above 3,500 meters the precipitation is scant, but cloud burst may take place. Absence of vegetation and ground covered with unconsolidated morainic materials aids in sliding.

Table III. 16 : LANDSLIDES IN THE HIMALAYAN REGION
(after V.K. Verma, 1985)

Name of the Slide	Location	Lithology	Causes and	Types	Remedial Measures
A. Slide affecting bed rocks					
Jammu and Kashmir					
Khuni Nala	Jammu-Srinagar National Highway between Ramban and Banihal	Salkhala gneiss	Sn	2	RB(i), T
Painthal Nala	Jammu-Srinagar National Highway near Ramsu	Salkhala gneiss	Se		Dw
Lolipura and Khairpura		Karewa formation	Sc	4	Ds, s's
Himachal Pradesh					
Chakki-Dalhousie Road	Near 42/1m on Pathankot-Dalhousie Road	Chail formation	Seb	1	Ds's
Dalhousie Cantonment		Chail formation		2,3	Ds,S's
Siul Hydel project	Baira Ballabha water conductor	Chail formation	Jt	2	Ds's
Baira Dam site	210m downstream of Baira Dam on Siul river	Chail formation	Js	1	Dw
Pandoh Dam	Right abutment	Chail formation	Pbs	1	Bs's
Giri Power house	Giri hydel project in Siram District	Lower Siwalik formation	Pb	3	Rw
Pong Dam intake	On Beas river	Upper Siwalik formation	G	2	Rw
Uttar Pradesh					
Kaliasor	located at 47 km of Haridwar and Badrinath Road	Quartzite and phyllite of Garhwal group	Jte	3	Rfe

Khari Dam	left abutment of Khari Dam on Tons river	Chandpur formation	Sr	3	Rfe
Pakhitangri	located at 252 km Haridwar-Badrinath Road	Dolomite and Phyllite, Garhwal group	Sf	3	Ra(r)
Balakuchi	located at 253 km Haridwar-Badrinath Road	Talc schist band		3	Ra(r)
Hydel project	A barrage near Rishikesh and power house	Upper Siwalik formation			
Kohna	East of Chamoli at a tributary of Alaknanda	Dolomite	Te	3	
Revi		Central crystalline gneiss	Te	2	
Nainital		Lower Krol shales		2	Ds,s's
Darjeeling subsidence	Hermitage adjacent to Raj Bhavan	Darjeeling gneiss	D4	D4	Ds
B. Slide affecting the overburden					
Jammu and Kashmir					
Thana Mandi	Located along Rajauri-Bafliaz Road	Salkhala formation	Se	3	Ds
Malori	Located at Batote - Bhadarwarh Road	-do-	Te	3	Rw
Chineni	Chineni Hydel Project area, Udhampur district	Murree formation		4	T
Nashri	Between Batote and Ramban, Jammu-Srinagar NH	Murree formation	Ste	3	Ra(r)
Pari and Tangri	Located at left bank of river Chenab	-do-	Ste	1	
Lower Jhelum hydel project	Located at right bank of river Jhelum	Quarternary formation		3	Rw

Himachal Pradesh					
Shimla	Northern slope of of Shimla ridge	Chail formation	Ste	1	Ds
Nanda Devi	At the foot of Naina Devi range, Bilaspur district	Lower Siwalik formation	Ste	1	
Uttar Pradesh					
Nandprayag	216 km of Haridwar-Badrinath Road	Gneiss and schist of Garhwal group	Rfe	1	Ds, Rw
Sial	249 km of Haridwar-Badrinath Road	Dolomites and phyllite	Ste	2	Ra(T)
Helang	13 km downstream of Joshimath	Slate and clayey matrix	Da	2	
Dhak	12.2 km from Joshimath, Joshimath-Malri Road		Te	2	Ra(T), Rb
Birehi bridge	Left bank of Birehi Ganga on Malri-Badrinath Road	Clayey matrix of Garhwal group		3,4	
West Bengal-Darjeeling					
Ambotia	2.5 km long and 600 m high slide at Ambotia	Darjeeling	Sr	3	Ds,S's,Rw
Hermitage	Adjacent to Raj Bhawan	-do-	J	2	Ds
Sikkim					
Berik	NH 31A, to Gangtok	Quartzite and phyllites, Dahing group	Te	3	Rfe
Likubir and Rangpo	26 km NH 31A, to Gangtok	Schist and phyllite		3	Ds,Rw
Teesta Bazar			misc	4	Ds,S's
Sweakhola	Quekhela		misc	4	
Vongmannal			misc	3	Ds,S's,Ra
Arunachal Pradesh					
Bomdila-Tawang	289-90 km Kiminzeo Road	Disang shale		3	Ra(T)

Abdhullah Nala	NH 39	Disang shale	misc	3
Nagaland	NH 20	Shale and sandstone	misc	3
Mizoram		Sedimentary rocks		
Manipur	milestone 23~ 36.8 kilometerstone of Lungleh-Tripaug Road	sand		4
Aizal-Lungleh	milestone 21~ 33.6 kilometerstone	Sandstone and shale	Tte	3

Symbols used : G-due to steep gorges; J-due to jointing, Jbs-Jointing in bedrock of Beas river; Js-Jointing and overburden; Jt-jointing and toe erosion; Sef-saturation of crushed and fragmented rocks; Se-saturation due to erosion; Sr-saturation by reservoir; Misc-other than above types of slides 1-Block slide; 2-Planar slide; 3-Debris slide; 4-Slump slide; Remedial Measure : Rb-Rock bolting, T-Tunnelling, Dw-Deflector wall, Ds,S's-Surface and subsurface drainage; Rw-Retaining wall; Rfe-Rain forced concrete; Rar-Road alignment.

A close glance on the roads in the region reveals that perhaps their improper alignments for construction of routes is an important cause of frequent landslides and blockades, resulting into repeated disruption of the traffic vital for people and communication. Many roads pass through old landslide zones, talus cover deposits, along the valley sides and, thus, are in constant danger. The Belakuchi disaster of 1970 was one such example (Nand and Prasad, 1972). The alignment of roads and National Highways ought to be geologically examined closely. In fact, where feasible, it is always better to construct the roads over the ridges, instead along the valleys.

MASS MOVEMENT PREDICTION

Mass movement prediction is based on an analysis of the landforms, geology and geomorphological processes operating in the area. Such investigations also necessarily include the past processes and the geological set up of the concerned areas. Solifiuction features produced under periglacial activity by seasonal thawing and saturation of the surface debris layers cause mass movements. Stability returns as the warmer climate sets in and when these debris layers are disturbed by human activity, instability again sets in. Also the glacial and fluvio-glacial sediments, abundant in the Himalayan region are as well susceptible to sliding.

Similary, soft rock formations like talcose schists are easily eroded triggering off landslides. Such formation/areas may be mapped in detail and landslide preventive measures taken up.

With the available present day technology, it is not possible to predict with reasonable confidence very large scale mass movement on regional basis. However, a new interesting Czechoslovak contribution referred to successful application of acoustic monitoring of small scale landslide (Tbilisi Symposium, 1979) with the help of geophones which produced noise due to movement along the planes of failures pin-pointed its source. When predicted in due time, appropriate remedial measures can be adopted.

CHAPTER IV

TECTONIC AND IGNEOUS LANDFORMS

The tectonic landforms are the resultant features developed in response to regional tectonic impulses reactivated or modified by the geomorphic processes which may vary in

(i) intensity
(ii) operational duration and
(iii) nature of rocks being sculptured.

These landforms, caused earlier by folding, faulting or due to igneous activity, may reflect the forces that initiated their developments and may be related to orogeny of a particular geological period or time.

LANDFORMS DUE TO FOLDING

Deformations usually buckle the rocks into folds comprising of anticlines and synclines. Most of the mountain ranges are related to anticlinal and synclinal structures often showing parallel or subparallel disposition. The anticlines generally correspond to upward positive physiography, called *anticlinal ridges* and the synclines to downward negative topography termed *synclinal valleys*. However, if the syncline has a hard resistant bed at the core, instead of a valley, it may form a hill on removal of relatively less resistant beds at the limb. The feature so formed is referred to as *synclinal hill*. Often on folding, the rocks weaken along its axis due to operating tensional forces, allowing rapid erosion, transforming the anticlinal ridge into *anticlinal valleys*. Divergent dips of the limbs away from the axis indicate that the nose is eroded away and a valley is carved out of an anticline.

Structural basins and domes are only modified synclines and anticlines. The beds dip from all sides towards a common centre in a basin and away from a common centre in case of a dome. However, the folded sedimentary rocks exhibit *linear homoclinal ridges* and *homoclinal valleys* on tilted beds of differing resistance. Homoclinal ridges converge in the direction of plunge of a syncline. Hogbacks and cuestas become particular instances of such features on simply inclined beds which have not undergone any folding. Alternate resistant and soft dipping beds, on erosion may give rise to a set of parallel *ridges* and *valleys*. Likewise monoclines of differing resistant beds, as also a lava flow in country leads to trap and/or *plateau* formation.

LANDFORMS DUE TO FAULTING

Variety of fault related landforms occur in nature. Faulting involves movement of two adjacent blocks past each other. Simplest landform produced by faulting is *fault scarp (scarp* is a short form of *escarpment,* a term used to connote a steep rather abrupt vertical to subvertical slope or cliff) formed either as

(i) a direct consequence of upward movement of a block, or

(ii) an effect of differential erosion of less resistant bed, now brought against a resistant block.

A range of 25^0 to 40^0 is typical angle of repose of scarps of practically any origin-tectonic (fault), erosional or depositional, and often they quickly slump down or collapse to a slope of gravitational stability.

A set of parallel faults dipping in the same direction gives rise to step fault. So arranged step faulting may develop into *fault-terraces*, a feature similar to other terraces, formed by river or marine action. When two similar sets of faults are so patterned that one set dips

(i) away from the other, a *horst* or hill like feature is developed

(ii) towards the other, a fault valley or *graben* is sculptured (if the convergence is deep enough, a *rift valley*, as that of East Africa is formed)

Much of our coal, particularly Gondwana coals are deposited into the graben, a sunken block in Archaean terrain.

LANDFORMS DUE TO IGNEOUS ACTIVITY

Landforms at times manifest igneous activity of a terrain. Such landscapes are constructional, but some secondary features developed on differential erosion may also be caused. Igneous landforms may be caused by

(i) volcanic activity, or

(ii) even by plutonic or hypabyssal injections, subsequently surfaced on erosion of the overlying overburden.

Volcanic Cones

More or less regular cone shaped central type of volcanoes are hills sloping around at 30^0 to 35^0, with a circular depression or crater developed on the top. They erupt

(i) gaseous material consisting predominantly of steam

(ii) hot molten lava flows, and

(iii) fragmented solids-the pyroclastic materials.

Nature of the material erupted-dominantly fragmented or lava flows determines the morphology of the cone. A *cinder cone* is formed when the dominant supply is solid/pyroclast.

Features like

(a) conical-shaped geometry
(b) steep-sided slopes varying from 30-40 degree to gentler angles
(c) symmetrical in shape
(d) built of angular large solid fragments to finely broken rocks and dust including ash, and
(e) formed due to violent explosions

are characteristics of such cones. Often cinder cones occur in clusters and are discerned by the presence of crater at the top on aerial photographs. In contrast, a *shield volcano* is practically made of lava flows. Some features like

(i) flat type of cones
(ii) formed essentially of lava flows
(iii) have gentle side slopes, and
(iv) formed by quiet type of eruptions

are characteristics of these volcanoes. They may feed parasitic cones as well. Yet another type showing characteristic of both the types-cinder cones and shield volcano, is developed in *composite cone* in which fragmental materials and lava flows are inter-layered with each other. They reflect slopes between the fundamental varieties and

(a) show crude stratification of pyroclasts and lava flows and therefore sometime
(b) slope intermediate between the cinder and lava types, and
(c) are of dual origin in which lava emission alternates with pyroclastic debris.

Slopes of such volcanoes are often furrowed by gullies radiating steeply in all directions from the crater and are termed *barrancos*. When dense, only narrow crest separates the barrancos giving a ribbed appearance. *Somma* (volcano-within-volcano), is a semi-destroyed old crater in which a younger and well preserved cone is developed by recent explosions. Volcanic cones are rather fragile and are rapidly destroyed by erosional agencies of which the testimony lies in the evidence of their existence limited to the feeding conduits when they are later buried below by younger rocks.

The volcanic mountains are crowned by a bowl-like depression termed the *crater*, and are the terminal part of the vent or volcanic pipe which in fact is a duct piercing across the host bed rocks through which the eruption products reach the surface. These craters measure hundreds to few thousands of meters in diameters which may be much larger, of even several kilometers in diameter specially when associated with explosive under cone type. These may be produced by

(i) explosion of the summit area
(ii) coning of the upper part of the volcanic cone or
(iii) even due to later erosion and enlargement of volcanic mouth.

Such *calderas*, as they are called, have

(i) steep inner walls and
(ii) a flat flow.

Laccolith

In this, the host strata are

(i) arched up like an inverted spoon or domes as concordant intrusion
(ii) with flat or concave base

These are developed on intrusion of relatively more viscous flow which limits the areal spread, consequently, they heap up around the eruption orifice arching up the overlying strata.

Plug Domes

These are rather uncommon type of features in which high viscosity lava ascends as a pasty mass confined in limited spread around its vent. Such features, termed *plug domes,* have usually steep margins and are composed of obsidian and pumice.

Dome Gneiss

North Bengal gneiss in some parts show peculiar weathering due to circular exfoliation of rocks giving rise to semi-circular, dome-like hills or ellipsoidal gneiss-masses. Bengal gneiss have been assigned the name of *Dome Gneiss* because of the shape of the resultant masses.

CHAPTER V

WEATHERING

The rocks are subjected to destruction by atmospheric/ geomorphic processes- physical and/or chemical-capable of securing and transporting earth materials such as those obtained from volcanism, seismicity, water, wind and ice *etc.* ever since the consolidation of the outer shell of the earth and the accumulation of water in depressed basin on sufficient cooling. Such affected rocks are said to have undergone *weathering.* Weathering is a group term for all those processes that produce an essentially irreversible change in rocks leading to new products and a residuum of original matter. It is one of the universal exodynamic processes, the principal supplier of plastic materials and dissolved substances of which sedimentary rocks are formed. It is a response by the rocks formed earlier to re-establish equilibrium in the new environment. *Weathering is defined as essentially the process of decomposition and disintegration of pre-existing rocks-igneous, sedimentary or metamorphic under the influence of certain physical and chemical agents like temperature, frost, wind, ice and water etc.* The process of destruction of rocks and their removal grouped together are designated by a general term *erosion* (derived from the Latin word *erodere* meaning to *gnaw* or to *eat away*) or *denudation* (derived from the Latin word *denudere* meaning *to strip bare*). These are by and large terrestrial processes operative on the land while those of reconstruction take place principally beneath water bodies such as sea bottom and lake beds. Weathering prepares the rock surface for erosion. But in some instances, solid rocks are eroded directly by scouring without the intermediate stage of loosening activities of weathering.

The land surface agencies *e.g.* wind, water and ice that are operative on the earth's crust are designated as *epigene* (as proposed by Lawson) or as *exogenous* (as proposed by Penck). Those geomorphic processes such as volcanism and diastrophism which have their origin within the earth's crust are classified as *endogenous* by Penck and *hypogene* by Lawson.

On weathering, the top layer of the earth's crust is broken down in size and/or altered chemically. Thus, a mantle of loose and partially decomposed rock referred to as *regolith* overlying the parent rock mass is formed. The existence of regolith is a pre-requisite for many of the gradational agents, facilitating loose materials of a suitable size to be picked up for transportation and ultimate removal. The depth of the regolith varies in part (i) according to the rate and type of weathering (ii) depending on the rate of removal by gradational agents.

Although often intact, *bedrock* is frequently exposed in hillsides, cliff faces and in desert or arctic regions, about 90 percent of the earth's surface is masked by regolith or other unconsolidated materials. This means that mostly the gradational agents are operative on the mantle of loose or weathered rock products.

There is general tendency of the geomorphic processes to bring the surface of the lithosphere to a common level. This action is termed as *gradation* (Chamberlain and Salisbury, 1904) and or *planation*, the process could be assigned to two categories

(i) those which level down-*degradation* or destruction

(ii) those which level up- *aggradation* or reconstruction.

In practice the term planation is, however, referred to erosion and not deposition. Likewise denudation is synonymously used with gradation, but this term usually connotes the removal of material keeping deposition still beyond its ambit. The two sets of processes—weathering and erosion—operate together in wearing away the land surface and their combined effects are described by the term denudation. In fact, denudation includes three-fold processes, namely

(a) weathering

(b) erosion, and

(c) transportation

The latter *i.e.* erosion, essentially of physical nature, is rather a broader term to denote more vigorous destructive effects of denudation followed by the transport of the resulting loosened material away from the source site for deposition elsewhere. The weathering may be

(a) physical or mechanical or physico-mechanical, and

(b) chemical

Generally, in all cases, the geomorphic landscape is a product of the combined action of

(a) mechanical weathering (disintegration) and

(b) chemical weathering (decomposition)

MECHANICAL WEATHERING

Variety of factors cause physical weathering. However, the most decisive and, therefore, significant role is played by those factors that bring about physical movement of rock particles destablizing the mechanical bonds between them. The nature of the processes of rock destruction by physical weathering differs depending upon the factors involved. The processes of physical weathering invariably involve application of forces, either from within such as volumetric changes in the rock constituents brought about by temperature variations or externally like crystal growth of substances foreign to rock bodies such as those of ice or salt. *Stricto senso* this is termed mechanical weathering. Accordingly, based on the application of forces involved, the following (Table V. 17) classification is erected :

(A) Stress applied from internal conditions
- a. Unloading effects (residual stress)
- b. Expansion/contraction of crystal lattice/ thermal stress

(B) Stress applied from external conditions
- a. Growth of foreign crystals - (i) ice crystals
(ii) salt crystals
- b. Wedgework action and soil stirring by organisms
- c. Electric storms
- d. Rain

UNLOADING EFFECTS

Even at shallow depths within the earth or beneath the sea, the pressure encountered is much more (about three to four meters of overlying rocks or ten meters of sea-water approximates to confining pressure of one atmosphere) than the usual atmospheric pressure at the surface. Since a rock is, generally, subjected to greater pressure at the time of its formation, the pressure is considerably released on exposure to the atmosphere. Thus, rocks expand from pressure release or unloading or on their removal.

The amount of expansive force in massive rocks is rather surprising. A three meter granitic block may expand to by about one eighth of an inch when cut across or blasted loose. The rocks adjust to the new pressure environment in some situation leading to rock- bursts due to pressure release. Likewise the rocks are severely fractured on their uplift preceding their eventual exposure. The pressure is relieved through open spaces on unloading by movement on these pre-existing rock surfaces.

However, unloading pressure release promotes rock mass disintegration as continuing series, for it helps air, water and even plant roots to penetrate through along the open space and, thus, prepares ground for further weathering and may not be a significantly major factor in itself.

THERMAL WEATHERING

Rocks consist of aggregate of minerals. Different mineral species respond to temperature fluctuation differently and at varying ratés, causing unequal expansion or contraction of the constituent mass on heating and cooling respectively. In addition, because of slow heat transfer from and to rockdepths,

TABLE V. 17 : CLASSIFICATION OF WEATHERING PROCESS

Type of Weathering		Processes	General Characateristics
Mechanical/disintegration	a.	Temperature variations (thermal expansion and contraction)	(i) reduction in particle size ;
			(ii) increase in surface area ;
	b.	Release of geodynamic stress(un/off loading)	(iii) no change in chemical composition
	c.	Hydration(volumetric expansion and contraction of rock-forming minerals)	
	d.	Crystallization processes (growth of foreign : ice/salt-crystals within a rock body or soil).	
	e.	Organic processes (root wedging and colloidal plucking)	
Chemical / decomposition	a	Oxidation	(i) complete change in chemical and mechanical properties
	b.	Hydrolysis	(ii) net increase in volume
	c.	Carbonation	(iii) formation of stable clay minerals
	d.	Solution	
	e.	Chelation	

differential surfacial expansion and contraction also takes place vertically, the exposed surface expanding and contracting faster, than those below. These temperature variations could be diurnal or seasonal. Rock contraction is caused by cooling in the night and/or winter against residual expansion from the daylight and/or summer heating.

Minerals of different composition tend to deform differently in response to temperature fluctuation owing to differential volumetric expansion factors. Accordingly, it is less to only marginally perceivable in monominerallic rocks (composed of a single species of mineral) than a polymineralic rocks (composed of two or more species). While monomineralic rocks essentially weather to disintegrate due to differential linear expansion factors along different directions of its axes or anistropic nature of the same mineral, in polymineralic rocks due to differences in expansion factors of the constituent minerals, the mutual cohesion of the individual mineral grains is dislodged on frequent and prolonged temperature fluctuations. (Table V. 18)

Table V. 18 : VOLUMETRIC AND LINEAR EXPANSION FACTOR OF CERTAIN MINERALS

(after A.L. Bloom 1969)

Mineral	ExpansionFactor t^0 C		
	(Volumetric)	(Linear)	t^0C
Quartz	0.000310		
ǁ L^3		7.5×10^{-6}	20
⊥ L^3		13.7×10^{-6}	20
Calcite	0. 000200		
ǁ L^3		25.6×10^{-6}	20
⊥ L^3		55×10^{-6}	20
Orthoclase	0.000170		
Hornblende	0.000284		

It may be observed from the table that the linear expansion factors in the direction perpendicular to the three fold axis is appreciably more, practically double than that in direction parallel to the axis. In monomineralic rocks, local stresses produced on temperature fluctuations, cause disintegration with time.

Rock colour also affects the thermal weathering intensity in as much as dark coloured rocks and minerals because of higher absorption capacity respond to sun-rays to heating much more strongly than light coloured rocks which, on the other hand, tend to reflect that heat. Obviously, therefore, for the reasons that the thermal weathering intensity responses with different colour vary, variegated rocks disintegrate more rapidly in relation to uniformly coloured rocks—light or dark.

Similarly, rock texture also contributes to thermal weathering. It is the function of the surface area available for thermal responses. As such, coarse-grained disintegrates faster than fine-grained rocks.

Relative to adjoining lowlands, at higher altitudes the

(i) slopes of high mountains are steeper

(ii) air more transparent and

(iii) insolation much stronger (L. word *insolation* meaning the irradiation of earth's surface by sun rays the magnitude of which is determined by the quantity of heat and light falling a unit surface in a unit time).

and as such thermal weathering is more intense allowing loosened weathered rock fragments to roll readily downslope under the influence of gravity, at or near the foot of the mountain slope. These ill-sorted or assorted angular fragments deposit to form *hillside waste* or *talus*, on accumulation and termed *colluvium* and, on the other hand, expose fresh slope surface for further weathering.

Diurnal temperature fluctuation seems to be the principal factor of thermal weathering as seasonal fluctuations are more effective. Observed practically in any climatic zone, thermal weathering is most intense in terrains/ regions of

(a) sharp temperature fluctuations

(b) dry air and

(c) no or poorly developed vegetative cover

as in arid and semi-arid regions. In such regions

(i) absence or negligible presence of moisture-relatively may drop down to 10% or less to 2 to 3%

(ii) insignificant precipitation

(iii) clear sky and

(iv) sharp diurnal temperature fluctuation amplitude, may be even of the order of 40 to 50^0 C.

Conditions are quite conducive for most pronounced insolation heating the rock far above the ambient air temperature by the sun rays in the day time and equally cooling down at night with the fall of temperature or sunset. Further, the precipitation in such areas is in the form of downpours for limited span of time, rapidly cooling the rocks causing nonuniform compression in the rock-mass dismembering the constituent minerals and accentuating rock disintegration in hot and dry climates.

Particularly under similar situation the outer or surfacial parts of rocks develop detached curved shells due to tensile and compressional stresses specially observable in dome shaped crystalline masses, like granites in semi-arid or arid regions, developed in response to differential expansion. Intensity of rock disintegration is a function of variation-extent. Such weathering is termed *exfoliation* in which curved shells peal off from the main rock body as illustrated by Madan Mahal granites near Jabalpur.

Scaly peeling off of rocks is also termed *desquamation* in which the shells are detached. Such a type of weathering is called *spheroidal weathering* and the inner rounded uneffected solid mass which yielded to desquamation is the main mass referred to as *residual boulder*. Temperature fluctuations produce concentric layers. However, recent expansion and contraction are generally feeble and are incapable to produce rock fracture or disintegration.

GROWTH OF ICE CRYSTALS OR FROST WEDGING

Destructive action of water is most intense and rapid. It is experienced in higher latitudes—polar and subpolar regions and at higher altitudes, where frequently temperature fluctuates about the freezing point of water. Water that enters into the rock fissures and pores and increases volumetrically by 10.11% on freezing. It develops enormous pressure, upto several hundreds of kilograms per square centimeter of area on the pores or fissure walls sufficient enough to overcome the tensile strength of rocks leading to the rock failure by way of disintegration into pieces. It takes place under the following conditions:

(a) in the presence of - (i) fissures and pores in rocks
(ii) water, and

(b) at appropriate temperature.

Such phenomenon of rock fragmentation by ice crystals growth is often termed *frost wedging* and the vast spaces covered by varying dimensions of such rock fragments produced, is called *stone seas* or *stone rivers* depending upon their geometry.

GROWTH OF SALT CRYSTALS

It is commonly observed in areas where evaporation exceeds precipitation, that is, in arid regions

(a) In such areas the groundwater in the soils rises towards the surface by virtue of capillary gradient developed due to excess of evaporation and when the water reaches the surface, dissolved salts are precipitated in the soils due to loss of water.

(b) Aerosolic atmospheric salts derived from evaporation of wave spray in distant oceans are leached away during occasional rain waters and drained to the ground or in rock pores or space.

On evaporation the salts precipitate out from solution, or just beneath the rock surface, and crystals grow in the pores and cracks and generate tensional force locally which may be in excess or rock tensile strengths that lead to granular disintegration, much the same way as that produced by frost wedging. Thus, salt weathering takes place. It is commonly associated with cavernous rotting or honey combing of rocks (tafoni) particularly along shaded cliff bases and on the undersides of boulders, where moisture is conserved longest in desert climates. Additional stresses are developed as changes in humidity or temperature favour alternating expansion and contraction of salts that encrust mineral grains.

Although the results demonstrate that the presence of salts in solution does not necessarily augment frost weathering, nor increase its efficiency, salt hydration and crystallization effects are capable for causing particle size reduction. Salt is apparently not uncommon in polar regions and salt crusts have been observed on pebble and granule sized materials in Alpine areas. Throughout the late spring, summer and early fall, Alpine areas experience substantial diurnal fluctuations in relative humidity and temperature, thus promoting salt hydration and crystallization pressures and leading to further particle breakdown and the production of fines, and affect that can also result from clay hydration, but to a lesser extent. Thus, while frost weathering may remain the principal cause of rock disintegration in cold climates, clay hydration and salt weathering can be expected to contribute to the disaggregation process, the former throughout the year, and the latter more especially during the spring, summer and the fall.

WEDGEWORK ACTION AND SOIL STIRRING BY ORGANISMS

Significance of organic/ biogenic weathering under arid and low temperature conditions is poor because the number and activities of soil organisms is reduced. The humidity assumes enhanced importance in as much as it supports the development of soil organisms reflected by the relationship between the nitrogen content, a product of the activity of organisms of soil and rainfall. Attributes of biogenic weathering are :

(a) rock splitting due to continued expansion of the root systems in joints

(b) some prying action particularly in lifting slabs

(c) in permitting water that percolates into the joints through the root system (seem to be the most important)

(d) soil stirring (as earthworms do) so that unaltered mineral matters (particles) are exposed to chemical attack, and

(e) breakage and alteration of soil particles moving through the digestive tracts of worms.

Organic acids released by bacterial as well as animal life tend to increase the H-ion concentration of the soil water. Much of the soil nitrogen is combined in compounds many of which affect the acidity of soil and hasten soil-forming processes

ELECTRIC STORMS

Electric storms at high altitude where the air is more transparent often produce lightning that shatter the rocks alongwith the insolation which is much stronger uplands. Lightning is natural cause of forest fires at higher altitudes, exposing fresh surfaces for rock shattering. Terrain attributes like mountain peaks and summits are locales of effective lightning attacks due to their build up charge and referred to as *coronal* discharge. Thunder cloud accentuates it.

Convex surfaces are prone to more direct lightning strokes than plane or concave slopes. Fractured and /or jointed rocks act as 'spark gaps' through which the body bridges for the current to jump, rendering shelter in narrow, vertical gullies or beneath rock overhangs more risky.

RAIN, SATURATION AND DESICCATION

The work of rain in (a) carving out gullies and small valleys and (b) transporting the eroded materials through them is of common observation particularly on slightly sloping and/or undulatory surface. The gullies develop and fan out in different directions from relatively higher elevations which helps in the formation of *earth pillars*, called *hoodoos*, in soft soil capped with rock blocks. The soft material is removed out by the rain, leaving behind the soil underneath the caps of boulders, which is also subsequently removed. These are seen near Billawar and in Ramnagar hills, Jammu.

The soil cover of gently sloping hilly surfaces are also brought down, of course, accentuated by gravity flow as well, to the lower parts by the rains

(i) exposing fresh surface for further weathering, and

(ii) depositing the angular rock fragments with soil forming rainwash.

Further, the rainwater is soaked in the rocks and on sunshine it is driven off. Recurrence of saturation and desiccation of rock materials loosens the constituent mineral particles often observed dotting mountainous terrain.

MAN

Man is the fastest weathering agent. He breaks the rocks for several geotechnical works—highways and tunnel construction *etc.* as also for use as road and building stones. Such activities of disintegration of rocks are

(a) even faster than those achieved by geological agencies and

(b) cause changes of the landforms leading to environmental consequences.

Table V. 19 provides summary of mechanical weathering in rocks.

CHEMICAL WEATHERING

The rocks formed at high

(a) thermal energy conditions, and

(b) pressure

when brought near or exposed at the surface, interact with the chemically potent constituents of the atmosphere, the hydrosphere and the biosphere in order to re-establish equilibrium with its new environment. On such interaction with the surfacial layers of lithosphere, the rocks tend to undergo exothermic chemical reaction, evolving heat. In this process, termed, *chemical weathering,* the chemical make up of the original material transforms by changing into new compounds/minerals of

(i) greater volume and

(ii) lower density

Oxidation of iron bearing minerals with the oxygen available in moisture in the wet atmosphere is a typical example of exothermic reaction. The chemical weathering processes are generally

(a) irreversible

(b) slow and

(c) deceptively simple

The following factors contribute to chemical weathering:

(i) *Surface area* : Chemical potency of minerals and rocks to react is higher along their surfaces, and thus, more pervasive is the chemical weathering on wider availability of surface area obtained on rock disintegration into fragments. Accordingly, mechanical weathering provides more surface area and thus promotes/facilitates chemical weathering.

(ii) *Moisture and warmth :* These two factors tend to increase the reaction rate, as is observed in the moist tropics and are thus directly related to chemical weathering.

Table V. 19 : SUMMARY THE VARIOUS PROCESSES, RELATED OPERATIONS, AND ATTENDIG EFFECTS/ RESULTS OF MECHANICAL WEATHERING

(after Karl W. Butzer, 1976)

Process	Operation		Effect/Result
Crystal growth (develops pressure)	a. as water freezes and ice forms (frost weathering)	(i) ice crystals grow and wedge apart rock fissures and joints (ii) freezing of capillary water (adhering to mineral grains) and gravitational water (in pores pace) exerts pressure on inter-mineral structures	Blocks wedged apart along vertical joints; slabs or slices spall off along cracks or bedding planes parallel to rock surfaces. Grain -by-grain disintegration of mineral aggregates at surface
	b. as salt crystallizes	Salt crystals grow as moisture evaporates creating pressures in pore spaces and cracks	Rock shell undermined, splices or spalls break off parallel to rock surface.
Thermal tension (generates differential stress)	a. as rock surfaces are heated by sun(or fire) b. as rock surface are cooled by re-radiation	Rapid expansion or contraction of surface rock weakens mineral bond between shell and rock interior, generating fissures vertical and parallel to surface (aided by hydration)	Segments of rock shell spall off as slabs or splices
	c. as minerals heat at different rates	Inter mineral stress as a result of differential expansion or contraction	Grain-by grain disintegration of surface particles
Pressure release (by unloading)	a. release of vertical pressure as surface rocks are eroded, sheeting, unloading, spalling.	Dilation of upper rock masses, facilitating development of weakening of surface shells.	Joint systems enlarge

	b. release of lateral pressure as valleys are cut and slopes undermined mechanical collapse	Dilation of exposed rock surfaces with growing weakness along joints and bedding planes	Splices or slabs irregular or blocks spall off
Others			
a. Abrasion	Friction of boulders or impact	Wearing away the weaker parts like corners and further fragmentation	
b. Colloid plucking	Clay film dries out	Plucking out grains	
c. Soil ripening	Evolution of fresh alluvium	Soil colloids lose water and volume, becoming sticky and then solid	
d. Moisture swelling		Changes in volume	

(iii) *Organisms* : They aid chemical weathering, by way of producing a host of organic compounds that react with rocks and minerals and transform them. Biogenic activity increases with higher moisture coupled with warmth which is responsible for further rapid rates of chemical weathering in moist tropics. New materials obtained from open spaces on root growth or upturn of the soils by earth worms become available for further reactions with the already developed organic acids and their deeper penetration through the space so developed.

(iv) *Composition of rock materials and their susceptibility* : Rocks are composed of minerals. Due to higher chemical gradient development in the new environment minerals formed at higher temperatures and pressures become more unstable and weather faster. Their susceptibility to weathering varies with the constituent compositional materials. As such olivine (FeMg) SiO_4 the first formed magmatic mineral chemically weathers much faster than the quartz (SiO_2) last formed at lower temperature and pressure, establishing low gradient.

WATER IN CHEMICAL WEATHERING

The role of water in chemical reactions in weathering the rocks is very important. It is a must in chemical weathering and acts as a

(i) catalyst
(ii) reactant or as the
(iii) carrier (solvent)

of the reaction products. It contributes in weathering of rocks through its following chemical and physical attributes :

(a) it is by far the most abundant material

$$W = 3AOS$$

where W = water and AOS = all other substance

(i) more than thrice of all other substances (AOS) in the outer five kilometers of the earth.

(ii) almost six times more than the next most common single substance, mineral feldspar F, W = 6F

where F= feldspar

(b) thus far, it is the only known natural compound at the earth's surface that occurs in all the three states of matter solid, liquid and gaseous state.

(c) its

(i) general solvent power and

(ii) surface tension

are greater than those of any other fluid

(d) of all known substances its heat of vapourization is highest

(e) possessing a strange and almost unique property in its maximum density at 4.0^0 C in the liquid phase with expansion both toward the

(i) freezing temperature by 9%, indeed a great volumetric expansion as well as

(ii) away from it.

The chemical potential of water is well demonstrated in its successive chemical analyses after its precipitation on the surface as rain or snow, through

(i) soaking into the soils (during which it is enriched with CO_2 due to decay of contained vegetal matter)

(ii) percolation into the openings, pores, joints, cracks *etc*, and

(iii) on reappearance at the surface in the form of springs.

In fact, rain-water charged with nitrogen, oxygen and carbon-di-oxide in its downward journey through the atmosphere-lithosphere interface helps largely in rocks accomplishing chemical weathering by alteration and solution. Marked difference, however, results, in the proportion of the dissolved gases in rain water and their concentration in the normal atmosphere (Table V. 20).

TABLE V. 20 : IMPORTANT GASES IN THE ATMOSPHERE AND IN RAIN WATER

Gases water million)	Non-variable Components of (parts per million at 20^0 C)	Dissolved Gas in Rain Atmosphere (parts per
Nitrogen	± 780, 084	± 636, 900
Oxygen	± 209, 460	± 341, 700
Argon	± 9, 340	not determined
Carbon-di-oxide	± 380	± 2, 140

Further, in addition ammonia, nitric acid, sulphuric acid, sodium chloride and chlorine are chemically very potent. Besides the above, but relatively in minor proportion are other substances. The degree or extent of the chemical activity is a function of

(i) composition and concentration of solutions so formed
(ii) temperature
(iii) presence of bacteria, and
(iv) the substances dissolved from the decomposing minerals.

Practically all natural waters are slightly dissociated into H^+ and $(OH)^-$ ions. The pH- acidity/alkalinity determines its nature. Natural water measures the pH to 7, above it is alkaline and below it the water is acidic. Rain water derives its acidity from dissolved CO_2, and measures pH lower than 7. The water on dissolution ionizes:

$$CO_2 + H_2O = H^+ (HCO_3)$$

Increase in acidity of water and, consequently, of increase in release of H^+ ions, increases the weathering rate of minerals and rocks.

SOLUTION

By and large natural solutions are rather dilute, although most solutes range from 1^0 A (Angstron) unit diameter of ions to 20^0 . A unit diameter of molecular structures (colloids) with consequent exception, like quartz which remain largely in the molecular state, are highly ionized and, therefore, reactive. At the border of the large particles, colloidal matter, the arrangement is rather very irregular facilitating free and more ionic mobility and in true solution, the regular arrangement of ions in larger, colloidal particles constantly changing their position to maintain a general state of random distribution. As such the non-ionized part slowly re-orients to assume into an orderly atomic arrangement of a stable crystal under formation when ionic concentration reaches the value of solubility product for a given temperature and some precipitation is about to occur.

The undersaturated rain somewhat acidic, 10^{-6} mole per liter of H-ion water in its downward journey through the atmosphere- lithosphere interphase comes in contact with reactive ions released from common minerals and dissolve them. Some ions are more soluble under slightly acidic conditions and other under basic conditions. For example, limestone and marble are essentially made of calcite ($CaCO_3$). The latter, *i.e.* $CaCO_3$ is susceptible to solution action in water that contains CO_2 and behaves as carbonic acid as

$$H_2O + CO_2 \text{----------}\rightarrow H_2\ CO_3$$

$$H_2CO_3 + CaCO_3 \text{-----------}\rightarrow Ca(HCO_3)_2$$

leading to the dissolution of limestone and marble. Bicarbonate is estimated almost 30 times more soluble than carbonate of calcium.

The dielectric constant of air is around 1 whereas that of water is of the order of 80 which obviously indicates that when water separates two oppositely charged plates, the tendency to neutralize or reduce the applied field acting between the plates is also 80 times that when air separates the plates. Dipole structured water molecule has two H-ions on the same side, thus the other side of the molecule exceeds with negative charge from the 0 atom, setting up an electrostatic field between the plates, causing partial orientation of water molecule and reduce the applied field. This electrical property of water is the key to the solution process.

The water molecule around the mineral enters the tunnel opening between the atoms in the lattice

(i) H^+ or OH^- ions of it attracts the loosely held ions present there

(ii) spread the lattice and separate it into units which become solute particles

causing

(a) total obliteration of crystal structure and

(b) consequent dissolution.

COLLOID FORMATION

Colloid (from the Greek word, *'kola',* meaning glue) particles are

(a) split off by ionization destruction, and

(b) whose dimensions are such as to possess at least partial lattice.

These particles are larger as compared to ions, in which molecular arrangements are such that many partially satisfied electrical bonds hold the surface atoms to the molecular structure. In such particles the outer borders are rimmed with several partially satisfied electrically charged atoms which constitute active 'handholds' for those H^+ and OH^- ions that tend to neutralize these charges and metal ions that tend to attach with such colloids. Generally, the ions are loosely held at colloid particles surface. In the lattice, they are spaced reasonably close, though still removable by other ions of stronger electrical field. They may as well penetrate into and occupy holes within the lattice. During the weathering of rocks silica, iron oxide and hydrous aluminium silicates (clay minerals) tend to form colloidal matter and often remain in soil due to their low solubility, slowly moving downward as the soil formation advances.

BASE OR CATION EXCHANGE

Base or cation exchange is a process in which

(i) the unsatisfied bonds of colloidal particles offer handholds to the ions oppositely charged and attach them on the outer surface of colloids—a form of *adsorption*

(ii) a strongly charged ion replaces ion of similar diameter but of lower electrical charge in a molecular or crystal structure. This is rather a stronger form of adsorption.

It is a reversible reaction the direction of which is controlled by the number of

(i) ions available for replacement and
(ii) replacing ions

In this mutual transfer of cation

(i) rich in an acqueous solution and
(ii) another rich or appreciably present in mineral, such as Ca^{++}, Mg^{++}, Na^{+} or K^{+} takes place. Such transfer rate depends on
    (a) the chemical activity and the abundance of various cations
    (b) acidity
    (c) temperature, and
    (d) other properties of solution.

Cation transfer in groundwater may expand or collapse mineral structure and release other chemical components. Such destruction of mineral grains eventually detaches to expose the neighbouring surrounding grains on which the same processes repeat. The principle of base exchange is used to

(i) improve soil fertility through fertilizer in solution adding the cation of the type required
(ii) 'soften' the 'hardwater' rich in calcium bicarbonate by replacing Na-ions for Ca-ions in the water.

CHELATION

Recent studies show the role of

(a) bacteria and
(b) organic activity, in chemical processes.

Chelation is derived from the word *chelate* meaning 'claw like' and refers to such bond and to co-ordination compound where in central metallic compound is attached to an organic molecule at two or more positions. It is a complex organic process in which metallic cations are trapped into ring structure of tightly imposed chemical bonds of hydrocarbon molecules. In their functions, several organic processes need metallic organic (metallo-organic) chelates. Plants chelate and absorb the metallic ions put into solution derived from minerals on hydrolysis.

The complex form of humic acid is produced under the stable conditions of

(i) moisture

(ii) temperature

(iii) length of microbiological activity and

(iv) neutral pH.

The type of weathering is influenced by the type of humic acid involved in a reaction.

The chelatory properties of *humus,* a decayed organic matter in or at the soil, which produces humic acid are plant-promoted. Rootlets of vascular plants maintain a negative charge on the surface and are surrounded by field of H^+ ions. The minerals migrate towards the roots along the gradient set by the two opposing electric fields. The organic compounds in the roots absorb and chelate the essential metallic cations put into solution derived from minerals on hydrolysis. The metabolic energy used to weather minerals underground in the process of chalation is derived directly from the sunlight falling on the plant leaves. Thus, the removal of metallic cations disrupts the mineral lattice. By this process, minerals many meters beneath the surface are weathered.

HYDROLYSIS

It is a process of weathering primarily limited to breaking down of complex silicates of hard igneous and metamorphic rocks into

(a) substances most stable on the earth's surface and

(b) of simple composition.

Water besides being

(i) a carrier of dissolved reactants,

(ii) is also one of the reactants

It occurs dissociated to some extent into its constituent ions as H^+ and OH^- and of these, the H^+ ion is most powerful in the chemical weathering of rocks. These ions which have rather small diameter characterized by strong electrical field moves, from the absorbed surface film, through the tunnel openings between O atoms of the crystal lattice, displaces the larger positive ion of relatively feebler electrical field and of the lattice, as an excess positive charge cannot exist, transforming the complex silicates into a variety resembling silicic acid. The degree of dissociation and, therefore, hydrogen ion concentration increase with

(a) increase of temperature (it doubles when the temperature rises from 0^0 to 30^0) and

(b) carbon-di-oxide saturation, (when the increase is 300 times or even more) :

$$CO_2 + H_2O \rightleftharpoons H_2\ CO_3 \rightleftharpoons H^+ + HCO_3^-$$

Carbon-di-oxide in water apart from dissolution, as above, also help in decomposition of minerals as illustrated by the decomposing weathering reaction of K-felspar (orthoclase) with carbonated waters

$$2KAlSi_3O_8 + 2H_2CO_3 + 9H_2O \text{--------}\rightarrow Al_2Si_2O_5(OH)_4 + 4H_4SiO_4 + 2K + 2HCO_3^-$$

orthoclase	carbonic acid	water	kaolinite-a clay mineral	silicic acid	bicarbonate

Thus, the frame structure of feldspars is transformed into layers of kaolinite and like minerals. In fact, plagioclases- Ca and Na feldspars hydrolyze even more readily in the acidic water

$$2NaAlSi_3O + 11H_2O + 2CO_2 \text{------}\rightarrow 2Na^+ + 2HCO_3 + Al_2Si_2(OH)_4 + 4H_4SiO_4$$

solid sodium plagioclase feldspar	liquid water	carbo- di- oxide gas	dissolved sodium (ions)	dissolved bi-carbonate ion	solid kalion clay	dissolved silicicacid

$$CaAl_2Si_2O_8 + 3H_2O + 2CO_2 \text{------}\rightarrow Ca^{++} + 2HCO^-_3 + Al_2Si_2(OH)_4$$

solid calcium plagioclase feldspar	liquid water	carbon di- oxide gas	dissolved calcium ion	dissolved bi-carbonate ion	solid kalion clay

On hydrolysis, feldspars yield three end products :

(i) clay mineral

(ii) silica in solution, and

(iii) a carbonate or bi-carbonate of K, Na or Ca in solution, which are transported by surface waters far from the place of their origin.

The clay minerals are indeed most stable and much of K-freed is

(a) incorporated into clay minerals other than kaolin or

(b) used by plants as essential nutrients

reducing its (K's) extent in river water or sea, than the theoretical value obtained from the abundance of feldspars in nature, in contrast to Na, Ca and bi-carbonate ions which are almost as common in river water. In the sea, the Na accumulates, but the Ca and bi-carbonate ions are absorbed by marine organisms in building their limely skeletal framework or shells. Under intensive and prolonged action, kaolinite also decomposes further to much more stabler substances produce 'aluminium hydroxide' (constituent of bauxite) and silica. The silica is also consumed by organisms, largely by the single-celled plants, diatoms. However, the bulk of silica, rather passes rapidly into colloidal state and is deposited locally as hydrous silica or opal. Thus, on the whole the process of complete hydrolysis of feldspars may be presented as follows :

			→ opal
Feldspars------	intermediate minerals------ chiefly hydro-mica	kaolin	→ bauxite
			→ soluble-carbonates and bi-carbonates of K, Na, and Ca

Basic and ultrabasic rocks are more susceptible to weathering. The ortho- and meta-silicates constituting ferro-magnesian minerals like olivine, hornblende and augite *etc.* decompose even more readily than alumosilicates.

$MgSiO_3$ +	$2CO_2$ +	$3H_2O$ ----------	Mg^{++} +	$2HCO_3^{--}$ +	H_4SiO_4
solid magnesium pyroxene	carbon-di-oxide gas	liquid water	dissolved magnesian ions	dissolved bi-carbonate ions	dissolved silicic acid

These ferromagnesian minerals constitute about 15% of the volume of crystalline rocks which are often capped with weathering products of varying stages of their decomposition. For instance, olivine weathers as below :

Mg_2SiO_4 +	$4H^+$ + $4OH^-$ ----------→	$2Mg^{++}$ + $4OH^-$ +	H_4SiO_4
olivine	ionized water molecules	ions in solutions	silicic acid

in which the mineral chemically breaks down in entirety, when excess water is available to transport the ions in solution.

In the course of decomposition of the primary minerals, silica is partly removed in solution and often re-deposited *in situ* in the form of hydrous silica or opal. With carbon-di-oxide, the water acts in the replacement of cations and formation of ferro-magnesin minerals to the point of formation of the ultimate residual products through number of intermediate products that may be presented as follows :

Ferromagnesian minerals (olivine, hornblende, augite) +water+carbon-di-oxide ----------→	intermediate minerals (ferruginous montmorrillonite, ferriuginous beidellite, chlorite and like-minerals)	→ Opal
		→ limonite
		→ soluble salts of Ca Mg and partly Fe

taking a course similar to that of alumosilicates, ending up with substances of most stable simple composition on the earth's surface.

HYDRATION

It is a process by which the entire water molucule is added on absorption to the mineral structure, such as the transformation of anhydrite to gypsum:

$$CaSO_4 + 2H_2O = CaSo_4 . 2H_2O$$

certain gypsum deposits are formed in this way.

Water of hydration causes minerals to expand/swell leading to local dislocations and mechanical disintegration on the enclosing rocks. This process, therefore, is regarded by some scientists as of physical weathering, whereas its chemical aspect produces, new relatively more stable mineral(s) in the new environment on the earth's surface. Of course, the water of hydration can be driven off on heating *i.e.* by re-altering the environment, but it is not possible in hydrolysis as water becomes an integral part of the new (clay) mineral. Many clay minerals are hydrated. Some of these minerals depending upon the availability of moisture alternately hydrate (say in rainy season) and dehydrate (in drier months), causing engineering concern on swelling and heaving the overlying rocks or soil and shrinking and cracking respectively.

Weathering of granite mainly involves the hydration and hydrolysis of feldspars, crumbling the mass into segregated grains of quartz and weathered feldspars, called *grus*.

"Water of crystallization" is held as water molucules in fixed number occupying space lattice openings in a complex mineral compound in the process of hydration as positive ions such as Fe^{++} tend to hold the oxygen side of the water molecules in a complex ion $Fe(OH_2)_6^{++}$. Such bonding of the water molecules may be rather strong and in some minerals is released only at above water boiling point temperature.

In the weathering of ferrous iron silicates, apparently the Fe^{++} ions leaves the lattice during solution and surrounds the water molecules to form such complex as Fe $(OH)_6^{++}$ or undergoes oxidation to form Fe^{++} which trips OH^- ions from water to form stable hydroxide $Fe(OH)_3$ which is characteristically transformed to hydrated form, limonite

$$Fe_2O_3 + nH_2O \text{ -------}\rightarrow Fe_2O_3.nH_2O$$

on aging in the dynamic weathering environment in which 'n' commonly has a value of two. Likewise, silica and alumina develop into hydrates, SiO_2. nH_2O and Al_2O_3.n H_2O respectively. Eventually, the anhydrous forms (Fe_2O_3 and SiO_2) are developed on further aging by loosing the water of crystallization.

CARBONATION

Carbonation, a common process of weathering, is the process of reaction between carbonic acid and minerals. Carbon-di-oxide is present in the atmosphere and all the waters in contact with the atmosphere do contain some dissolved carbon-di-oxide in it as the latter is readily soluble (in water). The dissolution of carbon-di-oxide in water increases with the increase in

(i) temperature, and

(ii) pressure.

Carbonic acid is a weak acid formed by the dissolution of carbon-di-oxide in water

$$CO_2 + H_2O \dashrightarrow H_2CO_3$$

Because of the presence of carbon-di-oxide in the atmosphere and its solubility characteristics in water, the rain precipitates, is always slightly acidic, with H^+ ion concentration 10^{-6} mole per litre. This rain water, while passing through the upper soil layer comes in contact with unusually carbon-di-oxide rich air trapped in the soil generated by the decaying plant tissue supported in the soil becomes more acidic. Carbonation of groundwater is of such organic origin.

Weathering of rocks like limestone or dolomite, involves the reaction of calcite or dolomite with carbon-di-oxide in solution involving ionic equilibria as below :

$$CaCO_3 \rightleftharpoons CaCO_3 \rightleftharpoons Ca^{++} + CO^-_3 \text{----------------- (a)}$$

Calcite--→ Calcite ion ion

solid in solution

$$H_2O + CO_2 \rightleftharpoons H_2CO_3 \rightleftharpoons H^+ + HCO^-_3$$

$$HCO^-_3 \rightleftharpoons H^+ + CO^-_3$$

In the above equilibria relation, the carbonate ion is depressed, but to maintain the solubility product of $CaCO_3$, more calcium is drawn into the solution, until in a closed system sufficient Ca- ions has accumulated to effectively satisfy the solubility product of equilibrium (a). Because, weathering is an open system, the ionic products on the chemical break off calcite leaves the site of reaction through the groundwater movement, the equilibria as at (a) above is not satisfied, with the effect continued solution of calcite as bi-carbonate goes on with the groundwater movement resulting into deeply grooved and furrowed limestone platforms or surfaces. The joints are opened into *grikes* with the intervening ribs or chutes of harsh undissolved bare rocks. Deep grooves like hemi-cylindrical pipes, termed *lapies* are developed by rain-water solution action along narrow vertical channels *e.g.* at joint intersections on steep or vertical limestone exposures. More acidic waters of sufficiently low pH, are reported to have developed on vertical surfaces of even basic rocks of composition other than the limestones.

Impurities like quartz and clays in the limestone practically remain undissolved. In fact, traces of iron minerals present in the limestone on weathering leaves insoluble iron hydroxides as residue giving rise to red-earths known as *terra-rossa.* These occur as red clays over grey or white limestone terrain on long accumulation of dispersed iron in it. As much as 3 to 6 meters of limestone on weathering produces about tens of centimeters thick residual soil, depending upon the extent of impurities the original rock had, at times lowering the topographic surface. Examples of such activity are offered by the

(a) limestone terrain around Mammoth Cave in Kentucky (USA) where the process of carbonation is estimated to have lowered the landscape by about 0.3 meter in 2000 years. and

(b) carbonate solution action in the Chalk Hills of England in lowering the land surface by 0.35 to 0.55 meter in 4,000 years.

Such solution action in carbonate region may also operate at depth particularly in regions of subterranean waters circulation, giving rise to limestone caverns developed along joints and bedding planes ultimately weathering the rocks into spongy network of caves and tubes, displaying a landscape of solution feature, called *karst topography* after the characteristic Karst region of Yugoslavia.

OXIDATION AND REDUCTION

The atmosphere contains about 21% of free oxygen and 30 to 35% of water (about 3% of the volume of air dissolved in water). Free oxygen, both of the atmosphere and of the air dissolved in water, is one of the most potent chemical reagent. The process of oxidation of minerals and rocks, therefore invloves, the reaction of the latter with oxygen, generally in the presence of moisture and is most intense in case of elements of different valencies and those appearing on the earth's surface in the protoxide from, which is particularly true of iron, a very widespread element. Low valency ferrous compounds pass into high valency ferric hydroxides on weathering. The Fe^{++} ion in the mineral loses electrons and is oxidized to ferric state. The latter when grouped with oxygen forms stable anhydrous iron oxide (Fe_2O_3) or gathers OH^{-} ions to constitute the relatively less stable hydroxide Fe $(OH)_3$.

Oxidation of iron silicate minerals seems to go on alongwith hydrolysis of the silicate structure, as in the case of biotite weathering

$(OH)_2K(MgFe)_3AlSiO_{10} + H^+ - e^- \text{--------}\rightarrow H_5AlSi_3O_{10} + Fe^{+++}\ Mg^{++} + K^+ +$

(biotite) (ion) (silicic acid like structure) (ions)

$Fe^{+++} + 3OH^- \rightleftharpoons Fe\ (OH)_3 \rightleftharpoons {}^1/_2\ Fe_2O_3.\ n\ H_2O$

(stable hydrate limonite)

$H_5AlSi_3O_{10} + OH \text{---------}\rightarrow (OH)_8Al_4O_{10} + SiO_2 + H^+$

(silicic acid like structure) rearrangement of lattice (clay mineral kaolin) (ion or colloid)

On availability of free oxygen of the air in the presence of moisture, the sulphides developed in reducing environment becomes unstable and gradually substituted to sulphates and Fe++ and proceed further to form carbonates and oxides schematically represented as --

$$FeS_2 + O_2 \text{--------}\rightarrow SO^-_4 + Fe^{+++} + e^{--}$$

$$F^{++} \rightleftharpoons Fe^{+++} + e^{--}$$

$$FeS_2 + nO_2 + nH_2O \rightleftharpoons FeSO_4 \text{--------}\rightarrow Fe_2(SO_4)_3 \text{------}\rightarrow Fe_2O_3.\ nH_2O$$

Ferrous sulphate ($FeSO_4$) formed at first on further oxidation is transformed into ferric sulphate [$Fe_2(SO_4)_3$] and subsequently into ferric hydroxide, limonite (Fe_2O_3. nH_2O), under the action of oxygen and water. Limonite is most stable iron mineral on the earth's surface condition.

Under certain special environmental condition where an excess of partially decaying vegetation is present

(a) in the upper part ('A' horizon) of the soil or

(b) in stagnant or foul smelling organic-rich mud or

(c) incorporated within a sediment during the deposition,

reduction of ferric iron to ferrous occurs, due to carbonaceous material. The anaerobic bacteria, a group of organisms that can thrive without free oxygen, reduce iron oxide completely to metallic iron, in an attempt to derive oxygen for their own metabolic processes. Such reaction deprives these soil of iron in as much as Fe^{++} is more soluble and, therefore, transported in solution.

Brown coloured 'iron-hats' derived from the oxidation of iron sulphides, occur in upper parts of many sulphide deposits. Ferrous iron bearing ferromagnesian minerals-meta and orthosilicate types (like augite, hornblende, olivine *etc.*) readily oxidize under surface conditions, producing a brown crust on the minerals. Likewise ferruginous minerals in sedimentary rocks, sands, sandstones, clays, marls *etc.* often show a brown or ochrous colouration, printing the oxidation process, the minerals have undergone and transforming the minerals into iron hydroxides. Brown to vermillion red colours in hydroxide, hydrate, or oxides is due to oxidation of the Fe^{++} and grey, green or blue tints characteristics of ferrous Fe^{++}.

The oxidation of minerals in chemical weathering is often controlled by the

(i) oxygenous surface - the upper boundary of distribution of free oxygen and

(ii) 'oxygen potential' *i.e.* oxygen content of a given environment, which is a function of

- (a) nature of rocks
- (b) their porosity
- (c) fracturing
- (d) depth of occurrence, and
- (e) intensity of circulation of waters and gases.

The oxygen penetrates to varying depths in the upper part of the lithosphere and may even coincide with the surface of the earth, such as in peat bogs or permafrost areas, but in some other instances may effect at a depth of 1 km or more. Generally, oxygen penetration under groundwater table is low, almost to the level of main drainage system through which the water reaches to the surface as springs. The oxygen capacity of the environment, referred to as *'oxygen potential'* is related to the quantitative content of active oxygen, varies from place to place which complicates the course of chemical weathering.

Aspects of chemical weathering are briefed in Table V. 21

BIOGENIC WEATHERING

Biogenic activities, a permanent geologic feature, play an appreciably significant role in the complex processes of chemical weathering like other exogenous agents leading to rock destruction and decomposition. *Biosphere* embracing a part of the troposphere, the world ocean and a certain part of lithosphere envelopes the earth and various organisms sustain in it. Maximum life activity of plants and animals is limited to the upper parts of the hydrosphere and the lithosphere-atmosphere interface with green plants which can live under the influence of the sun's rays and penetrate the lithosphere through their root-depths only, are limited to their activity zone. However, micro-organism, are numerous and of various kinds, operate to relatively greater depths, 10 to 20 meters. "*The role of organisms in chemical weathering is determined by their ability to assimilate various elements from the rock undergoing decomposition, and to evolve, in the course of their life activity*

Table V. 21 : SYNOPSIS OF CHEMICAL WEATHERING

(after Karl W. Butzer, 1976)

Process	← Operation	→	Effect/Result
Solution	a. in pure water	chlorides, bi-carbonates and some sulphates dissolved	salts removed in solution
	b. by dissolved acids	a. limestones and dolomites altered and dissolved except for impurities	carbonates and bases removed, leaving earthy residual of insoluble clays, oxides and quartz sand.
		b. bases produced during hydrolysis and altered and dissolved	Solutions may also lead to precipitation when saturated. Solute concentration in stream reflects rocks in basin and rates of percolation
Oxidation/ Reduction	minerals with iron or mangenese recombine with oxygen and water	silicates or carbonates of iron or mangenese, as well as ferrous form of iron, altered to oxides	a. various oxides that form part of an earthy residual product common forms of natural weathering via oxygen dissolved in water. Oxidation mostly in aerated zone above water table, also by bacterial action. b. reduction in waterlogged sites where red yellow oxides change to green/grey forms.
Carbonation			Reaction of carbontes, bi-carbonates with minerals often a step in weathering process including breakdown of feldspars.
Hydration	water molecules electrically attached to mineral grain	a. swelling leading to mechanical stresses within mineral matrix.	Abundance of CO_2 in soils aids process a) pre requisite of auxiliary process to hydrolysis carbonation,oxidation.
		b) formation of weaker or soluble hydrated minerals.	b) development of some clay minerals.
Chelation	chelating agents organic chemicals	extract metal ions form structures	assist leaching from humus

Hydrolysis	hydrogen ions from ionized water replace potassium, calcium, magnesium, or sodium ions in mineral combination	feldspars, ferromagnesian minerals and micas decompose.	fine residuals (including clay minerals hydroxides, carbonates and colloidal silicates as well as inert quartz sand and other unweathered minerals (released by rock deterioration), increased hydrogen ion concentration makes silica dissolve.

various chemical agents, including organic acids";

Penetration of plants through cracks and pores,

(i) on one hand mechanically disintegrate the rocks,

(ii) and on the other hand chemically decompose with the help of acids which are generated at the tips of their plant roots, by way of ionisation

and mechanically fragmented rock surfaces provide effective means of their etching followed by assimilation of various nutrients like K, Ca, SiO_2, Mg,Na,P,S,Al and Fe *etc.* elements of the rock. Micro-organisms are the first settlers on bare rock to initiate rock decomposition. A definite sequence of organic settlement is observed

(i) the first to take foothold are bacteria and blue-green algae - they destroy the rocks and prepare substratum of soil needed for the appearance and development of micro-flora-diatomaceous earth and the simplest fungi

(ii) followed by lithophyllic plants-lichens and mosses and

(iii) then development of higher plants and fauna, which carries the rock destruction a step further.

The strong bond between silica and alumina in the alumosilicate lattices is destroyed by lithophylic vegetation and later assimilated by the higher plants in the form of silica and alumina. This contention is supported by the chemical analysis of their ash contents. Organic humic acid is subsequently generated on the decay of vegetal matter which promotes further decomposition of silicates and alumo-silicates and remove the cations from them. Being very mobile colloids, the mobility of aluminium and trivalent iron increases in their presence and their precipitates obtained due to their own low mobility. These are carried in solution to a distant place, far from the site of decomposition. O_2 and CO_2 also evolved by the organic activity, play remarkably in the rock destruction. Free oxygen which owes its origin to autotropic vegetation in considerable amount, is a very active and potent chemical agent. Photosynthesis of organic matter is accomplished in the chlorophyll grains of green plants, and essentially consists in the development of carbohydrates out of CO_2 and water with the evolution of free oxygen that spreads for, both in air and water takes place on the land surface and the surface layers of oceanic waters, covered practically with green vegetation. Free oxygen also evolves by the reduction processes, the action of ultra-violet rays in the upper atmosphere on water vapour and possibly on carbon-di-oxide, the action of radioactive radiations on water *etc.*, but the evolution through photosynthesis is regarded as most significant.

Likewise, carbon-di-oxide in the atmosphere is derived from different sources, such as that obtained from

(i) the volcanic activity,

(ii) subterranean gas streams, as also from

(iii) the water of hydrosphere,

but again the living matters is its constant source

(i) through their combustion and decay and

(ii) through the animal respiratory action.

These facts constitute tremendous significance of organisms and their activities in the weathering of rocks by chemical processes.

A. Animals

1. Subsisting largely on plant materials : earthworms, insects (ants, beatles, grubs, *etc.*) multipedes, in rodents, snails, woodlice.

2. Largely predatory : centipeds, insects (ants, beetles *etc.*) moles, nematodes, protozoa, rootifers

B. Plants

1. Algae (green, blue-green, diatoms) actinomyees, bacteria (aerobic and anaerobic), fungi (mushrooms, yeasts, molds)

2. Roots of higher plants

TABLE V. 22 : SYNOPSIS OF BIOTIC WEATHERING
(source ?)

Process	Effects
Particles broken	Animal burrowing; growing root exerting pressure
Transfer, mixing	Animal movement, meterials moved to areas of different process
Simple chemical effects	Solution enhanced by respired carbon-di-oxide
Complex chemical effects	For instance chelation
Soil moisture effects	Roots, humus holding moisture; shading by plants
Ground temperature	Shade, fermentation increasing heat and activity rate
pH effect	Respiration and absorption by plants affecting pH
Erosional protection	Less exposure, lowered weathering rates.

FACTORS OF WEATHERING

Aforementioned discussions demonstrate that there are several factors—active and passive, geographical and geological that control weathering. These are

a. Passive factors -
   (i) Parent material
   (ii) Topography
b. Active factors -
   (iii) Climate
   (iv) Vegetation
   (v) Duration

Parent Material :- Composition and structures, such as jointing, bedding and porosity of the parent material have pronounced effect on weatherability. For instance, quartz does not readily respond to either physical or chemical processes, and thus rocks bearing them, such as sandstone and quartzite weather slowly in contrast to calcite bearing limestone or marble which is more susceptible to the solution action and thus to rapid weathering in wet chimates. Between the two, marble tends to be somewhat more resistant, presumably due to lower porosity. It is generally believed that minerals of metamorphic rocks are rather resistant to weathering such as garnet, talc and serpentine which are stabler than even quartz, chlorite may be comparable to the weatherability of potash feldspar (Tables V. 23 and Table V. 24)

Topography : From the viewpoint of weathering of rocks two topographic aspects

(a) altitude
(b) surface slope

contribute mechanical disintegration of rocks and is more dominant in high altitudes, say above timeberline, where the vegetation is sparse. Freezing and thawing is the characteristic in contrast to lower latitudes where chemical weathering under heavy forest cover may dominate (Table V. 25).

Steep slopes

(a) facilitate immediate runoff and
(b) do not support vegetation and

therefore, reduces the chemical weathering potential of rocks whereas the gentler slopes provide more water-rock interaction specially due to

(i) accumulation of already disintegrated rock fragments obtained from the higher reaches and
(ii) relatively denser vegetation cover over gentle slopes. Steep slopes or cliffs are barren.

Climate : Climate controls the nature of type dominance of weathering—physical or chemical

TABLE V. 23. ROCK CHARACTERISTICS AND WEATHERING
(after M. J. Bradshaw , A. J. Abutt and A.P., Gelsthrope 1979)

Rock type	Composition	Froms and lines of weakness
Unconsolidated sediments	Loose accumulations of sand, gravel and silt in alluvium, scree and moraines	Variable sorting and mineral composition, with few definite internal structures
Sedimentary rocks	Compacted and indurated sediments formed of clay minerals (shales), sands, calcium carbonate, pebbles with mixtures. These materials have been formed through weathering processes once.	Bedding and jointing, the latter emphasized during earth movements. Cross beddings form additional lines of weakness in some sedimentary rocks
Igneous rocks	Formed of a range of minerals including those highly susceptible to weathering. Extrusive lavas and intrustive bodies.	Lavas and intrusive sheets and massive bodies jointed; both cooling and unloading joints.
Metamorphic rocks	Altered forms of other groups may contain minerals more or less susceptible to weathering.	Often layered due to pressure cleavage in slates or to heat and pressure foliation (schistosity)

Table V. 24 : WEATHERING PRODUCTS OF COMMON MINERALS

(modified after M. J., Bradshaw A. J. Abutt and A. P. Gelsthrope , 1979 and E.C. Dapples, 1959, *combined*)

Mineral	Composition	Occurrence in	Weathering characteristics	Important Decomposition Products		
				Residual mineral products	Released ior.	Dissolved product
Quartz	SiO_2	Sandstones, granites, some metamorphics	Last mineral formed in cooling of the magma; no cleavage and high hardness renders resistance to physical weathering, stable at surface temperatures and, therefore, also resistant to chemical weathering. Quartz sets the standard against which other minerals are assessed.	Quartz	Same SiO_2 (colloidal)	Silicic acid
Feldspar group	Allumino-silicates	Igneous and metamorphic rocks, arkose sediments	Hardness close to quartz.. Disintegration and water penetration along well developed cleavages, facilitates physical and chemical weatherig. Ca-bearing feldspars weather much faster than Na and K-feldspars. Form clay mineral (kaolin) and sand.	Clay minerals quartz, calcite from Ca-feldspars	SiO_2(colloidal) +K, Nand Ca carbonates (soluble)	Silicic acid, potassium/ sodium calcium bicarabonates
Mica group	Aluminium silicates with sheet lattice	Common in pegmatites, granites and metamorphic rock also present in some sedi ments	Perfectly cleaved and alters to chlorite and clay minerals	(i) Muscovite tends to remain, eventually alter to clay minerals and quartz. (ii) Biotite alters to clay minerals, limonite and hematite	K, SiO_2(colloidal) K, Mg	

Olivine	Magnesium iron silicate	Basic and ultra basic igneous rocks	Formed at high temperatures; it is unstable at surface conditions and rapidly decomposes; highly irregular fractures facilitate weathering.	Clay minerals, limonite hematite	Mg, Fe^{++}	Silicic acid magnesium bi-carbonate
Pyroxene *e.g.* augite		Basic igneous rocks	Good cleavage helps in break-up. Other wise similar to olivine in many ways	Clay minerals, limonite, hematite.	Mg,Fe^{++}, Ca	Silicic acid magnesium bi-carbonates
Amphible e.g. hornblende		Igneous and metamorphic rocks	Relatively more resistant than pyroxenes and olivine, well developed cleavage aids susceptibility to weathering	Clay minerals limonite hematite	K,Ca, Mg Na	Silicic acid, magnesium bi- carbonate
Chlorite		Metamorphic rocks		Clay minerals	Mg, Fe^{++}	
Clay minerals				In tropics having high moisture content may alter to bauxite	SiO_2 (colloidal) Al_2O_5 nH_20	
Garnet				Garnet		
Gypsum				None for mineral, clay minerals and quartz from gypsiferous shale or sandstone,	Ca,SO_4	
Hematite, limonite pyrite and marcasite				Hematite, limonite Limonite, hematite		
Serpentine				Serpentine		
Talc				Talc		

Carbonates	Calcite; $CaCO_3$ Dolomite $MgCa(CO_3)_2$	Mainly sediments	Most soluble on weathering in presence of water and carbon-di- oxide	None for mineral, some quartz, clay minerals and hematite as impurities from limestone or dolomite rock.	Ca, Mg, HCO_3

TABLE V. 26 : WEATHERING ENVIRONMENT AND RELATED CHARACTERISTIC PROCESSES

(after M. J. Bradshaw, A. J.Abutt and A.P. Gelsthrope , 1979)

Weathering Environment	Characteristic Process
Humid tropical	High temperature and abundant rainfall helps pronounced chemical decomposition of rocks, accentuated further by decaying vegetation releasing acids for revitalizing chemical activity. Silica may as well dissolve, leading to the development of bauxite/ laterite. Tropical rivers charged more than ten times the solute load than temperate rivers, and thus, more polluted.
Seasonal tropics	Insolation weathering in summers, also salts brought near or at the surface by capillary action on evaporation impeding drainage and leaching out the salts in wet seasons. Laterite forms
Arid	With paucity of water, only little chemical processes operate, producing little clay or solution. Salt concentration by capillary action at the surface forming desert varnish, calcrete *etc*. more common
Humid temperate	Low evaporation consequently poor capillary action. Water available for some chemical reaction. Free thaw important in winters
Periglacial areas	Chemical activity limited to summers when water becomes available. Freezing through long winters and in springs and autumns free-thaw weathering.
Arctic	Mantle of snow, ice. Nunataks frost-river. When ice melts, carbonation beneath it.

(i) In areas of high rainfall evenly distributed through the year, the rocks are in contact with water, which promotes chemical weathering for longer time, decomposing the rocks deeper. This thickens the weathering zones. In the paucity of rains, water contact is limited, which results into surfacial weathering and development of only shallow weathering zone in depth. It's best example is limestone, which weathers rather rapidly in humid climate and which stands prominently as resistant ridges in arid regions.

(ii) Like rainfall, temperature also plays its role. Unequal distribution of freezing temperatures effect freezing and thawing on rock weathering differently from place to place (Table V. 26)

Vegetation : Dependent on the climate, vegetation plays a key role in the ground. Densely vegetated areas inhibit the runoff and promotes soaking of water for chemical weathering of rocks and soil development, whereas in regions of sparse vegetation, the runoff is more, reducing time of water rock interaction. The latter, in turn, has not much chemical effects on the rocks.

Duration : All other parameters being same, the degree or extent of weathering depends upon the process, operation time, those acted upon for longer time weather more and deeper than those suffered limited time operation. Such time differential operation leads to irregular surface exposures and several global phenomenon such as Pleistocene glaciation resulted in varying topographic expressions in different parts, due to uneven, both mass and time, ice action. An ancient weathering crust is characterized by

(i) great thickness

(ii) differentiation of eluvial formations and

(iii) their vertical zonality

in relation to the newer ones. The weathering crust on granites is in the follwing ascending order:

a. grus zone (of disintegrated granites occurring in pieces)

b. hydromicaceous zone

c. kaolinite zone

However, the weathering crust developed on ultrabasic rocks (say serpentinites) is composed, in contrast from bottom upwards of

(i) leached serpentinites

(ii) nontronitized serpentinites

(iii) nontronites and

(iv) red and brown striated ochres.

RATES OF WEATHERING/WEATHERABILITY OF SILICATE MINERALS, STABILITY OF MINERALS/WEATHERING SERIES

Field and laboratory investigations on minerals, especially the silicates indicate that different species have different susceptibility to chemical weathering and led to the following generalizations :

a. Quartz is the most resistant mineral.

b. Amongst feldspars, K-feldspars weather more slowly than the plagioclases (Na - and Ca - varieties) and within the latter the sodic ones are more resistant than those which are calcic.

c. Biotite weather faster than other dark coloured minerals, and muscovite is more resistant than biotite.

d. Amphiboles *e.g.* hornblende, are generally more resistant to weathering than the pyroxenes, such as augite.

e. Olivines seem to weather much faster than other ferromagnesian minerals.

The above when arranged according to the susceptibility of minerals to chemical weathering particularly of silicate composition which constitute the major bulk of the earth's crust, results into a pattern, similar to Bowen reaction series of crystallization from magma, in which the first formed minerals weathers first and the last formed is most resistant. The relative resistantce or stability of these minerals to decomposition is a function of prevailing

- (i) conditions under which they were formed, and
- (ii) surface condition to which they are exposed.

The difference between the above two conditions establishes a gradient which determines their weatherability, and thus, the rates of their weathering, and the minerals arranged in the sequence of weatherability led to the recognition of *weathering series.* The weathering order

- (a) may, thus, be considered to be the reverse of the Bowen's order of crystallization, and
- (b) with the variation that in weathering successive minerals formed do not react with one another as they do in a continuous reaction series.

Olivine	Calcic plagioclase	Rapid chemical weathering
Proxene		↑
Amphibole	Sodic plagioclase	↕
Biotite		
Muscovite	Potassic feldspar	↓
Quartz		Slow chemical weathering

Relatively silicates with higher metal oxides to silica ratio tend to decompose more rapidly than those of low ratio. The varying rates of chemical decomposition of these silicates may also be attributed to their structure:

- (i) unconnected silica tetrahedra, as in olivine, weathers faster and thus are uncommon in sedimentary rocks, and
- (ii) completely linked tetrahedra of which all the corners are shared, as in quartz which is extremely resistant to chemical weathering, and therefore quite common in sedimentary rocks.

Quartz is formed rather late in the crystallization series, more closer to the prevailing temperatures and pressures, and therefore, more resistant in relation to olivine which is formed at high temperature and pressure, and therefore, more easily weathered in response to sharp temperature and pressure gradient.

Chemical weathering develops a weathering profile in course of time, its depth indicating the progress of chemical weathering processes and is governed by

(i) climate,
(ii) geology and
(iii) relief *etc.*
(iv) the stage of landscape development or
(v) time lapse since the evolution of landforms.

The interface separating the weathered rock from the unweathered rock pavement is referred to as basal surface of weathering or *weathering front* which may or may not be always sharply defined. Often the transition of decomposed debris down to the undecomposed rock is gradual and the interface irregular. In this process, massive and impermeable parts of the rocks are left as unweathered boulders or *corestone.* However, deep weathering strips off the weathering residue, and decompose the rock-forming minerals to distinctive clay minerals. The latter aspect is of vital significance to the evolution of landscape morphology in different climatic settings.

The differing rates of weathering may be caused

(a) chiefly by the rock-compositional variations - the more resistant zones spur out as ridges, or ribs above the surfaces of easily weathered rocks,

(b) due to varying weathering intensities prevailing in different rock sections, and /or

(c) because of weathering operation time differential from one to another section in the same rock.

Weathering so accomplished in different sections of a rock is referred to as *differential weathering.*

STAGES AND ZONES OF WEATHERING

Two types of products are developed on weathering of rocks, *viz.*

(i) *mobile,* which are transported away, from the sites over varying distances, and

(ii) *residual,* products that remains on the site of rock destruction and/or decomposition, which constitute *eluvium.*

Depending upon the extent or stage of weathering, the composition of eluvium varies with each stage characterized by its own products. The stage of weathering character is most pronounced on the magmatic rocks. Four principal stages of weathering has been postulated by B. Polynov:

(a) fragmental
(b) siallitic calcitized
(c) acid siallitic, and
(d) allitic

Fragmental stage is the initiation of weathering in which the rocks are primarily subjected to physical weathering in areas of young mountainous terrains, in polar regions and in deserts. The rocks are fragmented into varying sizes, with no or very indistinct mineralogical changes and thus practically free from chemical weathering that brings about changes in mineralogical composition prevelent in warm and humid climate.

It is followed by the second, *siallitic calcitized* (the term 'siallite' is derived from elemental names Si and Al constituting mineral parts emerging at this stage) stage in which the alumo- silicates and silicates are decomposed by the chemical weathering with the leaching of cations. When alkali and alkali earth minerals pass into solution, the reaction environment becomes alkaline which results in the formation of intermediatae clay minerals of the montmorillonite group, and partly hydromicas also. Some $CaCO_3$, obtained on the interaction of Ca liberated in the weathering process with CO_2 of the soild and groundwater, is accumulated in this stage, as a result of which the eluvium is enriched with lime carbonate and therefore termed calcitized eluvium that often develops at the outcrops of magmatic and metamorphic rocks in a dry continental climate.

The third, *acid siallitic* stage is characterzed by

(i) further, leaching of cations

(ii) partial leaching of SiO_2 and

(iii) substitution of alkaline conditions by acid

In this stage, the earlier formed montmorillonite and hydromicas are destroyed and gives rise to new clayey minerals of kaolinite group. Most of the calcium is removed into solution from the mantle of waste and therefore the eluvium of the magmatic rocks are practically free from $CaCO_3$. Such processes of leaching of $CaCO_3$ and partly of SiO_2. are more rapid under humid and warm climatic conditions with abundant rainfall, which has intense leaching effect on the mantle of the waste. Under such condition, naturally second stage is rather short and poorly expressed.

In the *fourth, allitic* stage most simple compounds, very stable under the surface conditions, are formed on further decomposition of clayey minerals. Such simple compounds are hydroxides of Al, Fe and Si, which constitute typical colloidal minerals (constituents of bauxites, *etc.*). Accumulation of aluminium hydroxide in the weathering products of this stage, is most widespread and thickest in moist and humid regions of tropics and subtropics. Under such conditions, the eluvium in the presence of iron hydroxides is bright red coloured and thus termed as *laterite* (from the Latin word, *later* meaning bricks). Associated with ancient laterite zones of weathering are iron ore, and occasionally bauxite deposits.

Weathering processes are rather non-uniform and therefore, selective destroying less stable softer rocks more rapidly, than the more stable harder rocks, as a result of which the former has thicker eluvium than the latter. This is reflected in the uneven lower boundary of eluvium, which is expressed in the topography as well. For this reason, more dense and harder rocks constitute elevation, whereas the less stable rocks occupy depressions, so often exhibited by veins protruding over the country rocks.

CHARACTERISTIC WEATHERING FEATURES

Weathering develops varying types of features which are function of

(a) the nature of rocks and

(b) their disposition or attitude.

Badland : are seen in the areas washed by the river Yamuna, and in deforested terrain of Damodar valley. Badland, also called ravine land, are areas intricately traversed by irregular gullies often developed on argillacoues rocks in semi-arid regions due to selective removal of relatively softer constituents, by running water *e.g.* such as the terrain washed by the Yamuna in Agra, Mathura and Dholpur, in the foothill regions of the Siwalik rocks, and deforested terrain of Damodar valley.

Escarpment and Cuesta : Such features are developed on inclined beds of harder rocks overlying the relatively softer rock which are weathered more readily than the former giving rise to steep slope called *escarpment* on one side, with relatively gentler slope of harder rocks towards the dip, *dipslope*. These features are called *cuesta,* a term of Spanish origin.

Hogbacks : These are ridge-like features developed again on inclined beds, but characterized by high steep sides on two sides instead of one, as in the case of escarpment.

Mesa : Of Spanish origin, '*mesa*' means 'table'. It is developed as a table of the overlying relatively horizontal hard rock, on removal of softer underlying rocks that renders steep side slope.

Butte : it is a modified mesa with much reduced flat topped upper surface with terrace aspects.

PRODUCTS OF WEATHERING

Processes of weathering of rocks produce ill or poorly sorted loosend materials such as talus, residual boulders *etc.*

Scree : Scree is the name given to a mass/aggregate of varying sizes and shapes of rock fragments, collected at the toe or base of a cliff or steep slopes. These materials are derived on

(i) exfoliation,

(ii) freezing and thawing of water in cracks

breaking the host rock into pieces which fall, slide or roll down the cliff or steep edge or a slope until the slope angle at its bottom becomes rather too low to facilitate further movement. Such accumulations of materials, themselves also form slope of 25^0 to 40^0, with largest blocks generally towards the bottom. This is because the larger blocks due to their mass gain greater momentum and therefore move further and faster than smaller ones. These deposits may attain lengths and depths of tens or hundreds of meters in mountainous regions situated in severe climatic zones, characterized by large and rapid temperature-variation fluctuations. Their geometry generally assume a cone or fan at the mountain base, with apex upward indicating their direction of flow on mass failure along a plane of failure on the slope and therefore designated as *talus cone/fan* (refer p. 52 also). Occassionally the rock fragments may pile up parallel to the ridge on mass failure of the crest of the hill. Such parallel bodies with relátively planar (not necessarily smooth) surface are termed *talus terrace.*

Felsenmeere or Residual Boulders

Moderately sloping or flattish rock surfaces on high mountain summits or on subpolar islands are dotted with spheroidal boulders resulting from the exfoliation of massive rock blocks, attesting the rapidity of high altitude weathering processes. These boulders termed *felsenmeere* are believed to have developed by vigorous frost riving on massive rocks with widely spaced joints and fractures. Some of these boulders are formed *in situ,* but they could be glacially transported boulders as well. Similarly some may be stabilized in position, but others may move downslope by solifluction and concentrate into stream-like masses to develop *'stone-rivers'* (also refer p. 85).

Rock Mantle (Regolith) and Mantle of Waste

Rock mantle, as the name suggests, is the covering of bed rock of an area with loosend rock debris, which may have developed on the weathering of rocks below the cover, classified as *'residual mantle'* or may have been transported by geological agencies away from their source. In the latter case, the covering is termed *transported rock mantle.* The residual mantle can generally be traced down into unweathered country rock. The lithosphere's upper part composed of such loosened and friable materials obtained on weathering of different primary rocks is called the *mantle of waste* of which the thickness varying from a meter to hundred meters or even more, depends upon such factors as

(a) climatic condition

(b) topography

(c) the (i) nature
(ii) intensity and
(iii) duration
of the weathering processes, and

(d) the rock composition undergoing destruction.

In the tropics and the subtropics characterized by

(a) high temperatures
(b) abundant moisture and
(c) combined with luxuriant vegetation—a source of free oxygen and carbon-di-oxide

the mantle of waste developed generally over magmatic rocks is thickest. Opinions vary regarding the concept of locale of the rock mantle. Some workers visualize it as the upper part of the lithosphere above the ground water table level, a zone most suited for weathering because of

(i) intermittent moistening by percolating atmospheric water and
(ii) presence of carbon-di-oxide enriched air

circulating into the available spaces, cavities and pores, in the rocks.

Mantle of waste has assumed significance, both theoretically as also practically. Their study help in the reconstruction of

(i) weathering processes responsible for rock alteration
(ii) stages of weathering and
(iii) the prevailing climatic condition during which the weathering proceeded *etc.*

As to the applied aspects, the mantle of waste is

(a) a natural hoarder of economic minerals (iron, manganese, nickel, aluminium, lead-zinc ores, refractory clays *etc.*)
(b) the zone where placer deposits such as that of gold, or residual deposits are stored.
(c) the locale on which engineering projects, such as dams, canals *etc.* are generally developed.
(d) the zone where ground water occurs and new minerals-rocks appear,
(e) supporting a new natural body-the soil, so important for human existence.

Desert Varnish : Salts of iron, manganese and silica occur as an exceedingly thin film of hundredths of a millimeter to 1-2 mm. thick as brown or black and even on boulders and pebbles in deserts and semi-deserts and appears as a stone coat of brown or black colour varnish over the rocks or fragments. This coating is referred to as *desert varnish.* This film is more brilliant on harder and finer grained rocks than softer and coarser ones. The coating is invariably only on the upper sides and the lower sides are free from such cover. Such a cover reflects the sun rays like a mirror or like a tiny bluish sparklings from the rocks and boulders. The origin of desert varnish in not very clear. The two more often referred views are :

(i) The salts are derived from the rocks and the atmospheric dusts by the dew and coats the outer surface.

(ii) Sun's rays heat the rocks and helps in evaporating water containing dissolved salts in the pores. Insolation and moisture thus draws the dissolved salts to bring them to the surface by capillary action, which on evaporation deposits the salts on the pores and on the surface.

Weathering transforms the rocks through their constituent minerals. Such changes at times bring about manifestations of economic significance.

A. *Production of Soils* : Weathering processes ultimately lead to soil formation, vital for organic sustainment.

B. *Groundwater Storage* : Weathered zone acts as suitable receptacle for surface or subsurface water into its abundant pore space developed on loosening the hard and compact rock mass.

C. *Weathering of Ores and Secondary Enrichment* : Veins often bear gold, silver, copper, lead and zinc, pyrite (FeS_2) *etc*. Many changes are brought on weathering of such veins. Weathering processes may also lead to the concentration of the valuable minerals by

(i) leaving them where they are, and removing others by the subsurface ground water

(ii) detaching the minerals from the parent rock mass or veins and transporting the eroded fragment particles away from the site as placer deposits, *e.g.* gold in Kargil. platinum, diamonds and rubies also occur in placers.

(iii) selective dissolution, for example pyrite on oxidation yields iron sulphate and sulphuric acid. On further oxidation iron sulphates are converted into rusty stained insoluble iron hydroxides on the top as *gossan* or *iron hat*. The sulphuric acid may carry copper, present, in the presence of air and water.

(iv) dissolution or leaching of minerals in the oxidized zone and trickling down below under the influence of gravity to no-oxygen ground water level, where the dissolved compound is precipitated enriching the upper part of the primary sulphides with the addition of secondary sulphides. Thus:

FeS	$+ CuSO_4$	$= CuS$	$+$	$FeSO_4$	or
pyrrhotite	copper sulphate	copper sulphide		iron sulphate	
ZnS	$+ CuSO_4$	$= CuS$	$+$	$ZnSO_4$	
sphalerite	copper sulphate	copper sulphide		zinc sulphate	

This is the zone of *secondary/supergene enrichment.* where the primary ores are further concentrated by the addition of secondary sulphides brought in solution from the oxidation region down under the influence of gravity on surface weathering and precipitated in the no-oxygen reducing environment. The cycle repeats on the exposure of primary ores and thus depletes the copper content in the gossan. Just above the water table and below this, due to destruction of primary ore by oxygen and carbon-di-oxide, malachite or azurite (copper carbonates) is met. Next in sequence in a vertical section is present the water table fully saturating all the available space with water and thus delinking down below with free oxygen which facilitates the precipitation of copper ore, chalcocite (copper sulphide).

D. *Engineering significance of weathering* : Rocks and soils, on weathering loose their strength and become rather weak and therefore unsuitable to bear structures. Hence, it is always necessary to see

(a) the extent the rock or soil under consideration has been affected, and

(b) the possible weathering effects on the constructional materials in the areas and choose the material less susceptible to the operating processes.

CHAPTER VI

SOILS

The soil, which supports the vegetation, forests, grasslands and crops from which food is derived by the animal life of the land constitutes practically infinite simply thin layers, a few to several centimeters over the surface of the earth in relation to the dimensions of lithosphere and atmosphere at the contact of which it is situated. 'Pedology' is the formal name given to that branch of earth sciences which deals with the soils study though 'soil science' is more commonly in vogue. To a pedologist or soil scientist, soil may be characterized by distinct sub-layers or horizons. Soil layer is, in fact, a zone of complex processes—chemical, organic and physical in nature, constantly engaged to maintain its characteristic in tune with the locale, and is a function of

(a) climate
(b) topography
(c) vegetation and
(d) underlying rocks

and as such it is a dynamic layer capable of change or development of new characteristics in response to change in the above parameter(s).

Subsoil, underneath the soil layer is

(i) infertile
(ii) the original source rock from which it is derived and
(iii) the subsequent processes of weathering and/or transporation determines the transitory characters it acquires,

as after prolonged operation of the geologic/atmospheric processes it would also transform into soil, similar in character as the overlying true soil layer.

PHYSICAL AND CHEMICAL MAKE UP OF THE SOIL

Soil in nature, is composed of all the three states of matter-solid, liquid and gaseous and maintain some kind of balance, which when destablized deter plant growth.

The solid matter of the soil includes both organic and inorganic or minerals, the latter constituting major weight or bulk ranging in size from boulders down to minute clay particles of colloid size (less than over hundred thousandth of a centemeter across), derived from rock weathering. The organic constituents comprise both living and dead plant and animal life. Of these small tiny materials of partly decomposed plant matter of colloidal size, 'humus' share important chemical function with the inorganic partner colloidal mineral particles.

The soil liquid, aptly termed soil solution is charged with ions, of several chemical compounds in the soil water, including

(i) dissolved atmospheric gases

(ii) salts obtained on rock weathering and

(iii) produced by the bacterial action during vegetal decay.

The gaseous matter of the soil, almost similar to the free atmosphere but for much higher proportion of certain gases *e.g.* carbon-di-oxide liberated by the ongoing processes within the soil, occupies the pore spaces between the solid soil particles. This gaseous matter is normally close to the saturation point with water vapour.

SOIL COLOUR

Although, like colour of rocks or minerals, the colour of the soil also is not always very important property, but it is the property of the soil notable immediately on seeing and unlike minerals and often in rocks, it has been utilized in naming soil groups. The soil colour gives an idea of its

(i) composition as also

(ii) history.

Variation of colour, for instance depends upon the humus content in it and varies light to black on increase in its proportion. Humus content in the soil is a function of prevailing climate on which the vegetation luxuriance and bacterial action intensity depends. Accordingly, in humid regions the soil colour is often dark brown and grey in arid regions in response to decrease in vegetal matter from which humus is derived. Red and yellow colour stains are due to minor iron oxide and hydroxides produced on the oxidation of ferromagnesian minerals developed on drainage of which enables the atmospheric oxygen to enter the available pore space wet with moisture. In areas of poor drainage, such as in bogs, the soil colour, instead, is greyish or bluish, due to non-availability of oxygen, leaving iron in reduced state. To standardize the soil colour studies, it is preferred to compare it with the chips of defineable colours of the Munsell system of which there are three basic components : *Hue, value* and *chroma* :

Hue is the quality of *spectral colours* that distinguishes as far instance, red from yellow *etc.*

Value indicates the *colour brilliance* expressed as relative lightness in comparision to absolute whiteness and varies from white to black as light grey, dark grey.

Chroma refers to *gradations of purity* of colour or the apparent degree of departure from neutral greys or white, such as greyish brown or brownish grey.

In the Munsell chart, these three properties are codified with numbers arranged in the order hue, value and chroma. For instance in the Munsell notation 8YR 5/3,

8 YR	is the hue
5	is the value
3	is the chroma

In the Munsell system, the chip of all the colours of a given hue are mounted on a single sheet of cardboard in a book arranged in order of increasing or decreasing wavelength of the dominant colour. Each of these chips is referred to by a numerical system as above. A circular hole punched between two adjoining and similar colour chips permits to compare the soil sample colour by putting the soil ped next to the chips and the sample is referred to the number corresponding to the best match chip or by the number obtained on interpolation between two chips.

SOIL TEXTURE

The description of the particle sizes constituting the mineral matter is referred to as the soil textures. Particles may be described in decreasing order of size as

	fine gravel -	grains 1-2 mm. in diameter
	sand	grains 1-0.05 mm. in diameter
	silt -	grains 0.05-0.005 mm. in diameter
and	clay -	particles of < 0.005 mm. in diameter

Their varying proportions produce different types of textures such as

Sand - consisting of 80% or more of sand; 20% or less of silt and/or clay

Sandy loam - consisting of 20-25% of silt and clay; the remaining balance of sand

Loam - composed of 20% or less of clay; 30-50% of silt, 30-50% sand (if silt predominates the soil it is termed silt loam; if clay predominant, it is a clay loam)

Clay - containing more than 30% of clay; less than 70% of large size constituents.

Soil textures control

(i) the drainage rate of water through it and

(ii) the amount of water it can retain which may be used by the plants.

Sands allow quick drainage with relatively lesser amount in storage whereas clay impede water percolation and thus can store large quantities (Table VI. 26). Thus, a combination of sand and clay *i.e.* silt is better suited for cultivation.

TABLE VI. 26 : QUALITATIVE PHYSICAL PROPERTIES OF MAJOR SIZE GRADES IN AGGREGATES

(source ?)

	Property	*Pebbles*	*Sand*	*Silt*	*Clay*
(i)	Volume change from dry to wet condition	None	None	Slight	Large (marked swelling)
(ii)	Strength under tension :				
	when wet	Low	Low	Intermediate	High
	when dry	Low	Lower than when wet	Higher than when wet	Very high
(iii)	Volume change under compression :				
	when wet	Very low	Very low unless particles separated by water when change is high	Intermediate unless saturated with water when change is high	Very high
	when dry	Very low	Very high	Low	Intermediate to low
(iv)	Plasticity :				
	when wet	None	Slight	Intermediate	Very high
	when dry	None	None	None	None partial cementation
(v)	Porosity	Very high	High	High	Very high
(vi)	Permeability	Very high	High	Intermediate to low	Very high
(vii)	Size of pore openings	Large	Intermediate	Capillary	Subcapillary
(viii)	Shape of fragments	Rounded	Rounded to angular	Angular	Thin tabular
(ix)	Water interaction	Very low	Low	High	Very high
(x)	Water retention	Very low	Low	High	Very high
(xi)	Base exchange capacity	Absent	Absent	Low	High

DENSITY

Soil density is a function of

(a) its structure and

(b) mineralogical composition

and can be determined as *Particle density* or *Bulk density.* Weight per unit volume of solid matter constituting the soil is referred to as *particle density* whereas in *bulk density* the volume also includes that of the voids besides the particulate materials. Different types of soils do not show marked variation in particle density but the bulk density may vary considerably.

VOID-RATIO

It is the ratio of the volume of 'open' or 'void' spaces per unit volume of solid particles in a given soil mass. It varies with the soil density and it is seldom that component grains in the soils are jam-packed.

PERMEABILITY

It is the faculty of soil to allow water to pass through it and has a direct influence in such soil properties which have a bearing upon in moisture content. It is determined with the help of permeameters and its coefficient *i.e. coefficient of permeability* is measured in terms of Darcy. The Darcy's law constitute the basis of most permeability studies. The coefficient of permeability

(a) shows great variations for different types of soils and also for similar type of soils under different conditions, and

(b) is considerably influenced by the factors like

  (i) size

  (ii) shape and

  (iii) arrangement.

of the component grains *etc.* and therefore often the average of repeated tests is considered while preparing the design of any structure as earth dams, dykes and embankments *etc.*

SHEARING STRENGTH

Shearing strength is the resistance of soils to the shearing forces. It is influenced by its such characters as

(i) particulate size and shape,

(ii) structure,

(iii) density, and

(iv) water content.

SOIL COMPRESSIBILITY

Soils often yield on loading pressure *(e.g.* in foundations of structures) and this settlement of soils is ascribed to the consequent change in the volume. Some soils are more compressible than others and the controlling factors for such variations are:

(a) the water content

(b) permeability and

(c) texture of soils.

Soils with fine grained textures (*e.g.* clays) show relatively higher compressibility even with comparatively very low rate of consolidation whereas compact textured (*e.g.* dense sands) granular soils are least compressible with a rapid rate of consolidation. The *bearing capacity* of a soil is defined as the load that the soil can withstand without yielding to detrimental settlements.

SLAKING TEST

It is a test to investigate the effects of alternate wetting and drying of soils or rocks. Repeated saturation and drying weakens them and would require some pre-treatment before being used for foundation or abutments.

SOIL COLLOIDS

Colloidal particles are electrically charged tiny bits of mineral or organic matter that cannot be seen even through the ordinary optical microscope. They are important in soil formation - attract and hold ions, the dissolved parts of chemical compounds. On dissolution in soil water certain common salts liberate Ca^-, Na^-, and K^- ions, collectively referred to as 'bases' needed by the plants which they obtain from the soil colloids through 'base exchange' process for their own growth. Soils rich in such ions are referred to as *'alkaline'*, whereas those with larger hydrogen ions concentration in contrast termed *'acidic'*.

Colloids remain suspended in water practically indefinitely and hold surrounding water molecule layers, which do not evaporate under ordinary temperatures and in doing so renders the soil moist and soft. Colloid-rich soils swell when moist and shrink when dried. It is due to the fluctuations in the holding of water by the colloids.

These colloidal matter in soil can be in two states *viz.*

(a) *dispersed state,* wherein the individual particles are freely separated from one another. These particles often occupy the small pore spaces available between larger soil particles and tend to impede water drainage through the soil, which in turn increase, the soil stickiness and tillage resistance.

(b) *flocculated state,* wherein the individual particles, are tightly held together, with large pore spaces between the flocs facilitating free movement of water and air in the soil. In contrast to the dispersed state of colloids in its flocculated state is not so sticky and facilitate tillage.

However, chemical properties of soil governs the state of colloids-dispersed or flocculated.

SOIL WATER

Water occurs in soils in different ways. Water on precipitation moves down through the soil under the action of gravity, until all the available space or pores are filled with it. The upper most surface of such water saturated zone is termed the *water table* and the saturated area below constitute the *zone of saturation.* The water below in the zone of saturation is the ground, underground or subsurface water. Gravity cannot always transmit all the precipitate to the zone of saturation. Though, a sizeable amount percolates down since water envelopes the soil grains and form a film around which clings the particles presumably with force stronger than the force of gravity. Such water is termed *pellicular water.*

Just above the water table, the soils draw the water by 'capillary tension', acting upward against the force of gravity by a force from down below. The space between soil particulates may unite or branch off frequently. So drawn water is referred to as *capillary water.* The capillary rise is inversely proportional to the fine pore-space diameter. The upper end of such water surface has a *'capillary film'* or skin of surface water molecules that clings to one another, tending to creep upward along the walls of the pores, thereby drawing the water column upward. This zone of capillary water, lies above the zone of saturation. It measures a fraction of an inch in gravels and coarser sands and may be several feet in fine silt as smaller soil particles provide smaller diameters of opening between the particles consequently increased capillary action. This water, held in the *capillary moisture zone* in which the pore-spaces are filled with water, is available to the plants and on evaporation of water from its upper fringes, water immediately below moves upward in an effort to catch up with the lost position.

Above the capillary moisture zone, the moisture occurs as connected film of capillary water, but all the space or pores are not water-filled. In such a situation the soil air moves freely through the connected openings between soil grains and water-films. This is the *continuous-film zone* in which water is left behind after the soil drainage between heavy rains. At the tangential contact of the soil particles in the uppermost moisture zone of the soil, small fragments of capillary moisture films adhere to the soil grains. This zone is referred to as *discrete film zone,* as the films are detached with each other. Such tiny films bind together the silt-size soil particles and offer cohesion to the soil. In dry season due to water evaporation, the film shrinks and the soil becomes stronger and harder because their adhesion force becomes stronger as the individual films become smaller.

The ability of the soil to absorb water is conditioned by several factors:

(i) *Vegetation cover* - Woodland and tall grasses will slow down rain impact by interception, so that water drips off the leaves or runs down stems and tree trunks. A rooted sod is usually well aerated and rich in humus, all of which soak up and hold considerable quantities of water. By contrast, base cultivated soil that has not been maintained or rock surfaces have a low infiltration capacity.

(ii) *Soil moisture* - Whereas a moist soil initially absorbs water rapidly, a dry, baked soil with little organic matter will repel water. However, moist soil is soon saturated with water, while dry soil can potentially absorb a great deal, although at a slower rate.

(iii) *Soil texture and structure* - Porous and permeable soils, with a high proportion of sand or gravel, absorb water far more rapidly than dense, clayey soils. The loose structure of sandy soils and the crumb structure of agricultural or woodland soils also favour greater absorption.

(iv) *Parent material* - The nature of the subsurface lithology plays an obvious role for infiltrating soil water. Unconsolidated sands or gravels are optimal, while clays and most forms of compact bed rock are unfavourable, depending on their permeability.

(v) *Time* - Infiltration capacity in most soils rapidly decreases with time, even in moderately permeable material. Once the subsoil is saturated with water, subsequent percolation is limited to lateral seepage (through flow) in the topsoil and slow, downward flow of free water to the groundwater zone.

Hygroscopic moisture occurs in the upper most soil zone comprising soil colloids which holds the water molecules by attracting its unlike electrical charges to the surfaces of the colloidal particles. Both hygroscopic and discrete capillary films adhering to relatively larger particulates are not completely lost even under fairly strong evaporation. Such soil-moisture may be driven out only artificially at higher temperatures. It is this property, which enables silt and clay retain considerable amount of moisture even in adverse conditions of natural temperature onslaught than the coarse-grained sandy soils.

Soil water is chemically a complex solution of different dissolved compounds which exist as ions. It acts as a weak acid, carbonic on dissolution of atmospheric carbon-di-oxide (carbonic acid) and on decay of vegetal matter is abundant, is relatively more acidic in contrast to those of dry climates where the soil is somewhat saline (rich in salts) or alkaline.

WEATHERING STAGE AND SOILS

The minerals in soils - amorphous or crystalline are derivatives of

(i) physical
(ii) geochemical and
(iii) pedochemical

weathering:

(a) may be *primary,* if directly inherited from the parent rock itself, such as constitute the bulk of soils and

(b) *secondary,* when produced by chemical weathering of silicates-feldspars, micas, amphiboles and pyroxenes whence trace elements are liberated. The secondary minerals, often occur in clayey fractions of soils, include aluminium and iron hydroxides, carbonates and aluminous silicates. They are simpler in chemical composition. Crystalline aluminum-silicates are structurally layered as in chlorite, halloysite, kaolinite, montmorillonite (smectite) and vermiculite. These clay minerals are best identified on the bases of X-ray differentiation pattern. The clay group secondary minerals population distribution varies

 (i) in different soils and
 (ii) with depth

 and under well-oxidized environment on leaching undergo differential

 (i) susceptibility to decomposition
 (ii) transportation and
 (iii) disappearance

 from soil profile which provides a basis of arrangement of these minerals in order of increasing resistance in chemical weathering.

Three stages, with characteristic minerals have been identified on the basis of minerals present in the fine sand and silty fraction of the soil ; early weathering stage (near the top), intermediate weathering stage and advanced weathering stage (near the bottom).

Both, the initial and advanced stage of soil weathering show rather poor fertility in relation to the intermediate weathering stage (Table VI. 27).

TABLE VI. 27 : MINERALS IN SOILS - WEATHERING STAGES
(source ?)

Weathering Stage	Representative Minerals	Climatic Zone and Fertility
	Early Weathering stage	
1.	Gypsum (also, halite, sodium nitrate)	These are typically desert soil
2.	Calcite (also, dolomite, apatite)	characterized by low chemical weathering
3.	Olivine-hornblende (also pyroxenes)	owing to limited water availability as in
4.	Biotite (also glauconite, nontronite)	arid conditions.
5.	Albite (also anorthite, microcline, orthoclase)	
	Intermediate Weathering Stage	
6.	Quartz	Wheat and corn belts and are developed
7.	Muscovite (also illite, sericite)	in temperate regions.
8.	2:1 layer silicates (including vermiculite, expanded hydrous mica)	
9.	Montmorillonite	
	Advanced Weathering Stage	
10.	Kaolinite, allophane	Typically intensely weathered soils of
11.	Gibbsite, boehmite	the warm and humid equatorial regions
12.	Hematite (also goethite, limonite)	dominated by these minerals. Generally
13.	Anatase, rutile, zircon	such soils are infertile.

SOIL PROFILE

In a vertical section many soils display/exhibit distinctive layers, differing in texture, structure, consistency and in colour. The arrangement and properties of the layers limits the characteristic horizons of the *soil profile* which helps in distinguishing and defining their nature. These soil horizons are best developed in cold humid climatic zones of middle and high latitudes which demonstrate complete ideal soil profile with practically all the horizons *viz.* A, B and C and the underlying unweathered country-bed rock, sometimes referred to as D horizon, although *stricto-senso* it does not constitute the soil.

The topmost horizon is designated A-horizon from which soluble substances and colloids are washed and transported downward. This process of removal of material is termed *eluvial* or leaching and leached out material is called *eluvium.* Because of removal, the eluvium is often light coloured, porous and open textured. Below A, is the B-horizon in which the material removed from A-horizon is brought and accumulated. The process of leaching in contrast to leaching out from A to B horizon is referred to as *illuvial* and the material comprising the transported in and the *insitu* weathered products together is termed *illuvium.* The illuvium of B-horizon is generally darker than eluvium of A-horizon and relatively less porous, rather somewhat dense and compact. A- and B- horizons together constitute the *true* soil or *solum* also referred to as *topsoil* in contrast to C-horizon consisting of weathered fragments of the parent rock called *subsoil* or *gley* layer. This horizon indicates the beginning of rock transformation into soil. Gleying is soil mottling caused by partial oxidation and reduction of its constituent ferric iron compounds due to conditions of intermittent water saturation. This process is also called *gleization.* Obviously the topsoil is more fertile than the subsoil. These A, B and C horizons are further classified (reference Table VI. 28). Underneath C is the D-horizon which is the unweathered bed rock.

FORMATION OF SOILS

The soil formation is regarded as the function of the following five major factors grouped under two head : active and passive.

I Active

A. *Climate* - particularly (i) temperature and (ii) precipitation—amount and kind spread period.

B. *Soil biota* - including both (i) vegetative cover and (ii) animal life within the soils.

The remains of green chlorophyll containing plants-angiospermous, gymnospermous and pteridospermaphytes constitute the bulk of the organic matter in soil. Very important are the remains of the root system, especially in the zone of development of grassy vegetation. The plant debris in soils interaction enters into a cyclic nature, in as much as on one hand,

(p) the plants extract various mineral substances and water, from soils *i.e.* essential nutritive elements for their own sustenance

and on the other hand

(q) when they wither away their remains contribute towards the accumulation of organic matter in the soil.

So in nature soil witnesses continuous movement of substances from plants. The accumulated mass of the dead organic matter undergoes complex biochemical reworking and decomposes.

Taking part in decomposition of organic matter are various soil organisms. Various burrowing animals-

(i) losen the soil

(ii) mix mineral particles with altered vegetal remains and

TABLE VI. 28 : GENERALIZED COMPOSITE SOIL PROFILE AND RELATED PROCESSES

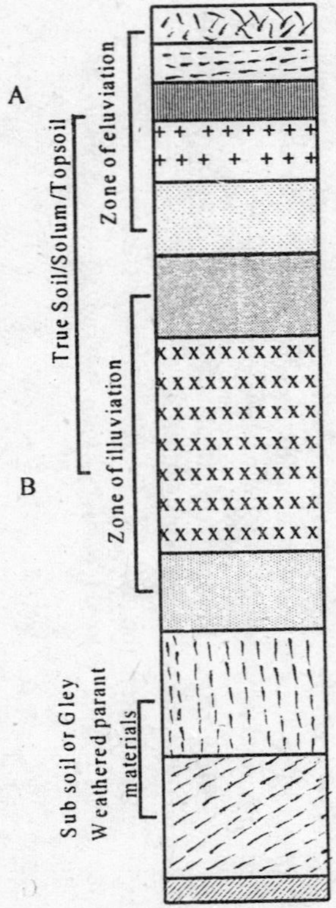

A
B
True Soil/Solum/Topsoil
Zone of eluviation
Zone of illuviation
Sub soil or Gley
Weathered parant materials

	Materials	Physical Processes	Chemical Processes
A_{00}	Loose leaves and organic debris		
A_0	Partly decomposed organic debris		Bacterial decomposition of vegetation
	Dark colour, rich in organic matter	Accumulation of humus	Solution of silica and
A_1	Light colour, zone of leaching	Downward movement	Removal of colloids
A_2	of colloids and organic matter and breakdown of clay minerals	of water removing colloids	
A_3	Transitional to B, but more like A than B		
B_1	Transitional to A, but more like B than A, finely divided humus silt and clay. Calcium carbonate or gypsum	Slow downward movement of water	Precipitation of colloidal Fe and Al hydroxides
B_2	Deeper coloured zone of maximum accumulation of colloids, clay Fe/Al oxide and hydroxide and some humus (more or less absent) prismatic blocky soil structure	Addition to colloids	Decomposition of most silicate minerals
	In dry climates $CaCO_3$ in nodules		In dry climates precipation of $CaCO_3$ from evaporation of water
B_3	Transitional to C, few pebble		
Upper	Accumulation of Calcium carbonate, No roots; altered bed rock pebbles in sandy or clayey matrix mottled soil material, primatic structure	Faster downward movement of water Than in B-horizon	Solution of carbonates, hydrolysis and oxidation of most silicate
Lower	Broken and partly/slightly altered bed rock and oxidized soil material	Downward and lateral movement of water	Oxidation and partial hydrolysis of silicate minerals
	Unaltered underlying parent rock		

(iii) make it more open to the action of air and water through their mechanical work.

This facilitates and intensifies chemical process and accelerates the decomposition of organic matter. Other biochemical changes are produced by animals which feed on vegetal remains. The work of various micro-organisms (bacteria and fungi), the quantity of which in

(i) the air
(ii) the water and
(iii) the soil

is most important in this respect : They exist under most diverse conditions.

Decomposition of organic matter gives rise to compounds of two types:

(p) simple compunds of CO_2, H_2O, NH_3 and salts of nitric, sulphuric and other acids (in the course of mineralization), and

(q) new more stable brown or black coloured organic compounds.

This relatively stable, very complex sum of organic compounds, accumulated in the soil on decomposition of primary organic matter by micro-organisms is called *humus* (from the Latin *gumus*, soil). Two types of humus may be distinguished

(x) wild humus (i) dark in colour
(ii) rich in humic acid and
(iii) serves to stablize clay and

(y) raw humus (i) more red in colour
(ii) less basic
(iii) rich in fulvic acids and
(iv) favours clay disperson

It is the principal factor of soil fertility

(a) Rapid and complete decomposition of organic remains results in the formation of only the simplest products of decay (complete mineralization).

(b) The humification, on the other hand, requires slow and incomplete decomposition of organic matter.

The decomposition intensity depends on

(i) temperature
(ii) humidity
(iii) air penetration
(iv) the chemical nature of the environment and other factors. Humus is composed of 85 to 90 percent of substances of high molecular

compounds specific to humus alkali in soluble humic and fulvic (crenic and apocrenic) acids. The remainder is composed of proteins, carbohydrates, organic acids, fats, wax resins and other organic compounds abundant in plant and animal tissues and in the products of their decay together constitutes the remaining part.

C. *Time* - The extent or length of time the soil forming pedologic process have been operating.

The quality and quantity of humus depends on the
(a) vegetation from which it has been formed, and
(b) the direction of the biochemical processes involved.

The role of climate through
(i) atmospheric precipitation and
(ii) temperature

determines the intensity of weathering that produces the mineral bases of soil on which some particular type of plant cover develops.

Water, too, plays a great role particularly in the formation of vertical soil profile by dissolving and moving downwards some soil substances together. They are redeposited either through

- crystallization or
- coagulation

down below where conditions are favourable, enriching the horizon. Contrarily upward movement may also take place through capillaries depleting the soil horizons of dissolved substances. Several genetic horizons may be developed in the vertical soil profile on the downward movement of substances. These horizons are expressed differently in different kinds of soil, depending on the stage of development of the soil forming process.

II. Passive

D. *Parent material* - including
(i) chemical and mineralogic make up and
(ii) texture and structure of material

E. *Topography* - especially as it affects, both, external and internal drainage.

Of these at least the parent material, topography and time are geologic in nature. Although the climate is more geographical in character but is considerably influenced by the topographic expression of the region.

It is commonly believed, as it often does, that the soils possess some such characteristics which is reflective of the parent materials from the weathering and decomposition of which they are derived. The residual soils are accordingly classified *e.g.* granitic soils, sandy soils *etc.* Likewise transported soils are classified on the bases of processes of their derivation, such as glacial, alluvial, marine, lacustrine, colluvial soils *etc.* Genetically such classification appears to be good, but for the fact that soils are considerably modified by other factors. It has been established that "*regardless of what the parent material was, with similar topographic, climatic and vegetative conditions and after equal intervals the soils produced were essentially the same*". This is, though, not strictly true, but is not too far from the truth than many geoscientists realize.

Time or duration affect the

(i) thickness and

(ii) degree of maturity

of a soil profile. A relatively young, not subjected to soil forming processes for long time, is immature and the various layers or horizons of its soil profile are rather poor, indistinct or even lacking in contrast to mature soil which exhibits rather well developed distinct horizons. Similarly, longer operation of soil forming processes enables deeper penetration in the parent material resulting in increased thickness of soil column and thus helps in the estimation of the ages of the soils.

Topography affects

(i) the soil profile in as much as a steep slope does not support mature soil profile as the transported weathered material under the action of gravity limits the operation of soil forming processes and often truncates soil profiles in which the upper layer (s) or horizon (s) if ever present, is (are) removed. Also it influences

(ii) the water table position and thus

(iii) the depth of penetration of the soil forming processes as well.

The relief largely controls the water circulation through the soil and parent material and therefore, the development of soil profile attests the good or poor drainage conditions as may be the case.

Current interpretation and classification of soils is primarily based on the fact that

(i) everywhere soils are the product of certain chemical and physical processes

(ii) the soils differ from place to place, particularly in their stage of development, varying in responses to climate, soil biota, parent material, duration(time) and topography.

(iii) the original differences largely on account of different parentage gradually dwindles and becomes obscure over which the affects of climate, which assumes dominance in differentiating soils, is printed afresh that helps soil differentiation on its newly acquired charateristics. Climate acts directly, through

(i) temperature and

(ii) rainfall variations

and indirectly through

(iii) its influence upon the density (around) and

(iv) kind of vegetation.

In similar climatic areas, the mature and old soils are strikingly similar and classifiable in soil groups developed under similar climatic conditions. So grouped soils are referred to as *'zonal soils'* . Within the broad climatic influence, pockets of zonal soil may show some minor variations as to the effect(s) of local conditions. Such soils are classed as *'intrazonal soils' e.g.* soils of bog areas and of alkali flats. Soils may locally lack well defined profiles and are termed *azonal* soils such as those of recent developments, *e.g.* stream alluvium, dry sands and colluvium *etc.*

SOIL SERIES

Consists of soil members, individually referred to as soil types derived from subordinate lithological variation in the parent material and topographic conditions possessing the same colour, structure and drainage conditions, but may differ in the texture of the surface soil and upper subsoil

SOIL CATENA

It consists of a group of soils of a region derived from similar parent material but develop dissimilar soil profile for reasons of differences in

(i) topographic and

(ii) prevailing drainage conditions

Varying degrees of slope—slight, moderate or steep, flat uplands, may be with shallow to deep depressions are the topographic settings that affects the development and characters of soil profile because of consequent variations in surface and drainage conditions. Thus, a soil catena is an expression of both

(i) lithology and

(ii) topographic implications

of interest to geomorphologists and geologists.

COMPLEX OR POLYGENETIC SOILS

Some vertical sections of soil profiles show soils that do not match with the prevailing soil-forming conditions overlain by those types, which show its development to the existing topographic and climatic conditions. Obviously the two sets of soil lower older and the upper younger, are formed under different conditions produced in response to

(i) climatic change with attendant variation in vegetative cover and

(ii) change in topographic and consequent drainage conditions

Soils which show such genetic complexities are referred to as *polygenetic* or *complex soils*. Since the recession of last ice sheet world experienced significant climatic variation and produced polygenetic soils.

Dissection by streams are affected by the topographic changes and may appreciably result in modifications in external as also internal drainage conditions under which a soil profile is developed. Complex soil profiles are noticed in response to improved drainage conditions causing dissection of till plain and show *siltil* characteristics above weathered till called *gumbotil* (derived from the popular name *gumbo*) profiles developed on flat and relatively poorly drained area. Advanced chemical weathering produces sticky clayey mass in the B-horizon of a *gumbotil profile*. Gumbotil is often light dull grey in colour when dry, at times mottled with brownish till consisting of a few but small smooth pebbles of quartz, chert and occasionally of quartzite and of few crystalline rocks, in otherwise clay material which shows shrinkage cracks on drying. A series of distinct horizons characterize below gumbotil which grades downwards into yellowish or chocolate colour till with relatively more but still subordinate pebbles and merges further below into yellow coloured till containing calcareous pebbles and further below passes into unaltered dark grey till which may be strong calcareous. Some of these gradations are more rapid than others and are, therefore, unequal. With good drainage, the B-horizon is composed of loose silt-like materials and the profile so developed is termed *siltil profile*. Intermediate profiles showing part characteristics of both gumbotil and siltil are termed *mesotil profiles*. Polygenetic or complex soil profiles reflect the recent geomorphic history of the area.

FOSSIL SOILS AND COMPOSITE SOIL PROFILES

Composite soil profiles, occasionally found in glaciated areas, may develop through two distinct parent materials, such as one developed in glacial till is subsequently burried with a thin veneer of wind blown materials—loess or sand or outwash sand or gravel. The old profile is called *fossil* or *burried soil*. When subsequently the new profile developing downward through the covermass reaches the fossil soil, the profile aptly becomes composite, in nature and termed *composite soil profile*. Ordinarily, the two parts of the composite soil profile are distinguishable except in such circumstance wherein till is overlain by loess because of similarity of the soils formed from the two materials. Soil developed over the till in this case is the fossil soil and can be distinguished from loess soil by careful petrographic studies. Fossil soils alongwith associated organic deposits help in deciphering the climatic and topographic conditions that prevailed during the time of their formation.

CLASSIFICATION OF SOILS

Soils are classified in different branches of sciences and engineering in various ways depending upon the primary objective borne in such classification. The first classification was given by Dockuchaiev (1882) in Russia, who grouped the soils into three categories

(i) *normal* soils of the dryland vegetative zones

(ii) and more *transitional* soils of washed or dryland sediments, and

(iii) *abnormal* soils

Soils may also be classified into three categories as

(i) *young* soils - show their genetic relationship to the parent material and typically associated with flood plains or hill-slopes where the soil surfaces are constantly replenished or disturbed,

(ii) *mature* soils - typical of flat lands, little erosion, possess good drainage. Often show only little relationship with the parent rock. Climatic and biogenic factors are more dominant, and

(iii) *old* soils - associated with old flattish surface with practically no erosion.

A classification of soils may be based on

(i) climatic conditions,

(ii) humus content

(iii) geology *etc.*

A. One such classification is the geological classification which is rather simple and genetic in nature. Two major classes of soils have been identified:

1. Residual soils and
2. Transported soils

A. *RESIDUAL SOILS*

These soils are practically developed above the parent rocks as cover on the weathering of latter and have undergone very little or practically no transportation. They are, therefore, *in situ* soils and their important characters are as below :

a. *Thickness* : Depending upon the

(i) climate

(ii) lithology of the parent material and

(iii) the extent of the operation of soil forming processes,

the thickness of residual soils vary from place to place. Development of thick soil column over the rock is attributable to

(i) a very rapid rate of soil formation and /or

(ii) the prolong span of time, may be without any break of soil forming operation.

b. *Stratification* : The successive penetration of weathering of the parent material, generally assumes stratification in residual soils.

c. *Chemical Composition* : As the residual soils are derived from the disintegration and decomposition of the underlying parent rock without any visible transportation, they tend to inherit gross parental chemical composition. However, in older soils the genetic inheritance of inorganic mineral content may become rather obscure because of their advance chemical decay. In addition, it also becomes enriched with humus derived from vegetal matter decay.

d. *Leaching* : Leaching out of soluble soil materials in humid regions depletes the plant food in them leading to the concentration of infertile salts, resulting in the soils infertility. In the paucity of moisture in arid regions, the desert soils are not leached and therefore the soils are generally fertile.

B. *TRANSPORTED SOIL*

In contrast to the residual soils, transported soils are those soils which have been removed away from the place of their formation and deposited elsewhere bearing no relation with the underlying rocks. Often during their transportation, they mix up with different types of soils derived from other places, rendering the solid heterogeneous, bringing in mineralogical complexity and adding to the difficulty in discerning the extent estimation of individual soil types. Such soils may have travelled short or long distances over hundreds of kilometers.

Transported soils are further classified on the basis of transporting agents as below :

(a) *Colluvial Soils* : Poorly to not at all sorted, these soils are accumulation, of short distance travelled, material moved down from the high relief under the influence of gravity, at or near the base of a steep-sided mountain. They support rather very little vegetation.

(b) *Alluvial Soils* : Stream-borne, fairly well to well sorted sediments are arranged in stratified form at the river banks and in flood plains. They are very fertile as in the Sindhu (Indo-) Ganga plains.

(c) *Glacial Soils* : Heterogeneous in size angular in shape and containing undecomposed feldspars, these soils of glacial origin lack stratification and are generally infertile, but boulder clay or till, rock flour, moraines and drift boulders at times constitute good soil.

(d) *Aeolian Soils* : These are wind blown soils *e.g.* Loess or Dune soils. The former consists mainly of silt and clay fraction and therefore are fine sized in contrast to sand soils referred to as Dune soils. They are poor in organic matter and incapabale of supporting vegetation because of lack of moisture.

(e) *Lacustrine Soils* : Rich in vegetal matter, these soils have their origin in lakes. They are well stratified and are quite fertile.

(f) *Marine Soils* : Mainly sand and coarse grained, these well stratified soils occur in the coastal plain or on the uplifted continental shelf zone.

(g) *Volcanic Soils* : Derived from the pyroclastic material and lavas in volcanic eruptions. The fertility of these soils is high.

Soils are also classified as Sandy soil, Loamy soil and Clayey soil and reflect textural responses in soil classification (see p. 125 also).

C. There are several other types of soils as classified below keeping in view besides texture and composition, porosity, permeability, moisture retaining capacity, organic content and agricultural use.

(a) *Black Soil or Chernozem* (Russian Black earth) : Generally formed from the decomposition of basic rocks, such as basalts, they own their black colour to humus (*e.g.* Russian black soil). They have high proportions of oxides of Al, Ca, Mg and others with variable humus content. It

(i) has a high capacity of retaining water and

(ii) is highly fertile and most suitable for cotton plant.

Indian block soil is iron rich.

(b) *Peat Soils* : Rich in decomposed vegetable product and containing very little clay, this soil too has good water-retaining capacity. It is good for production of onion *etc.* when well drained and properly manned.

(c) *Podzol/ Podsol* (Russian salting, saltiness) : Grey in colour and of acidic nature it is rather infertile soil and supports only tree growth. Characteristic of northern, humid regions of coniferous and hardwood forests, it rests over dark subsoil. Greybrown colour soil, termed *Forest* soil is of medium fertility. This together with red and yellow soil of warm temperate lands are regarded as special cases of podolization.

(d) *Laterite* (Latin-later, brick) : Laterization, like podzolisation is also a soil forming process in tropical countries characterized by hot but moist climate, where the decomposed products lie in layers over red coloured soils. Such soils are brownish in colour and chemically composed of the hydroxides of iron, aluminium and subordinately of hydrated oxide of manganese and in some instances with considerable titanium oxide percentage, but with very little silica, magnesium, calcium and alkali-materials that are flushed away from the source material by leaching. These soils are rather poor in humus content and are not of much value for agricultural purposes.

Tropical monsoon- climate with seasonal rainfall and alternating drought, lateritic soils are generally derived from basalts but occasionally granites and gneisses may as well yield such soils.

SOILS IN ARID ENVIRONMENT

Soils of arid and semi-arid environments are commonly prone to carbonate enrichment rather than leaching and eluviation. This process of *calcification* produces the class of soils known as *pedocals.* It involves high or complete base saturation, in combination with net accumulation of carbonates or water soluble salts. Several factors favour calcification.

(i) Under conditions of a generally dry season, the subsoil is seasonally or permanently dry, while rainwater percolation is limited. Intensive drought periods draw up capillary moisture from the subsoil or from a water table charged with time. As a result, some of the solubles washed into the subsoil by percolating rainwaters are carried upward until precipitated in a zone of lime or salt accumulation.

(ii) Nutrient-demanding vegetation types, particularly deep and densely rooted grass, also promote calcification. A plant cover that supplies abundant organic matter, rich in mineral nutrients, favours high base saturation of clay-humus molecules. Leaching is, therefore, impeded, since leached bases are rapidly replaced, due to healthy nutrient cycle. Grasses each year provide more organic matter than trees. Much of it is added directly to the soil through slow decomposition and replacement of the intricate root network. Burrowing rodents also aid in humus dispersal from O_2 into the A-horizon. In cooler semiarid environments, the result is intensive and deep humification

(iii) Nutrient-rich parent material is another factor. Bedrock that contains carbonates (*e.g.* limestone) or that weathers to a residual product rich in mineral bases (*e.g.* basalt) also favours a healthy nutrient cycle with huge base saturation and limiting leaching.

Duricrusts are concentrated surface encrustations of minerals dissolved in wet seasons and later brought up by the action of capillarity. These are characteristic of semi arid climate with alternating wet and dry seasons. The upper part of such profile is composed of concentrated aluminous, siliceous, ferruginous or calcareous materials.

Relative aridity, causes the development of *caliche* in the soil profiles which may exhibit differences in depth and maturity on different topographic surfaces. These variations in them may be referrable to the variations in ages in development of topographic expressions and therefore, may also offer clue to the climatic changes in the area. The caliche may show solution cavities, as also solution pits in limestone pebbles indicating effects of water period.

D. Also, soils are grouped into ten oxides based on the degree of development of various horizons in soil profiles and further classified into several suborders, great groups, subgroups, families and soil series.

TABLE VI. 29. SOIL CLASSIFICATION : ORDERS AND GROUPS

Azonal order	Arid semi-arid and subhumid climate		↑ Humid climate	
Lithosoils	Interzonal order ↑	Zonal order	Zonal order	Interzonal order
(Mountain soils)	(i) Saline soils	(i) Chernozem soils	1. Podzol soils	1. Bog soils
Regosols	(ii) Alkaline soils	(ii) Chestunt soils	2. Grey-brown Podzlic soils	2. Meadow soil
(alluvial soils, dunes)		(iii) Brown soils	3. Red yellow Podzolic soils	3. Plain soils
		(iv) Grey desert soils		
	↓	(v) Red desert soils	4. Latosols	
			↓ 5. Tundra soils	

E. The Soil Survey staff of the U S Department of Agriculture, has adopted a soil classification (Table VI. 30) referred to as Sêventh Approximation (as it is the seventh revision of the classification of soils).

SOIL GROUPS OF INDIA

Deduced from the nature of the foundation rocks, including

(i) composition

(ii) depth and

TABLE VI. 30 : SOIL ORDERS OF THE 'SEVENTH APPROXIMATION'

Name of order	*Derivation*	*Soil Characters*
Entisol	Nonsense, syllable ent from recent	Frozen ground, desert sand *etc.* in all climates, poorly developed horizons in alluvium
Ventisol	L. *verto,* turn, invert	Mostly in subhumid to arid terrain/regions, clay-rich soils that swell when hydrated and crack when dehydrated/dried
Inceptisol	*L. inceptum,* beginning	Tundra soils, soils on new volcanic deposits and recently deglaciated areas. Soils in these areas show slight development of soil horizon.
Aridisol	L. *aridus,* dry	Accumulations of carbonates, salt and gypsum in dry soils.
Mollisol	L. *mollis,* soft	Organic rich thick, dark surface layered soft soils of temperate grasslands.
Spodosol	Gr. *spodos* wood ash	Diagnostic iron or organic enriched B-horizon and leached out ashy-grey A-horizon, mostly occurring under conifers as humid forests soil.
Alfisol	Syllables from the chemical symbols *Al, Fe*	Common under deciduous forest young soil with clay enriched B-horizon.
Ultisol	L. *ultimus,* last	Deeply weathered red and yellow clay enriched soils on old land surfaces in humid temperate to tropical climate.
Oxisol	F. *oxido,* oxide	Almost horizonless, intensely weathered old soils such as tropical and sub-tropical lateritic and bauxitic soils.
Histosol	Gr. *histos,* tissue	Not distinguishable on climatic basis. These are bog soils, organic soils, peat and muck.

(iii) the profile of their soil caps

the soils of India have been grouped as under :

RED SOILS

Red soils including its several minor types constitute the largest Indian soil group and encompasses an area of 20,00,00 sq km covering the Archaean basement of the Peninsular India, embracing Bundelkhand in parts of Uttar Pradesh and Madhya Pradesh, Bihar, South Bengal, Orissa, East Andhra Pradesh, southeast of Maharashtra, Mysore and a sizeable part of Tamil Nadu.

These soils are derived mainly from the acid granites and gneisses quartzitic and feldspathic, and subordinately from iron and magnesium-rich other rocks including ancient sedimentary sandstones and clays-shale of Cuddapah and Vindhyan Supergroups. The main characteristics of these soils include—

(i) higher texture, porous and friable structure,

(ii) absence of lime, kankar and free carbonates, and

(iii) soluble salts in a small quantity.

They owe their red colouration to the wide diffusion of iron, and often grades into chocolate, brown, yellow, grey or even black and therefore not all red soils of south India are red in colour and on the contrary at times soils of quite different constitution, such as that which occurs as surface capping over lateritic rocks or soil weathering of impurities-clayey and sandy in nature, in the calcareous country rock which is dissolved on the limestone terrain, are red coloured.

A variety of subgroups of red soils are met. These variations may be in texture, depth, porosity, humus, presence/absence of calcareous seggregations, called *kankar* or ferruginous layers (iron pan), or soluble free salts. It may as well be brought also by the free or drainage conditions on which depends the nature and extent of leaching out of salts from the accumulation zone of soils profile.

The red soils chemically are siliceous and aluminous with free quartz as sand, and poor in lime, magnesia, phosphates, nitrogen and humus, particularly in drier regions. They are fairly rich in potash and other alkali contents derived from the muscovite and orthoclase, but the iron percentage is not particularly high.

Generally red soil group is deficient in soluble exchangeable bases and their total base exchange capacity for K, Ca etc. is rather low. The clay fractions usually are mainly kaolinitic.

BLACK SOILS

Included in it several distinct sub-varieties deep, medium and shallow black as also black soils undifferentiates, are yet another large groups of Indian soils. It commonly occurs as discontinous bodies capping the basic and ultrabasic igneous rocks in Gujarat (Surat, Bharuch and Vadodara districts), and Kathiwar, western Madhya Pradesh (along Narmada river and level portions of the Vindhya and Satpura plateau), Rajasthan (Kota, Bundi, Jhalawar districts), Uttar Pradesh (Jalaun, Hamirpur, Banda and Jhansi districts) Andhra Pradesh (Adilabad, Warangal, Khamam, Mahbubnagar and Kurnool districts), Maharashtra (along the Tapi, Godavari, Bhima and Krishna rivers courses).

Karnataka (Bijapur, Gulbarga, Bidar, Belgaum, Dharwar, Raichur districts), and in some parts of Tamil Nadu (Ramnathpuram, Tirunelvelli in South Arcot districts). It often occurs over the Deccan Traps as a mantle of rich soil of no great thickness, but extends also beyond, over granitic and basic gneisses, and shales and sandstones of Cuddapath or Vindhyan age. Transported black soils seem to be limited in areal extent. Black soils cover an area of low to medium, rainfall (45 cm to 100 cm). In some sections of river valleys-like those of Tapti, Narmada, Krishna and Godavari, these soils often are as thick as 8 meters.

Black soils are generally derived from the decomposition and decay of

(i) basalts of the Deccan Trap region, and

(ii) feruginous schists and gneisses, such as those of Tamil Nadu. These soils are rich in iron, aluminium, lime and magnesia, but the Indian black soil is rather poor in organic matter (humus) in contrast to Russian black soils, phosphorous and nitrogen. The black colouration of the soil may be due to the presence of titanium, iron or some organic compounds. These are fine grained and highly argillaceous. They contain high proportion of carbonates and magnesium and often characterized by the layers of calcareous nodules, referred to as *kankar,* particularly in those derived from ferromagnesian schists.

(a) When wet the black soils are very sticky and considerably show contraction on drying with development of large and deep cracks. This property of moisture sensitivity responsible for swelling and shrinkage of these soils is troublesome from engineering consideration. Their moisture retentivity is rather high and when wet, these soils are extremely compact and tenacious, making it rather unworkable because ploughs get struck in them.

(b) However, the base exchange capacity of the black soils including regur is generally high, and therefore, favourable for crop production.

The black soils may be *shallow* or *deep* types. The deep type of black soil is highly productive for cotton cultivation. They are largely clayey with 62% or more clay without coarse sand or gravel and very tenacious of moisture. The clay fraction of *regur* broadly comprises 60% of silica, 25% of alumina and 15% of ferric oxide and the black cotton soil are characterized by 6-8% of carbonate of calcium and magnesium, 9-10% of iron oxide and fairly constant, 10% of alumina. Potash content is usually less than 0.5%.

LATERITIC SOIL

Laterites are derived from the underlying basalts on flushing out silica under typical monsoon climatic conditions and may be of great thickness extending to 33 to 66 meters. These laterites are further modified by the biologic and pedognic processes forming about 0.3 to 1 meter lithomarge, are red coloured soil profile over it. Such soils are depleted of exchangeable bases on vigorous leaching in short-span high precipitation conditions rendering infertility. However, it improves to some extent on tillage and secondary changes brought about such as acquiring enough potash, nitrogen and phosphates. Lateritic soils of variable thickness are developed in the summits of the plateaus of Malwa, Malabar, Madhya Pradesh, the Eastern Ghats, Rajmahal hills, Orissa, East of Bengal and Assam.

ALKALINE SOILS (*Reh, Usar, Kallar* soils)

This is yet another, of course, negatively important, group of soils of India occurring in relatively drier regions in parts of north Bihar, Uttar Pradesh, Panjab and Rajasthan, containing many-still undecomposed rock and mineral fragments, which liberate Na, Mg and Ca salts and sulphurous acid on weathering. These soils are particularly impervious because of the salts occupying available pore and other spaces and therefore have impeded drainage. Once impregnated with salts many fertile grounds have lost productivity from agriculture viewpoint. These salts are brought up to the surface from the lower levels, may be called *root-zones,* under the capillary action and this process of upward transmission received acceleration in the last four-five decades on development of canal irrigation. Such salts are high in alkali content with large excess of free salts. There is hardly nitrogen or organic plant food material present in them. Such lands are variously referred to as *usar* or *reh* in north, *kallar* in Sind and *chopan* in Maharashtra. Sodium carbonate and sulphate, as also calcium and magnesium compounds constitute the most common salts in the reh and there reclamation would add millions of acres to cultivable lands.

The *nitre-impregnated* soils of Bihar and Panjab are the variant of the alkaline soil wherein the potassium nitrate salt has been introduced by the anthropogenic activity from the above into the soil-cap in warm humid climate.

Irrigation facilities develop these days, help

(i) water table rise and

(ii) compounds drainage congestion

leading to salinity problem in agricultural lands. About 7 million hectares of land is salt-affected in India. Percentages of cultivable land affected by salinity for some areas are-

Delhi region	> 21%
Panjab region	> 16%
West Bengal and Haryana	> 1.3%

Sufficient quantity of water with adequate attendant drainage alone can help reclamation of alkaline soils. Water may dissolve the salts present in the soils which tend to precipitate later unless washed away by the adequate drainage. It is further desireable to lower the groundwater level rather, below the zone of root penetration, say below 2 to 3 meters and efforts not to ever allow to rise nearer than 1.5 to 2 meters from the surface is most desireable. Conjunctive use of groundwater with surface water may keep groundwater lands at safe limits for avoiding rise of salts under the influence of capillary action. Major irrigation canals may be lined up with concrete slabs to prevent water seepage from the canal to the ground and avoid addition of water to raise undergound water. Leaching out of the impregnated salts is rather slow in fine-textured soil. Addition of chemicals and/or tillage is not always helpful. However, with

(i) sufficient water and

(ii) adequate drainage

the base exchange capacity of the soil tend to increase on addition of proper chemicals involving replacement of Na by Ca and conversion to replace Na salts into neutral Na_2SO_4. Finely ground $CaSO_4$ (gypsum) or even sulphur or Fe/Al sulphate may help in bringing out the desired changes. The supply of Ca^{2+} improves the soil fertility.

SINDHU (INDO)-GANGA ALLUVIUM

It is the most important and largest group of soils of India which supports the most thickly populated region of the country. This is transported soil, brought by large number of tributaries of the Ganga and Brahamaputra river systems that drains through the Himalaya. However, since their deposition in the valley, the Sindhu (Indo)-Ganga alluvium have only little, if any, effected by the subsequent pedogenic processes and in this sense, these soils may be regarded as immature. On the basis of kankar concentration in them, the alluvium are classified into two sub-groups :

(i) the *Khadar* - the newer younger light coloured deposit, less kankary, but more siliceous in nature

(ii) the *Bhangar* - the older dark coloured clayey deposit with high kankar concentration.

Assam, Bengal, Bihar, Maharashtra, Tamil Nadu, Orissa, Panjab and Uttar Pradesh are the regions where alluvial deposits are developed alongside the river systems draining through the area. Within the group variation with respect to the depositional pattern, texture, colour and composition of the alluvial soils are met at different places. For example, Assam soil is typically acidic and the Panjab alluvium mostly comprises loam or sandy loam.

FOREST AND HILL SOILS

As the name suggests these are typical soils of forests and mountains and their genesis and characters depend on geology, topography, climate and the type of vegetation. They may occur on the slopes, in the depressions or in the valleys and generally show a great variation in their chemical and mechanical composition, but usually charged with high organic content and nitrogen. Forest and hill soils are well developed in Assam, Panjab, Himachal Pradesh, Uttar Pradesh Tarai and Nilambur teak forest of Tamil Nadu.

DESERT SOILS

These soils contain high percentage of soluble salts, but are poor in organic content. They have virtually blanketed the entire region of Western Rajasthan, called the Thar deserts, and southern Panjab, encompassing an area of about 10,00,00 sq km Because of extreme paucity of water, the area has a brown look, with scant vegetation of xerophytic type.

PEAT AND MARSHY SOILS/CUMULOSE SOILS

In contrast to the above desert soils peat and marshy soils are residual types and possess high content of partially or wholly decomposed organic matter such as peat. They are also rich in soluble salts. Charged with high content of ferrous salts, these soils sometimes bluish in colour are often developed in the depression of dried river basins or lakes in coastal areas, as in some parts of Kerala, Orissa, Sunderbans in Bengal and in Almora district of Uttar Pradesh.

SOIL EROSION AND IT'S CONTROL

Removal of soil particles by the geological agents like wind, water and ice from its depositional place to elsewhere is termed *soil erosion.* Factors affecting soil erosion include-

(i) rainfall concentration

(ii) ground relief

(iii) nature of the soil

(iv) carrying capacity of running water and

(v) nature of vegetation cover.

It may manifest itself in either

(a) Sheet erosion,

(b) Rill erosion

(c) Gullying

Sheet Erosion

In this process, during the initial stages of rainfall

(i) the particles of the soils are loosend to float and then later with more water

(ii) pouring drain away out of the field.

Such type of erosion is more or less universal on

(a) all bare fallow land

(b) thinned plant cover uncultivated land, and

(c) sloping (i) cultivated fields, and

(ii) forests, scrub jungles

where natural soil porosity has been removed by heavy grazing, felling of trees *etc*. With this process ongoing, layers of soils are uniformly skimmed off, one by one may be for years, without any visible trace of damage and thus the owner is robbed unnoticed. It is common in the foothills of Himalaya, Assam, Western and Eastern Ghats. It depends on

(i) the velocity and

(ii) quantity

of pronounced surface runoff and

(iii) the erodibility of the soil itself

Rill Erosion

The runoff water in this case tend to accumulate in small streamlets or rivulets on heavy precipitation. With the increase in volume, velocity of water too increases which cuts into narrow trenches or *rills*. Such erosion removes large quantities of soil and rapidly lowers the soil fertility, a detrimental posture. It is active over large stretches of land in Bihar, Uttar Pradesh, Madhya Pradesh, as also in semi-arid tracts of Maharashtra, Karnataka, Andhra Pradesh and Tamil Nadu.

Gully Erosion

Continued sheet erosion over an area for a long period tend to selective concentration of removal along certain path more than others, resulting in greater erosion of soil along even very slightly depressed, tract than the others. Eventually this and similar tract form gullies of varying depth and extent. This gives rise to gully formation or gully and erosion so caused is termed gully erosion.

Gully erosion is common in areas of

(i) stream-dissected upland plateaus where feeders develop intricate gully patterns, and

(ii) relatively levelled country where field run off concentration takes place.

PREVENTION AND CONTROL

Soil conservation aims at the preventive measures to control soil erosion, may broadly be grouped as

(a) agronomic practices and

(b) engineering practices.

Agronomic practices of soil protection is based on the scheme of crop cultivation. These are mainly of two types :

(i) *Crop rotation* scheme practices a definite and planned sequence of crop rotation for cultivation *e.g.* a small grain and then grass and so on in rotation

(ii) *Strip cropping* include alternate crop cultivation by sowing the cover crop in strips generally parallel to the local contour.

Most widely used engineering practices include :

(i) different types and sizes of *ditch* excavations for the diversion of excess water from the fields. It is particularly more suitable in areas of steep slopes. Two types of common ditches are -

Diversion ditches which are excavated above the cultivated portion of the sloping area to divert away the runoff.

Interception ditches are excavated at suitable regular intervals covering the entire cultivated fields, which help in the prevention of sheet erosion by retarding the water velocity carrying soil grains.

These ditches are so designed as to ensure maximum expectable flow at a non-eroding velocity

(ii) *Terraces* are constructions across the hill slope sides to collect and divert the run-off to an erosion-proof outlet. They are so constructed as to prevent the downward flowing water and direct the flow along their lengths until the water is drained off the field. Such terraces are built at carefully worked out intervals depending upon slope characteristics with the first one close to upper limits of the slopes in an effort to prevent the initiation of or reduce erosion. These terraces when uniformly graded prevents ponding of water or its reaching erosional velocities that would facilitate soil removal instead prevention.

(iii) *Dams* are particularly useful in combating gully erosion. such check dam, could be made of stones, timber and steel *etc*. with the idea of reducing run off velocity that causes the load deposited which may ultimately support vegetation which further holds the soil and prevents gully erosion.

Vegetation- Tree plants often successfully helps in prevention of soil erosion on hill slope side, because of their root system which offer a powerful grip on the soils preventing their further loss.

CHAPTER VII

GLACIERS

Glaciers are river-like masses of ice and granular snow of size varying from a fraction of a square kilometer to hundreds of square kilometers, formed by compaction and recrystallization, lying largely or wholly in the mountains at high latitudes or even at high elevation of lower latitudes and showing evidence of past or present spreading flowing down the hill under its own weight from its place of accumulation. They move out like a tongue and proceed to a distance if the supply of the ice is greater at its head than the amount of wastage on heat generated due to friction offered by the underlying surface on the way and while in motion it assumes the form of the area through which it passess. A combination of low temperature with abundant snowfall is the essential condition for the maintenance of thick snow for a long time These conditions are satisfied in colder terrains.

In regions of exceptionally cold climate -

(i) at high latitudes

(ii) even at low latitudes but at high altitudes,

the precipitation takes place as snow in place of rain water. Some of this snow melts in summers or in hot weather, appreciable quantity of snow remain unaffected in its state (a form) in some places all through the year. The permanent areas of snow formed on account of such accumulation in excess over the waste, by evaporation and melting are called *snowfields.* These snowfields are the feeders of all the glaciers. The work of snow and frost beyond the true glacial limits is termed *nivation,* a process common in Tundra and Arctic regions. *Snowline* is an imaginary line which is the lowest limit of the area of perpetual snow above which the snow does not melt even in summers or hot climate. The snowline differs according to (i) latitude (ii) amount of snowfall (iii) direction of the wind and (iv) physical features, particularly the altitude and the surrounding environment such as nearness or otherwise to the standing water bodies-lakes or seas. Snowline is endowed with the following general characteristics :

(i) it lowers down from the equator towards the poles,

(ii) in regions of dry winds the snowline is higher and lower in regions of wet winds, and

(iii) it is higher on steeper slopes than on gentler slopes.

Snow is abundant at high latitude and high altitude. In the eastern Himalaya the snowline descends down to about 4575m and stays back at an elevation of about 5,200 to 5,800 meters in the western Himalaya.

ORIGIN OF GLACIERS

The snow, as it falls, is powdery, porous with low density ranging between 0.07 and 0.18 grams per cubic centimeter (however exceptionally as low as 0.004 grams per centimeter and as high as 0.5 gram per centimeter have also been recorded) substance like in the form of cotton. Belonging to the hexagonal system, these lacy snow crystals exhibit hexagonal, prismatic or peculiar star shaped forms. The powdery substances may be tiny micro-perciptible grains in a severe frost region or may assume larger crystals as is noticed in relatively warmer climate. Such accumulations in the supply regions are subjected to complex transformation

(i) under the action of the sun,

(ii) as a result of sublimation, and

(iii) due to increasing pressure.

The upper surface of the freshly precipitated snow begins to melt under sun's heat particularly in summer. Rounding of the edges of precipitated snow results and at night in cold, it assumes the shape of grains. Successive snowfalls generate pressure on the earlier snow which assumes a flaky structure and as a result air is squeezed out and partially melts the lower snow. A part of this meltwater percolates/seeps into the loose snow and melts the snow crystals further down and thus the process goes on and gradually the whole mass grows granular. The process is repeated with each new snowfall and like this the initial loose cotton-like substance gradually passes through a more closely packed flaky structured mass into granular mass, under the influence of moisture, called *ne've* or *firn,* which has an intermediate character between snow and compact ice and is the main source of glacier-nourishment, Firn density varies between 0.4 and 0.8. It is opaque, though the individual granules are transparent, due to the pressure of air bubbles. Sublimation, vapourization of ice and subsequent recrystallization is the most dominant process in the transformation of snow into *ne've* or *firn.* Ice surface vapour pressure is a function of (i) the temperature (ii) the size and (iii) the shape. Because of the more surface area, such vapour pressure is higher over the smaller crystals as compared to the larger ones resulting into pressure (vapour) gradient from the former to the latter which enables larger crystals grow. Coalescence of individual crystal body into larger crystalline aggregation is thus achieved on release of heat on vapourization (or sublimation) gradually leading the mass transformation into relatively more compacted clear ice (due to escape of trapped air) and often characterized by conspicuous layerings separated by the dust particles settled during the intervening successive snow falls. So formed ice takes a motion under the action of

(a) increased pressure generated due to its own weight on successive snow falls

(b) shape, and

(c) the pull of gravity

and flow like streams. Such moving mass of ice is known as *glacier.*

Thus *"a glacier is a natural moving body of great dimension composed of crystalline ice (ne've in the upper layers) formed on the earth's surface as a result of accumulation and subsequent transformation of solid atmospheric precipitation snow".*

Conversion of *firn* into ice may take wider range often varying between only one year in temperate climate to even more than one hundred years in colder, say polar regions. Such a vast variations depends upon (i) temperature, (ii) rate of snow accumulation and (iii) gravity pull *etc.* Out of 10-11m^3 of snow, on compaction constitutes about 1m^3 of ice.

Under certain definite conditions solid bodies including glaciers undergo plastic deformation without any discontinuity or development of cracks. These conditions are temperature and pressure controlled. This is demonstrated experimentally by a mid sag of 2mm a day on ice bars placed on two supports at a temperature of -12° to -3/.5° C. Sagging is accelerated under pressure which causes plastic deformation in the glaciers, a property most pronounced into thick blue masses of glctchuri ice (that constitutes the principal/main body of glacier) in ne've basins. Under tremendous pressure of the overlying mass, the lower layers which become plastic set to flow when the underlying surface relief is favourable. Thus, a snowfield may show a basal campact ice overlain by the ne've in the middle, and fresh unconsolidated snow over it at the top. Earlier formed snow, occurring on sheltered mountain slopes, texturally is like very coarse sand.

The process of converting to glacier, thus includes

(i) sublimation

(ii) melting and refreezing, and

(iii) plastic deformation

Glaciology is the science of the study of ice crystals in high clouds, hail and snow frozen lake, river and ocean water, and glacier ice and glaciologists are a group of workers in itself, may be meteorologists, physicists and geologists. Ice can be studied in different ways such as

(i) an easily deformed crystalline solid with a structure to be analyzed under microscope, or

(ii) a geomorphic agent that erodes valleys and transports massive loads of rocky debris.

Glaciers study include field and experimental determinations. Glacial geomorphology became a subject of almost general fascination during the *nineteenth* century, when it was first discovered that much of Europe and North America had been glaciated during the Pleistocene Ice age.

GEOGRAPHIC DISTRIBUTION OF CONTEMPORARY GLACIERS

Over 10 percent of dryland is contemporary glaciers which encompasses about 16.3 million km^2 of area distributed as below-

	Km^2
Arctic regions	2,100,000
Moderate regions of the Northern Hemisphere	100,000
Tropical regions	100
Moderate regions of the Southern Hemisphere	21,000
Antarctic regions	14,100,000

The above areal statistics indicates maximum ice coverage in the polar regions, especially at the South Pole.

TYPES OR CLASSIFICATION OF GLACIERS

Glaciers have been classified in several ways :-

I. The forms, size and origin of glaciers vary and accordingly named differently. Ahlmann (in Thornbury 1985) brought out a morphological classification as follows:

A. Continuous sheets of glaciers from which ice may radiate in all directions

1. Continental glaciers or inland ice covering large areas
2. Glacier caps, relatively smaller in size than the continental glaciers
3. Highland glaciers, those cover the highest and central portion of a mountain area.

B. Confined glaciers limited to more or less marked courses guiding their main movement. This group includes both, independent glaciers and outlets of ice from glaciers of group A.

4. Valley glaciers of the Alpine type
5. Transection glaciers, occupying more or less a whole valley system
6. Cirque glaciers, which occupy localized niches on sides of mountains

7. Wall-sided glaciers, covering a side part of a valley

8. Glacier tongues afloat.

C. Glacier ice fan out in large or small cake-like sheets over level ground at the foot of the glaciated areas. None of these are independent, but on the other hand they are all connected with some other types of glaciers

9. Piedmont glaciers formed by the union of lower parts 4, 5 or 7 above

10. Foot glaciers, which are the lower ends and more extended portions of types 4, 5 or 7 above

11. Shelf ice.

II. Glaciers are also classifiable according to their mechanics of motion into two groups :

(a) ice streams and

(b) ice caps

While ice cap motion is produced largely

(i) by differential pressures within an ice mass

(ii) is multidirectional, and

(iii) is not controlled by the bottom topography although it may be influenced by it,

the flow in ice streams is largely influenced by the gravity, and is,

(i) unidirectional , and

(ii) controlled by underlying topography.

III. Yet, another classification of glaciers is on the basis of stages of their development, size, slope, form, mode of occurrence and the relationship between the supply and wastage areas as

(a) Mountain or Alpine or valley glaciers

(b) Continental ice sheets

(c) Intermediate or Scandinavian glaciers

Mountain or Alpine glaciers are commonly met in young mountain regions formed by tectonic movements. They are also designated as 'Alpine type', so called, after the area where they were first studied. They are ice streams flowing in valleys and have fewer tributaries in relation to those in streams. Valley glaciers are the most abundant type commonly of a few to 300 meters or more thick with widths of 0.4 to several kilometers and range in length upto about 13-14 km in Alps to about 115 kilometers on southern Alaska. Most valley glaciers occupy mountain valleys.

Valley glaciers have a distinctly expressed supply region (a ne've basin within which snow accumulates and is transformed into ne've and ice) and wastage area.

The supply region abounds the snowline in various basins and depressions incised into mountain masses and surrounded by a amphitheatre of high ridges and peaks. Ice flows or tongues advance in linear motion over the steep mountain valleys slopes.

A close relationship exists between the ne've basin and the glacier tongue, more abundant the ice supply in the ne've basin, the longer is the glacier tongue in the mountain valley.

Valley glaciers may be "simple" or "complex" (or "*polysynthetic*") depending upon the specific structural features

(i) *Simple* glaciers are isolated in a single flow without any tributaries (*e.g.* Alps and Caucasus glaciers)

(ii) *Complex* or *polysynthetic* glaciers, on the other hand, are a group of several coalescing glaciers resembling a river with tributaries. The homeland of such glaciers are transverse valleys or those longitudinal valleys which separate mountain ridges. On powerful glaciation the longitudinal valley possesses powerful ice flow to which the flows of transverse valley are also discharged. The process continues and ultimately a tree-like pattern in plan, called "dendritic" are developed, as we see in the Himalaya. When two opposite slopes of a ridge supporting ne've basins coalesce, a new gathering single basin is formed from which two glaciers are fed and are referred to as "*diffluent*" or "*through*" glaciers.

In addition to these well-developed valley glaciers, two more varieties are recognized

(i) "corrie" and

(ii) "hanging" glaciers, (these are relatively small, isolated, imperfectly developed, as it were or relict mountain glaciers).

Corrie glaciers are yet another type which emerges from corries (a horse shoe depression). Corries are incision into steep slopes of high mountains. They are often located on the slopes of trough valleys and cirques and are characterized by small thickness of ice and therefore lack flowage. On high steep mountain a valley glacier may be perched in narrow depression leading short tongues of ice masses virtually hang over the main valley, from which the ice breaks off and tumbles down. Such mass of ice is called "*hanging glacier*" and the valley it occupies, is called "*hanging valley*". Like tributaries, they are different from the main valley glaciers in the sense, they are smaller in size and unlike river they never meet at the same level. Some of the

Sikkim and Kashmir glaciers are of this type. In nature these glaciers are in between a "valley glacier" and a "cliff glacier" which occupies the cliff face. The separated ice blocks of the hanging and /or cliff glaciers that fall on the slope toe tend to freeze again and form a *"reconstructed glacier"*.

Continental Ice Sheets or Continental Glaciers - These types of glaciers typically have enormous areal coverage, especially in the polar regions and envelope the entire islands and continents below it and possess such distinctive features from mountain glaciers as

(i) huge ice thickness

(ii) lack of periglacial relief influence on their distribution

(iii) coincidence of the supply and wastage areas

(iv) radial nature of ice flowage towards the margins of the ice-sheet, and

(v) plane-convex shape of the glacier surface resembling a shield

Such glacial shields are thickest (~ 2400m) in the middle and thins out marginally from which island-like rocky projections of mountain peaks or even individual cliff called "*nunataks*" project out like islands in oceans. Small plateau are covered by relatively small ice- sheets, better termed as *"ice caps"* and are characterized by outward radial movement. Greenland's ice cap is one such good example. It does not reach the sea, but moving through mountain passes it gives rise to great mountain glaciers, some of which reach the sea where the ice breaks off a process referred to as "*calving*", producing *icebergs*, which floats, being lighter in density, on the sea water. In contrast in "shelf ice" the glaciers over run the coast without melting and spread over the ocean as great masses at one end to the continental ice. These are vast ice-blocks floating in the ocean. Shelf ice usually ends in shear cliff, rising 30-50m above the sea from which huge mass of ice called *"tabular icebergs"* often breaks off. The Antarctica is an ice shield at sea level, and only at certain "oases" the land stands up without any ice over it.

Intermediate or Scadinavian-type Glaciers - Plateau glaciers or ice caps and piedmont glaciers are included into it.

"Plateau glaciers" cover flattened surfaces of the summits of ancient mountains as solid ice caps for hundreds of square kilometers. They combine features of continental and valley glaciers in as much as they emerge through river valleys, descending as isolated flows or tongues moving from the centre to the margins. The summit plateau is covered as an ice cap which resembles continental ice sheets and those that descend through isolated channels, resemble valley glaciers. Such glaciers are extensively developed in the Scandinavian mountains.

"Piedmont glaciers" are characteristic of abundant snow fall area situated

in highly dissected mountains where both replenishment and wastage follow the pattern of typical valley glaciers. However, because of excess supply in the ne've basins, the valley glaciers emerge into the piedmont valley where they fan out into a shape like river delta. These spreadovers coalesce with each other and form a solid ice sheet at the mountain foot, referred to as "piedmont glacier".

Thus, a definite sequence- primary glaciers (*corries*, hanging glaciers *etc*) valley and piedmont glaciers,, continental ice sheets *etc.* are notable in the development of glaciers. However, their sensitiveness to the climate is evident from the fact that the lowering temperatures, snowfall becomes heavier, whence from the simpler forms may be successively replaced by more complex ones upto and including the formation of uninterrupted continental sheets.

IV. So far the classifications discussed above, are basically descriptive. To understand the geomorphic work done by the glaciers, another classification based on temperature has been found more useful. The following three properties peculiar to ice is of significance in this connection :

(i) Latent heat of fusion of ice is 80 calories per gram which means transfer of 80 calories of heat per gram to ice from the environment by absorbing when a gram of ice freezes without temperature change.

(ii) The temperature of co-existence of ice and water is constant at a given pressure. Ice water remains at $0^0C/32^0F$ under one atmospheric pressure, until either all water is frozen to solid state or the ice is entirely melted without any regard to the amount of heat added or removed from the mass during the co-existence of the two phases.

(iii) Ice has a low thermal conductivity that is comparable to or lower than the conductivity of many rocks.

In the early formational stage, the glacial ice has entrapped air too making the mass a poor conductor which helps the ice in maintaining the atmospheric temperature during snow accumulation period. Generally the temperature of each individual layer is a function of the particular time of the year of each snowfall. On an average about 100 meters thick accumulation is reached, the seasonal variation averages the temperature of ice or firn to the mean annual air temperature, if that temperature is below freezing point.

Two basic types of glaciers have been visualized on the basis of thermal properties of ice :

(i) the *cold* or *polar* glacier which is always below the ice melting temperature and thus maintain a solid state althrough, and

(ii) the *warm* or *temperate* glaciers which is characterized by temperature hovering around the ice melting at depth much of the year and interstitial water saturates the ice.

Snowfalls when the ambient temperature around is lower than the freezing point. Thus, the glacier, under this condition is thoroughly refrigerated at the surface as also the temperature distribution in a thick ice sheet too attains some kind of near uniformity at considerably lower than the freezing point. The geothermal heat of about 40 calories per square centimeter flows into the base of a cold glacier, but simply conducted upwards "through the ice toward the cold-surface, where it radiates in to the still colder atmosphere". Although with this geothermal heat the glacier does become somewhat warmer at depth, but because it simultaneously looses heat into the colder atmosphere above, and therefore the entire mass of a glacier remains frozen firmly to the cold rocks beneath as solid mass.

Thermal conditions are more complex in termperate glaciers. These galicers are met outside the polar region where the temperature fluctuates such that the ice experiences as summer melting season when it is likely to be warmed to the melting temperature down to a considerable depth inspite of its low conductivity. This melts the surface ice and would drain down under the action of gravity through crevasses in the ice, throughout its downward journey the meltwater practicaly remains at 0°C, there is change of state and consequent thereof transfer of heat in as much as each one gram of melt carries 80 calories of latent heat into the glacial mass. Now if the ice at depth is colder, the melt water would refreeze releasing 80 calories per unit gram of water, which is estimated to raise the temperature of 160 grams of ice by 1°C (specific heat of ice = 0.5). Thus a thick water saturated ice-zone at the melting point is produced not always by conduction, but by refreezing of the penetrating meltwater as well.

Warm glaciers become colder with depth, inspite of ice and water co-existence. The melting water of ice is somewhat depressed because of the pressure due to its own overlying weight. As a follow up, it is evident that a small amount of melting occurs when ice at 0°C is buried at great depth in a glacier and this change of state draws heat from the remaining ice which causes the sinking of the temperature of co-existing ice and water below 0°C proportionate to the overlying pressure. Accordingly, with some melt-water near the surface thermal condition is so established to maintain the pressure-melting temperature prevailing throughout the glacier. A mixture of ice and water is nevertheless slightly colder with increasing depth and pressure.

Temperate glaciers indeed prevent the geothermal heat to escape and therefore, act as perfect insulators to the geothermal heat. It's property being colder at the bottom than at the surface establishes a gradient and "heat will flow upward through them". The average annual heat flow of 40 calories per square centimeter frøm the earth under a temperate glacier would be expanded by melting a layer of ice about 0.5 centimeter thick at the glacier bottom enabling the temperate glaciers to move-over their rock floors on a film of water.

MECHANISM OF GLACIER MOVEMENT

The "flow" of glaciers is due to a combination of

(i) internal deformation in ice crystals,
(ii) melting and refreezing,
(iii) basal slip of the ice over rock, and
(iv) brittle fracture and faulting in the ice.

In polar glacier the movement is primarily by

(i) internal plastic deformation of ice crystals, and
(ii) a minor amount of brittle fracture near the surface.

It is demonstrated experimentally that the bond between frozen ice and rock forms is stronger than the internal strength of individual ice crystals. Therefore atleast theoretically a cold glacier is believed not to move by sliding over its bed. A consequence of the motion due to plastic deformation is that polar glaciers are rather poor agents of erosion and transportation as also evident from almost free of rock fragments in the Antarctic ice sheet. However, when a rocky obstruction is broken away by the flow of cold ice, the rock is fragmented and transported until the ice wasted by melting or it breaks off into the sea.

Forward movement of temperate glacier is

(i) primarily due to basal slip over the rock (it practically accounts for about half of the total motion of many valley glaciers), which seems to be determined by the bed roughness in a critical size range and in addition is also due to
(ii) melting and refreezing, besides some accounted by
(iii) crystalline deformation of ice, and
(iv) brittle fracture and faulting in the ice mass.

The temperate glaciers generally move faster than polar glaciers, perhaps because of the more options available including basal slipping accentuated by gravity as well. Melting and refreezing in some instances particularly under differential pressure is not so easily discernible from slip, because if a rock obstacle projects in to the bottom of a temperate glacier, the excess pressure exercised to ride over its melting the ice on the upstream side, and the melt-water migrates downstream to the leeside of the obstacle where it refreezes in response to continued lower temperature. The glacier advances by "slipping" on a water film over the obstacle, but in details the individual grains of ice melts and refreezes to permit the slip.

The rate at which temperate valley glaciers move downhill is rather variable may be between some centimeters to two meters, and in some cases may be even over 30 meters. In the case of the Himalayan glaciers the movement varies from 2.5 cm to 1 meter per day approximately. Commonly the velocity of glacier motion increases with

(i) the increase in the slope of its bed,

(ii) ice thickness,

(iii) course straightness, and

(iv) temperature.

Glaciers sometimes can transport as big as one storeyed building for considerable distance. The motion is appreciably faster

(i) on very steep slopes, or

(ii) during the meltwater season when the bed is lubricated with more water to overcome the frictional resistances.

Rarely, one part of a glacier surges forward for a few weeks or months at a rate several meters per day.

Glacial surges, *surging glaciers* are the waves of rapid motion that move down. Sudden increase in ice accumulation on the upper reaches of the glacier, possibly caused by snow avalanches or exceptionally heavy snowfall trigger glacial surge which generate bulges like waves some meters high that move through the glacier upto five times the actual ice movement velocity (prior to surge movement), similar to ocean wave movement through water. It is believed that normal glaicer motion before and after the surge passage is in response to a balanced set of forces because of the ease with which ice is deformed and the surge is the response to a single additional variable such as an abrupt increase in accumulated snow.

Because of the too low confining pressures and the stresses too abrupt the deformation in the surface layers of glacier ice is not in a plastic fashion. Ice to a depth of about 30 meters is brittle and cracks into a maze of crevasses as it carried along on the more plastic ice at depth. The surface crevasses extend as deep into the ice that a basal slab of ice may become detached and caused to move, and active ice will ride forward over it on a "thrust fault" (Figure 9)

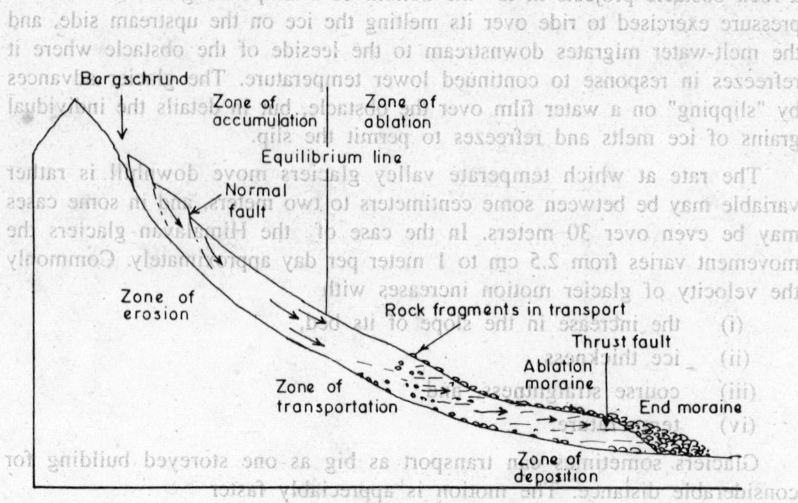

Figure 9. Glaciation Zones

The movement of a glacier is serpentinous and is caused by the stress generated by the overlying mass ne've and snow. The glacier ice comprises two parts:

(a) one, viscous due to pressure, moving below, and

(b) the other, solidified part which is caused by the flowing lower mass.

Thus cracks, or *"crevasses"*, as they are called, are developed in the upper part of the glacier. These crevasses attest the crystalline and solid nature of the glacier ice, at least on the upper part. They are developed over an inclined surface where a glacier bends low. Where motion is initiated in a glacier mass crack (or a series of them), parallel to the rocky wall behind is formed. This crack is referred by the German name "*bergschrund*". This forms where the thick body of ice, *ne've* and more compacted snow of the snowfield draw away from the thinner, less compacted snow of the upper margin of the snowfield as the head of a glaicer tends to detach from the snow and ne've of the upper slopes, leaving a deep crevasse (*bergschrund*) in which the bed rock is more or less exposed when the temperature during the summer days, water fills the joint cracks and crevasses in the rocks down the bergschrund and in much colder nights this water freezes developing internal force which pulls and loosens some of the joint blocks and in the process the disintegrated torn out rock pieces that are produced are enveloped in ice and in ne've formed from new snowfalls and thus got frozen and transported alongwith. This quarrying or excavating operation, which is repeated season after season, is most effective towards the bottom of the bergschrund, and so side of the valley head are cut back and greatly steepened, forming a *cirque.*

Glacial Cirques - These great bowl shaped singly or group of basins in the form of amphitheatre surrounded by high nearly shear walls and only on one side they are open and gradually pass into their valleys greatly widened and reworked by the glaciers, are limited to the upper reach of mountain valley glaciers and their development is a function of

(i) the action of glacier itself,

(ii) weathering, and

(iii) the action of melt waters.

Although cirque morphology varies considerably and consequently many gradations are possible, the principal features include the following distinctive elements:

(a) the head and side, steep may be deeper than even 45 meters and nearly vertical and often scattered walls

(b) the rock basin or circular hollow with a concave floor meeting the headwall in a sharp slope break and showing smoothing and polishing evidences

(c) the lip or threshold of the basin where the basin floor rises and is sometime capped with moraines. It is convexly rounded often shattered on the down valley side

(d) the lengths of cirques are usually three times their heights *i.e.* length/ height ratio is about 3:1.

Occasionally they may be several meters in width.

According to the position at which they are formed, crevasses are of several types :

(i) *Transverse crevasses* - These crevasses are developed where the slope over which the glaicer is moving is sudden and more than that the glacial body can mould itself over without rupture. Thus transverse crevasses develop across the glacier. A rapid change of slope even of the order of few degrees is usually enough to develop tensile stresses in the upper part more and cause transverse crevasses. Because of the forward motion of the ice, old crevasses often close up, and new ones develop.

(ii) *Marginal crevasses* - Since the central portion of a valley glaicer is free from side and bottom rocky friction, it moves relatively faster setting up stresses and strains in the marginal portions which often cause "marginal crevasses" to develop.

Marginal crevasses of the ne've field or basin, separating the ne've mass from the surrounding walls play a special role in the widening of cirques, where conditions are most favourable for physical weathering and subsequent removal of the loose weathering products by the glacier with the walls gradually retreating and the area of cirque increasing.

(iii) *Longitudinal crevasses* - These crevasses are almost parallel to the ice flow direction and are developed where the glacier moves on to relatively broader portion of a valley or where it terminates, when spreads on practically flat surface. The compacted ice mass, under such condition, in an effort to expand develop cracks aligned to the valley direction along which the glacier traverses. Crevasses do not extend far down in thick glaciers because the ice, under much pressure at depth, is more subject to flowage.

In course of time the crevasses tend to be filled up. Sometimes, of course, they may be covered by dangerous icebridges. Not much weight is required for such bridges to collapse.

(iv) *Splaying crevasses* - Limited to areas of intense compressive flow, they are also parallel to the long axis in their upglacier end, but curve towards the margin of their downglacier end.

(v) *Ticktacktoe crevasses* - They are rather uncommon groupings of crevasses which form criss-cross patterns and probably form over a bedrock knob on the valley floor.

ADVANCE AND RETREAT OF GLACIERS

Advance and retreat of glaciers respond to the climatic conditions, and is primarily dependent on temperature ; they retreat or thin out on temperature rise, by snow melting and the loss is replenished by the supply from the snowfield and consequently two opposing factors operate in the glacier movement :

(i) supply of snow and consequent pressure from uphill region

(ii) melting of ice tending to cause shrinkage in volume of the glacier and its consequent retreat.

The snout of front portion of a glacier moves on below the snow until the melting and evaporation gain upper hand and stop the movement. Besides the temperature other factors in the forward movement of a glacier are its

(i) volume, and

(ii) velocity.

Forward movement of glaciers is inevitable when the volume is large and the velocity considerable for it would compensate more than the wastage. Glaciers shrink in warmer months and advance in moist months.

LOWEST LIMIT OF HEIGHT

The lowest limit upto which glaciers descend varies. In the eastern part of the Himalaya, in Sikkim, they descend down to as low as 4,000 meters against about 2,500 meters in Kashmir, western Himalaya. This is because of latitudinal trend (variation) of the mountains (the Himalayas). At present most of the glaciers are generally retreating, but for the occasional temporary advances. In Pleistocene times they were numerous and advancing.

PERMAFROST

Near the existing ice-caps or continental glaciers as well as in some high mountain regions rocks are filled with perennial frost known as "*permafrost*".

PEAT RINGS

Soil materials are kneaded by alternating freezing and melting of water which may drive off much of it (water) from partly or boggy soils resulting in the caving in or down thrusting of the surface into which chunks of soils may creep and flow down. Such "cavings" are termed "peat rings".

LAWS OF GLACIER MOTION

The glacier motion is rather complex and involves "differential motion" in a complicated manner. The "laws of glacier motion" which is better applicable for valley glaciers also serves to indicate the complicated nature of the movement. These laws (Miller, 1949) are as follows :

(i) A glacier, to a greater or lesser extent, actually glides or slides over the earth's surface. This is abundantly proved by the eroded, and often polished and striated, rock surfaces left by glaciers.

(ii) The top portion of a glacier moves faster than the bottom, because of friction of the glaciers on its bed. This has been proved by observing the change in position of a vertical line of pegs driven into the steep side of a valley glacier.

(iii) The middle portion moves faster than the sides because of the glacier against its containing banks. This is easily proved by observing the changing position of a row of marked objects placed across a valley glacier.

(iv) The velocity increases with steepness of slope of the bed. This has been proved particularly for certain glaciers in the Alps. It must be so because gravity is the ultimate force which causes the motion.

(v) The velocity increases with the thickness of ice. This again is due to the fact that the force of gravity is more effective in causing movement of a body of glacial ice on a slope is relatively thick.

(vi) The velocity increases with temperature. In warm weather glacier moves faster than in cooler weather, that is, faster when it is melting and contains more water.

(vii) Velocity increases with straightness of course. A glacier flows less rapidly through a crooked valley because the friction is greater as the ice current rounds the curves.

(viii) Velocity diminishes with roughness of bed. The motion of the glacier is retarded by being forced over inequalities of obstacles in its bed.

(ix) Velocity diminishes with amount of load in the basal portion. This is because of increased friction on the glacier on its bed.

(x) The line of greatest velocity is more winding than that of the glacial channel. Just as in a river, the tendency also in a winding glacier is for the line of greatest current to swing back and forth from one side to the other.

(xi) A stream of ice does not conform to minor irregularities of the sides of the channel. A glacier several hundred feet thick may move past the end of a tributary valley without flowing into the latter to seek the general ice level. In this respect, glacier movement is very much different from that of water.

Two schools of thought have developed in respect of the nature of glacial motion :

(a) One school is led by Hugi, Forbes and Agassiz (in Thornbury 1985). This group holds the concept of the flow by continuous plastic yield, which recognizes that a glacier is *"feuve de glace"* termed by Bishop Rendu which exhibits the characteristic flow of a highly viscous fluid on the basis of the analogy drawn out from many measurements made on Alpine glaciers which indicated differential velocities within a valley glacier essentially similar to those of a stream but for the fact that they are also slow. This viscous theory flow is at least apparently substantiated out by flowing of the crushed ice, through an orifice placed in pressure chambers in laboratory.

(b) The group headed by Tyndall, TC Chamberlin and R. T. Chamberlin opined that fractures and shears accompanied by regelation were largely responsible for glacial motion. In this connection crevasses in the ice mass and brittleness are cited as evidences of the rigid nature of the glacier. It is assumed that motion is affected by individual particles. Shear planes, overthrusts and "blue bands" which are common in the lower parts of glaciers cited as illustration in this connection and were interpreted as slippage planes as further evidence of flow by fracture and shearing.

It has become increasingly evident that neither theory adequately explains glacial motion as there are evidences of both types of motion. Demorest (1942, 43) proposed four types of flows -

(i) gravity flow,

(ii) extrusion flow,

(iii) obstructed gravity flow, and

(iv) obstructed extrusion flow.

REGIMEN OF GLACIERS

The regimen of a glacier embraces variations in

(i) nourishment,

(ii) movement, and

(iii) wastage

and any change in it determines a state of glacial advance, recession or stagnation. Accordingly, the concept of glacial regimen include :

(a) the supply of a glacier (the quantity of precipitation on the ne've basin- "accumulation")

(b) the movement of the glacier (or flowage), and

(c) the change in the glacier mass through melting and evaporation (wastage- "ablation").

and is determined by

(i) the inter-relationships between the above components, and

(ii) their variation in time.

"*Accumulation*" is the process by which glacier is constantly replenished. A glacier is nourished by

(i) precipitation in the form of snow and rain,

(ii) avalanches of snow from the valley sides above the glacier, and

(iii) wind drifted snow.

Most snowfalls are due to the rising and cooling of moist air in mountain terrains. Because of the necessity of the moisture content, the rate of accumulation of snow and ice tends to be more in mountainous areas adjacent to oceanic regions.

"*Ablation*" is the process in which ice and snow are lost from a glacier may be effected by

(i) melting,

(ii) sublimation, or

(iii) calving.

Melting often caused by

(i) solar radiation,

(ii) warm air, and

(iii) rain

is by far the most prevalent of these in most glaciers. Rock debris on a glacier tends to become warm relatively faster than the snow and ice because of better absorption of radiation and where the rock debris layer is not very thick, it would help deeper penetration into the underlying ice and forms a hole whereas on the other hand too thick layers of rock debris would act as an insulator and protect the underlying ice, and as the adjacent ice melts down, a rock table results. Hot springs and heat supplied by the geothermal gradient of the rocks beneath the ice may also act as secondary source.

Practically very little ablation, may be less than 1 percent is accounted by evaporation and sublimation, believed to be essentially due to the large latent heat of vapourization (40 calories per gram). Also capacity to absorb moisture in cold air is limited.

The relationship between the processes of accumulation and ablation is referred to as "*regimen*". When accumulation of ice and snow is more than the amount lost by ablation "*zone of accumulation*" is developed at the glacier-source and in the "*zone of ablation*" loss of ice and snow exceeds accumulation towards the lower part of a glacier. "*Firn line*" separates the zones of accumulation and ablation. It does not necessarily mean that accumulation does not occur in the zone of ablation; it only refers to the balance between the two processes.

In the zone of accumulation, the amount of ice passing through any given cross-section equals the amount passing through a cross-section upvalley *plus* the amount gained by net acccumulation between the sections. Thus discharge increases from the head of a glacier to the firn line. In the zone of ablation, the amount of ice passes through a cross-section upvalley *minus* the amount lost by net ablation in between. Thus discharge decreases from the firn line to terminus.

The position of the front of a glaicer is governed by the relationship between accumulation and ablation. It is stationary; if the relationship is balanced over a period of time, but when the accumulation exceeds ablation, the glacier has a positive economy, and the terminus advances in contrast the glacier has a negative economy when the ablation exceeds accumulation, and the terminus retreats. Even when its terminus is retreating upvalley, a glacier continues to flow down valley unless the ice is stagnant.

AVALANCHE

Thermodynamic changes induced by

- (i) diurnal temperature rise or rise in spring,
- (ii) increase in wind velocity,
- (iii) steepening of slopes in response to progressive snow accumulation, and
- (iv) seismicity in the terrain

bring about disequilibrium resulting in dismembering of superimposed different snowlayers leading to the topographic drop of snow-ice fragments entrapping in its body a part of the surrounding air and water vapour. These visco-elastic bodies termed *avalanche* travel downward aided with gravity at varying speed of 30 to 300 km/h.

Avalanche-prone segments in the Himalaya indicate that they

- (a) orginate predominantly at elevation of 3500-4000 meters and
- (b) are most vulnerable when the slope
  - (i) is convex,
  - (ii) is smooth,
  - (iii) gradient is between 30^0 - 45^0 (however wet avalanches are recorded from slopes less than 15^0) and
  - (iv) bereft of vegetal cover or large rock mass.

The diurnal temperature variation of 5 - 10^0C is found to be most conducive for generation of avalanche (particularly of the delayed-slab type). Slopes more than 60^0 do not allow snow accumulation and therefore do not constitute avalanche-risk surface.

Two broad types (Agarwal, 1979) of avalanches recognized in the Himalaya are:

- (a) *storm induced*, loose fresh snow (powder) *avalanche* (admixed with air, gathering speed upto 300km/h) develops strong pressure that creates shock waves in the valley atmosphere below
- (b) *delayed action slab avalanche,* also called *delayed avalanche* are triggered by seismicity explosion or even induced by vehicular traffic that leads to the dismemberment of differently cohesive layers developed due to load metamorphism with newer layers

depositing above and lower layers under their own pile of weight getting firmly fixed to the slope in order of their deposition. Sunny south facing convex slopes of Ladakh frequently experiences this type of avalanche (Raina *et al.*, 1982).

Dirty-snow Avalanche - Snow avalanches can also build up a peculiar variety of rock deposits termed "*dirty snow*" that includes quantities of rock waste of varying size which picks up large masses of weathered rock and talus during which avalanche cascades down. When the snow melts away, waste in the form of low irregular ridges, parallel to the cliff faces separated by a depression earlier filled with clean snow, known as *protalns ramparts* or *nivation ridges*, appear.

Avalanches are also classified as given in Table VII. 31

TABLE VII. 31 : CLASSIFICATION OF AVALANCHES
(after B. A. Bolt, ; W. L. Horn, ; G. A. Macdonald, ; and P. F. Scott).

Movement	Snow type	Conditions for occurrence	Scale	Character
Primary (during snow fall)	Dry	Slope steeper than friction angle of snow	Small	Surface runs off snow
Secondary (after snow fall)	New snow on crust	External trigger	Small to intermediate	Snow block debris
	Crusted surface	Settling of under-lying snow	Intermediate	Large slabs
	Wet surface layer	Solar radiation warm wind, rain	Small	Wet mass; slow
	Depth hoar	Constructive metamorphism	Intermediate to large	Large blocks entire layer
	Wet snow	Warm weather weak base	Intermediate to very large	Large blocks; slow to medium velocity

The Snow and Avalanche Study Establishment of the Defence Research Development Organization in their investigation computed that on an average 30 persons die every year due to this menace in the Himalaya. The avalanches in this region engulf valleys for several kilometers and destroy anything they encounter. Avalanche threatened places include body shrines of Hemkund, Badrinath, Kedarnath, Gangotri and Yamnotri in Garhwal (U.P.), Trilokinath (H.P.), and Amarnath (Kashmir). South facing slopes in the Himalaya receiving relatively larger solar rediations are more avalanche prone particularly in the afternoons of early spring (Valdiya, 1987). It is estimated that:

	type	*caused by*
72% of the avalanche	mainly loose snow slab avalanche	very heavy snow fall over peaks
15%		temperature and rain wind
13%		
and		

Date	*Destruction*	*Area*
March, 31, 1970	25,000 people died	Areas Mts. North-Central Peru. one of the worst natural disasters of the country avalanche caused by severe seismicity
March, 1979	237 people died in a day	Spiti and Lahaul valley
1982-83	100 people in a day	Jammu and Kashmir
December 1984	27 soldiers	Siachin glacial area in Ladakh-avalanche generated by artillary fire

Mitigation (Alleviation) of Avalanche Hazards

Minute microscopic fractures in ice and snow, if present, emit supersonic signals when stresses are monitered for warning (Cupper, 1982) and the Snow and Avalanche Study Establishment of the Defence Research and Development Organization has successfully launched avalanche forecasting on selected and strategic alingnments. Some techniques utilized to alleviate avalanche hazards include-

A. hurling explosive projectiles using Artillery recoilless launchers and airguns on the snowy slopes, blowing off the overhaning cornices and releasing potential avalanche detering their growth to unmanageable size.

B. erecting avalanche preventing structures such as

(i) retaining structures which may prevent build up stresses in the upper part of the slope by checking snow deposition

(ii) slope terracing which also reduces the gliding of snow substantially

(iii) snow drift control structure

(iv) *snow bridge* or *snow rakes* of timber/aluminium or /steel tubes and pre-stressed concrete of *snow nets* of wires or nylon ropes which help in changing the wind-flow characteristics and thus preventing snow accumulation concentration on vulnerable slopes.

C. diversionary structures

(i) *snow galleries* protect Srinagar-Leh highway by conducting avalanche over them within guide walls limiting the flow-width (in addition dams and mounds are also built to combat avalanche).

(ii) *jet roof* causes the wind to blow swiftly and wipe off cornices and overhangs.

(iii) *avalanche ramps* allows passage over installations and building. Avalanche wedges deflect the snow, break the speed of descending masses of snow and ice and thus provide protection to buildings and installations. This is being planned to protect the Badrinath temples.

GEOLOGICAL WORK

Glaciers like other geological agencies, also act as agents of erosion (Table VII. 32), transportation and deposition. Ice is considerably powerful to erode relatively hard rock when adequately charged with tools.

TABLE VII. 32 : ICE EROSION CONTROLLED LARGELY BY THE THERMAL CHARACTERISTIC OF THE ICE

(after M. J. Bradshaw. A. T. Abbott. and A. P. Gelsthrope. 1978)

Thermal zone	Subglacial erosion	Transportation	Deposition
Zone A net basal melting	Subglacial streams may be very important, also ice with embedded particles crushing and ice striations	Melting and refreezing at base, giving debris charged regelation ice, also melt water	Subglacial lodgement till when frictional drag on particles moved over bed tractional force exerted by glacier ice, more under greater ice thickness with rough bed, giving drumlinoid and filling hollows
Zone B balance melting freezing	Similar to zone A, but reduced unless down stream of Zone A	Similar to Zone A, but basal debris layer of constant thickness on smooth surface	Rates higher than Zone A; no water
Zone C water freezes to moving glaciér sole	Sole slips over as meltwater freezes to it Crushing, polishing and plucking	Water freezing in lower section will add debris rich layers to ice base.	Net freezing tends to re-incorporate material dropped giving net erosion and englacial debris
Zone D Subglacial materials frozen	No slip differential movement above ice/rock junction. Erosion where rock protrudes into glacier. No crushing or polishing	Frozen material gained by plucking. Thrusting in Zone D below Zone C gives upward movement of englacial material	Less subglacial deposition than ablation till. Some drumlins where rock projects

A. EROSION

Glaciers accomplish erosion in two ways :

a. *Plucking* or *quarrying* is the process by which ice after making its way into cracks, joints, bedding planes and other fractures holds a firm grip over the associated rocks surrounding these cracks and because of the glacier body motion the surrounding ice exerts sufficient pressure over the block under the grip to yield by being torn out block by block and removed away by it. Thus, a rough surface left behind the process and the process continues till the irregularities are all *"plucked off"* to make a smooth plane.

   This process is a unique to temperate ice. Heavily jointed crystalline rocks and thin bedded sedimentary rocks are particularly affected by plucking. Further, it has been shown/ proved that quantitatively glacial plucking is more important than glacial abrasion on suitable rock types.

b. *"Abrasion"* is an important process of glacial erosion. It is the rubbing, grinding, scratching and scrapping action of the rock fragments either frozen into the bottom and sides of the glaciers or situated underneath it. Much of the erosion actually achieved through the rasping, grinding and rubbing action of the rock fragments acquired during plucking and transported by the glacier. They act as "tools" of abrasion against rock surface when concentrated towards the sides and bottom of a glacier. The longer and relatively harder fragments groove or cut into the rocks in contrast to the finer materials which act like sand paper in polishing and smoothening the surface.

   Ice tends to hold rock particles in contact with each other or with the rock floor because of its slow rate of plastic deformation. Rock debris at the glacier base cannot easily be pressed upward into the ice and is therefore, forced scratch a long groove in the underlying rock and in turn also get ground flat in the process. The most distinctive features of glacial erosion and transportation are:

   (i) the scratches (called "striations", if they are fine, sharp and indicating the direction of glacier movement)

   (ii) grooves made on the glaciated rock surface, and

   (iii) the flattened or "facetted" polished surface, on ice transported stones.

The glacier erosion depends upon several factors. In general, erosion would be higher with

(i) increase in glacier thickness,
(ii) more amount of rock material carried by it,
(iii) higher velocity, and
(iv) the relatively softer nature of bed rock.

Features of Glacial Erosion

a. *Striations* (discussed above p. 178)

b. *Glacial scars* - Glaciers heavily loaded with rock debris abrade, polish, and groove the underlying hard bed rock. Quartz, sand grains, gravel and big jigged stones have on the surface scratches, scars and furrows, often referred to as "glacial scars", usually one to few meters long, several centimeters wide, and a few millimeters (rarely a tens centimeters) deep. In some sections, the scars run parallel to each other in a particular direction which offers clue to indicate the direction of ice movement in the given place. Sometimes, the scars intersect. Intersecting scars are suggestive of a change in the direction of movement of ice.

c. *Roche moutonnee* - Glaciers smoothen, round, polish and scour the rocks small elevations composed of bed rocks and encountered in their way producing a peculiar, somewhat elongated "sheep back" forms (viewed from a distance, these features appear like sheep lying down). They are longitudinally asymmetrical in which the side (stoss) facing ice movement direction is gentle, usually well polished and striated whereas the opposite slope (lee side) is steep and less affected by polishing. A combination of such forms produce a series of smoothened asymmetrical bosses and depressions called *roche moutonnee,* occasionally of considerable dimensions.

d. *Crag and tail* - When a highly resistant rock (like an igneous body) stands in the course of glacier as an obstruction, it greatly retards the erosive action of ice on the soft bed rocks sheltered behind it, resulting in a peculiar feature in which the obstructing block, boldly faces the ice direction. Such features are called "crag" and the sheltered rock as "tail". The entire structure is termed "crag and tail".

e. *Fiords (Fjords)* - These are overdeepened, narrow channels carved out by the glacial erosion between high walled rocks of which several of them reach the coastline below the sea level through which arms of sea stretch into islands. At least three conditions are necessary for fiords development :

(i) presence of mountainous or plateau topography of appropriate structure

(ii) formation of deep valleys by streams through their selective erosion (*i.e.* eroding weak rocks), and

(iii) intense glaciation of so formed stream valleys. Subsequently, the so developed valleys along the coast are occupied by the invaded sea inlets due either to

(i) the rise of the sea level or

(ii) by the sinking of the landmass

in response to isostatic readjustment following Pleistocene deglaciation. Such features are abundant in high latitudes, as seen along Norway coast.

f. *Corries* - A number of characteristic forms in mountain regions are associated with the work of ice and snow. They are : corries, glacial cirques, glacial troughs. "Corries" are spoon shaped depressions surrounded by high shear walls on the valley in a slope ledge in mountain slopes from which originate the corrie glaciers. Its formation may be initiated by snow accumulations in

(i) small erosion rills, or

(ii) other chance depressions

in the mountain slope and subjected to continued

(i) daytime melting and refreezing with the falling temperature through night, and consequent there to

(ii) intense frost weathering, disintegrating more and more the underlying and surrounding rocks, and

(iii) meltwaters washing so loosened fragments.

Thus corrie depressions enlarge and the corrie glaciers increased.

g. *Glacial Cirques* refer pp 167-168

h. *Glacial valleys or troughs* - Emerging from a ne've field, mountain glaciers descend down the pre-existing stream incised erosion valley and subsequently rework them considerably to the extent that the earlier V-shaped (stream) valleys in most instances, under the action of the moving glaciers are widened and deepened into a U- shaped trough form of which the upper part is open, with slopes of medium (40-50^0) steepness and the lower trough part with steep sides passing into a broad floor. High level benches which meet the walls as prominent "*trough shoulders*" are sometimes observed where the trough passes into the upper wider part of the valley.

The longitudinal profiles of trough valleys are also characterized by some including

(i) extreme irregularity of the floor, and

(ii) the existence of numerous steps called *"riegels"* or rock bars, alternating with "troughs".

Latter's formation is aided by the

(i) differences in the resistance to abrasion of the rocks composing the valley floor, and

(ii) variations in the ice thickness *etc.*

In some sections of the trough valley, occasionally the benches are inclined backwards, most often directly before the ledges.

Characteristic of Glaciated Valleys - Usually glaciated valleys in mountains are relatively recent. They show certain unmistakable evidences of the ice action -

(i) U-shaped cross-sectional profile of its valley

(ii) Many of the glacial valleys (or canyons) in the mountains are relatively much straighter and more open for long distances in contrast to those eroded by streams.

(iii) The tributary valleys in a glacial valley show typically a "discordance" of position, that is, they join the main valley much above its bottom and are, therefore, called "hanging valleys".

(iv) the heads of glacial valleys are very commonly characterized by big, steep-sides, amphitheatre-like basins known as "cirques". Two cirque walls may be cut toward each other from opposite sides of mountain mass until only a very sharp divide known as "knife-edge" ridge is left between the cirques. A knife- edge ridge may also develop where glaciers in two parallel valleys erode and steepen the valley sides until only a very sharp divide separates the valleys. If three or more heads of glaciers cut cirques into a mountain mass from several sides at once, a high pyramid-shaped rock mass, commonly known as *matterhorn* may result.

(v) A less common feature of glacial valleys is that large glaciers entering the sea may erode their valleys tens of meters below the sea level. This is because the moving ice is able to displace the water until its depth becomes so great that ice is buoyed up and broken off. Rivers, too, can cut their channels but very little below the sea level.

EFFECTIVENESS OF GLACIAL EROSION

Areas covered by an ice sheet commonly display three distinct zones :

(i) a central protected area where both erosion and deposition have been relatively less,

(ii) an intermediate zone where ice scouring is pronounced, and

(iii) a peripheral area in which depositional features are prominent.

Other indicators include

(i) abundant striated, polished, and grooved surfaces which give evidence of effective erosion

(ii) lack or scarcity of residual soils over much of the glaciated area

(iii) unweathered glacial deposits (including undecomposed feldspars)

(iv) unassorted deposit and tillite

(v) U-shaped valleys

(vi) hanging valleys

(vii) cirques

(viii) absence of red ferruginous and black carbonaceous matter in the drift.

Those who doubted the effectiveness of glacial erosion cited in support of their view that

(i) striated, polished and grooved surfaces are scarce or lacking in many areas,

(ii) in many places pre-glacially weathered rock underlies glacial deposits above it,

(iii) in many areas glacial deposits of one age are found overlying old soils developed upon older glacial materials, and

(iv) many of the features along glaciated valleys which were attributed to glacial erosion were formed by streams prior to glaciation and only modified slightly by glacial erosion.

Most of these arguments can be equally well cited as evidences for differential erosion. On attempting to visualize the movement of ice within and beneath an ice cap, it seems that one expects localization of flow and erosion as determined by variations in ice pressure and influenced and directed by subglacial topography. It is not always incompatible to find evidence of effective erosion near areas showing little erosion, for glacial erosion like other types of erosion is differential.

Melting of ice and related features- Specially in summer times, glaciers experience melting and sometimes evaporation too due to temperature rise throughout its entire surface. The product is melt water. On favourable slopes this melt water forms streams supported by glaciers on its upper surface; where the slope is opposing, lakes are formed. Emerging streams often fall into the surface fissures (the crevasses) and enlarge the opening to cylindrical hollows called "*moulines*" or "glacier mills" by the whirling action of stone carried by such waters. These hollows sometimes reach the bed rock but generally they merge into the inside caves in the glacier ice and form *"subglacial"* streams. These streams emerge out charged with a good quantity of suspended fine rock materials which turn the water milky, termed "*glacial milk*" in front of a glacier from underneath.

Streams flowing within the glacier body itself are termed "*englacial*" streams and those supported on the upper part of a glacier "*superglacial*" streams. These are frequent along the lateral margins of glacier valleys. Streams occurring at or near the bottom of a glacier are called "*subglacial*" streams. The terms subglacial, englacial and superglacial also refer to the rock materials, moraines respectively associated therewith.

The dust particles on the surface of a glacier are likely to absorb more solar heat in relation to the surrounding snow and melt the snow beneath it forming small basins, called *"dust wells"*. But if the rock fragments are large and considerably thick the situation would be other way round. The rocky substance would serve as an insulating cap and prevent the snow below from melting. Sometimes, huge blocks stand on a pedestal of snow under such condition. They are known as *"glacier tables"*.

TRANSPORTATION

Glaciers carry enormous bulk of load including rock fragments of varying sizes and are transported by glacial ice and water to their final destination where they are deposited.

DEPOSITION

Glaciers pick up every thing that comes in their way and transport the acquired materials embedded on or trapped in their body during their development and movement. The material, termed *"load"* of the glacier is solid and has no opportunity of sorting and therefore, they are generally heterogenous in character and may be comprised of fine clay, rock fragments, boulders and even bigger blocks. There is end to the gathering of materials when the glacier moving downslope either

(i) becomes "overloaded" with debris, or

(ii) even when at a farther end wastage of ice by melting increases greatly.

These factors tend to reduce the transporting ability of a glacier, initiating the process of dropping down, partly or wholly, the debris and deposition begins, in some cases only temporarily, for often they are lifted again by the melt water streams emerging from the glacier mouth and transported further. Glacial deposits, irrespective of the way they developed, are collectively known as "*drift*" which may be of two types :

(a) *Glacial Drift* or *Till*- These are deposited directly by the glaciers, and therefore are unstratified in structure and heterogenous in composition, material derived from local bed rock. Moraines and drumlins are typical examples.

(b) *Fluvio-glacial* or *Glacio-fluvial drift* - These are deposits formed by the combined action of water and ice, stratified and somewhat sorted.

MORAINES OR TILL OR GLACIAL DRIFT

Moraines are abraded, gathered, transported and finally deposited rock materials. Ancient moraines often consolidated are called *tillites*. They include mixed and varied lithologies of heterogenous, non-stratified to stratified sediments, within intermediate type, both stratified and unstratifed nature (Table VII. 33). All such non-stratified and stratified deposits of glacial activity are collectively referred to as *glacial drifts*. Poorly sorted and non-stratfied deposits of glacial origin are designated as *"till"*.

Some advance more together with the ice and others are deposited on glacier melting. Accordingly, they may be "moving" and"deposited" types.

"Moving moraines" - In every mountain glaciers several varieties of moving moraines are identifiable. Depending on their location, moving moraines may be further classified :

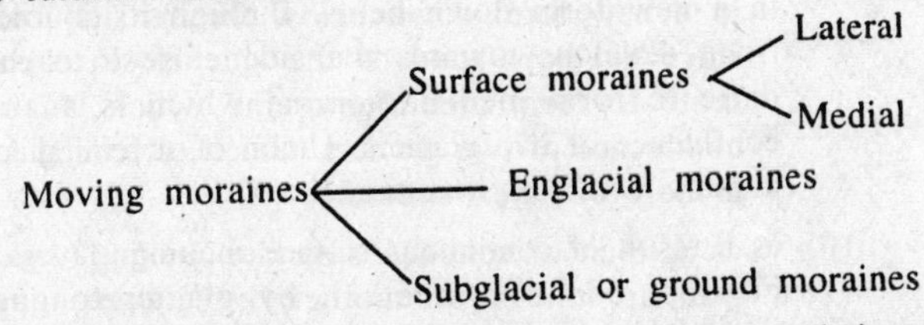

"Surface lateral moraines" are longitudinal ridges - like accumulations of rock debris liberated by weathering from the steep valley slopes along the sides rising above the glacier tongues upto or more than 30 meters often persisting for sometime even after the disappearance of the glacier. They are formed mostly of rock fragments. Debris carried by streams along lateral gorges, and in part derive from hillside creep add to the glacially gathering mass.

TABLE VII. 33 : CLASSIFICATION OF GLACIAL DRIFT AND LANDFORMS

(after V. K. Sharma, 1986)

Locus	Depositional Environment	Dominant material	process	Major morphological features
At or inside glacial margin	Active ice	Nonstratified drift (till)	Subglacial	Ground moraines, flute ground moraines
			Superglacial	Moraines due to dumping, pushing, crushing and squeezing (numerous types) Drumlinoid ridges
	Stagnant ice	Till and stratified drift	Ice disintegration	Hummocky dead ice moraine (an assorted feature)
			Ice contact	Crevasse filling, Linear disintegration Ridges, Eskers, Kame, Kame terraces
Outside glacier margin	Glaciofluvial and Glaciolacustrine	Stratified drift	Proglacial	Outwash plain (included pitted outwash), valley train, lacustrine plain, marine delta.

"Surface medial moraines" are developed either

(i) by coalescence of lateral moraines of two glaciers which result in a new form down below through its journey where a long ridge develops towards the side close to each other and they unite to form medial moraine which is an indication of the confluence of two glaciers. Union of several glaciers may present a more complex structure,

(ii) as a result of continuous surface melting layers after layers may pile up, one over another, by glacier tongues. Under these conditions even rare inclusions of rock fragments form noticeable accumulations in the form of a sheath on the surface lowered by melting.

"Englacial moraines" may accumulate both in the ne've field and in the area of the glacier snout. Debris tumbling on to the ne've glacier field enter into the ice crevasses and further trapped by the fresh abundant supply, and become firmly embedded in the ice and thus engulfed in the moving ice until they reach the wastage area, where they are dropped.

The "subglacial " or "ground" moraines are debris, derived from

(i) periglacial weathering,

(ii) subglacial weathering, and

(iii) destruction of the glacier floor by the moving ice and frozen into the sole of the glacier. As a result of periglacial weathering at the beginning of glacier formation, the snow falls on a surface already envelope the pre-existing surface debris. These entrained clastic materials are enriched on the way by the products of scrapping and plucking of the rock of the glacier flow and subglacial weathering on flowing out of the ne've field. Subglacial and englacial moraines are most pronounced in the continental type of glaciers.

"Deposited moraines" - The detrital materials engulfed in or lying on the glacier surface begins to accumulate on the ice melting, a process already in progress with the glacier movement and it is most intense at the time when the glacier retreats. Moraines of the various types are deposited on the surface consequent upon complete disappearance of the glacier flow, some of them retaining features they had when the glacier was moving (ridges of lateral and medial moraines) even after deposition. Deposited moraines may be

(i) terminal, or

(ii) basal

Terminal moraines - Enormous quantity of detrital material carried by the glacier through its journey from supply region to the glacier periphery are laid off at its tongue front of ice sheet, where the ice is thinnest and begins to melt. In a relatively stationary glacier snout the entire ridges of diverse detrital material, termed "terminal moraines" are formed in its front. All types of moving moraine-surface, englacial and subglacial-take part in the deposition of terminal moraines.

The terminal morainic ridges fringing the glaciers reflect the shape of their front parts. They are 30-40 meters high, and arcuate shape with varying curvature [in the case of mountain glaciers, but they follow the outline of the ice sheet in case of continental ice sheets]. Sometimes they assume a complex festooned shape and achieve great height and width.

These moraines are valued for clearly defining the prolonged stationary position of the glacier front edge reflected in frontal ridges and several of them are indicative of intermittent glacier retreat. No such typical terminal moraines are formed of continuous glacier retreat.

Basal moraines - This is a composite term for no distinctive morphological features as they are formed mainly out of the subglacial and englacial moraines, also on the thawing of the surface moraine which develops on the glacier surface. The basal moraines are possibly laid down only on complete glacier disappearance. Due to the changes in the supply conditions ice becomes inactive, its degradation intensifies, thus causes the ice melting and growing thinner, lowering down the surface and englacial moraines to sink below until they finally reach the glacier flow, coalescing with the subglacial moraine. The regions of development of basal moraines are characterized by sections of different relief. The predominant types are:

(i) smoothed plain morainic, and
(ii) undulatory morainic

reliefs. The non-uniform concentration of the detrital material of the englacial moraines within the glacier apparently is the reason for development of the undulatory morainic relief.

The lateral and medial moraines are projected onto the surface of the basal moraine on ice melting in mountainous terrains, that is often clearly expressed as longitudinal ridges in the relief. The lateral moraines thus deposited are then referred to as *"flank moraines"*, while the medial ones as *"longitudinal moraines"*. Sometimes, however, moraines of either type are designated as longitudinal. In contrast, longitudinal ridges do not develop in terrains that formerly experienced continental glaciation, and instead the plain or undulatory swell and swale morainic reliefs as well complex series of terminal morainic ridges are developed.

"Morainic structure" - Moraines are composed of most diverse detrital materials-fine clays, loams, clayey sands, gravels, shingles and boulders. The boulders range may be a few centimeters to 2-3 meters or more in diameter. The relative distribution of the component elements of a moraine may be most diverse, depending on many factors, such as

(i) the proximity to glaciation centres,
(ii) the intensity of supply of detrital material,
(iii) the composition of rocks of the glacier flow, and
(iv) the length of the route covered by the glacier and its thickness.

The variations in moraines are indeed very large, at some places they may be predominantly composed of clays or loams with inclusions of larger detritus, such as gravel, broken stone and boulders and other places, it may be just the other way, composed by detrital material with varying clayey sand grain size.

Thus, the features distinguishing glacial moraines from other continental deposits

(i) are non-uniform

(ii) lack the sorting of the detrital materials and

(iii) show no stratification.

DRUMLINS

In addition to irregular, chaotically scattered swells produced by glacial accumulation, certain distinctly regular features relatively low elongated hills usually gently sloping with long axis parallel to the direction of the glacier movement, called *drumlins,* are also identified. They are heterogenously composed of dense clay with boulders above a polished ledge of bed rock, the core of which is located either at the middle of the drumlin or at one of its ends. In the latter case the ledge is often exposed and, in fact, constitutes a roche moutonnee of which the drumlin is the continuation. These features vary in length from a few hundred meters to 1-2 km and are 2-3 times as long as they are wide, but generally are a few meters high. They often occur in groups separated by intervening depression. Drumlins generally occur in areas of very thick glaciers of the continental type.

Drumlin formation theories may be grouped into two categories :

(i) erosional, and

(ii) depositional.

The drumlin dynamics is, however, inadequately understood:

(a) The advocates of erosional hypothesis are of the view that drumlins are formed of a pre-existing drift. Such a postulate necessitates two clearly defined ice advances :

(i) the first ice advance provides an irregular till cover which

(ii) during a second advance is eroded into drumlins.

This theory suffers as much as instance(s) of a second advance is (are) not forthcoming.

(b) The depositional hypothesis assumes a progressive deposition of drift. The manner of drift deposition at present is only conjectural. It is believed that the following factors may initiate deposition :

(i) deceleration of iceflow due to a large debris load in transport,

(ii) presence of rock cores and/or knobs of frozen till around which deposition occurs, and

(iii) subglacial accumulation of till into streamlined forms of deglaciation.

This postulate also suffers from limitation, such as

(i) a core is not always present in the drumlin,

(ii) many drumlins are rich in sand or sandy loam, and only small amount of clay, and

(iii) the presence of stratified drift in drumlins cannot be explained by deposition in a moving glacier.

Gravenor (1953) proposed a modified erosional hypothesis. It is based on the observation that drumlin eroded from a pre-existing drift occur in the same drumlin fields as those made of till. It is suggested that during a temporary halt of an advancing ice, stratified and non-stratified drifts accumulate at the margin of the glacier.

Kupsch (1955) recognized both erosional and depositional environments in the formation of drumlins. Melt water selectively removed finer fraction, leaving boulders as lag deposits. The drumlin evolved from the kame following the advance of the ice. The boulders were fractured by the pressure of the moving ice and stratified drift got deposited in the leeside of the original kame.

A model of drumlin formation in terms of dilatancy of material has been discussed by Smalley and Unwin (1968). By definition, dilatancy is a property of granular mass. Granular material on deformation expands and becomes more resistant to shear stress. According to the model, till is continually deformed by the movements of the glacier. The deformed till layer experiences varying level of stress. When the stress drops below a certain critical level, the expanded material collapses into static stable forms. Once a stable form is achieved, further stress pląy cannot make it unstable. The heap of collapsed material is shaped into drumlins by the ice. The model best describes the drumlins formed by basal till only, but, fails to account for the rock-cored drumlins and drumlins formed of stratified drift.

ERRATICS

These are glacial boulders varying in size from pebbles to masses as big as small houses, strewn irregularly over the country on high mountain as left over by the melting of the ice. They comprise of hard rock, the softer material are generally ground up soon by the glacier action. Some erratics are reported to have been moved for short distances from their ledges; many have been transported at least some kilometers; while some have been carried over hundreds of kilometers. Erratics weighing from 5 to 20 tons have sometimes been left in such remarkably balanced positions on bedrock that they can be made to swing back and forth slightly by hand pressure. Such boulders are sometime called *"rocking stones"*. Erratics occur in abundance in Malla-Johar area in trans-Himalaya.

GLACIO-FLUVIAL/FLUVIO-GLACIAL DEPOSITS

These deposits are laid down by combined action of water and ice and are commonly stratified.

"Valley trains" - Waters charged with rock debris emerging from the ice flow down a gently sloping valley away from the glacier end and on the way down lay off some or most of the load on the valley floor, often for kilometers beyond the ice front, forming a crudely stratified, valley train.

"Outwash plains/Frontal aprons" - These plains are situated immediately in front of the outward edge of terminal moraines and are undulatory plains located in nature, composed of stratified sand, gravel and shingle. They constitute very gentle and broad coalesced alluvial fans of subglacial streams formed over a smoothed relief. The streams emerging from below the glacier onto such smooth surfaces sloping insignificantly 3-4^0 to 8-10^0 only become sluggish because of reduced flow velocity of the streams, spread out widely, break into a network of streams and deposit the detrital material they brought, coarser detritus, such as gravel and shingle, coarse sand usually deposited near the edge of the terminal moraines and the finer ones predominantly particles composed of sand followed by fine silt and clay fans over a wider area. Thus, outwash plains are for the most part composed of sand.

"Kettle/Kettle holes" - 3 to 30 meters deep irregular depressions without any outlet and with steep sides are often formed in outwash plains resulting from the melting of blocks of ice detached from a retreating glacier and trapped in the outwash material. Kettle holes may also develop in glacial lake deposits where ice-bergs are stranded, buried under sediment and later melted.

"Kames" - Jamieson in 1874 introduced the term "kame" for randomly scattered hills, isolated or in small groups, averaging 10-12 meters high. It resembles morainic hummocks in form but for its inner structure. Deep depressions occur between the hills when they are grouped, giving rise to *knob and kettle topography*. Most common locale for their occurrence is valley bottoms, but sometimes they may be on hillsides or even on hilltops. They occur in association with terminal and recessional moraines. Sometimes they form so called "kame moraine" ridges composed of stratified and somewhat sorted well graded sands at times with gravels and shingles; fine clays in places resembling varve clays and boulder material.

Presence of fine clays is suggestive of the fact that they had formed in stagnant, rather than moving water (it is believed sometimes that kames form at the margins of glaciers by debris loaded streams which heap up the materials, usually sand and gravel, as they emerge from the ice, sometimes the debris charged water rises as great fountains). It is for this reason the kame formation is linked with the association of stagnant ice, that stopped moving on the surface of which on rapid thawing various and numerous depressions are developed; the ice are filled with water and turn into small lakes. A large number of big and small streams empty alongwith detrital material of different composition into such lakes from elevated section and deposit them. The presence of boulder material may be due to its direct thawing out of ice. As the ice mass melts, the materials accumulated in the lakes gradually sinks down and imposed on the surface the basal moraines as a cluster of low hills.

"Varves" - In closed glacial lakes with no outlets, the glacier melt water deposits thin alternating sand and clay srata 0.5 to 5 cms. thick, presumably in response to seasonal variations and this water level increases and decreases, depositing the sand strata in spring and summer (on snow melting) and the clay strata in winter (on water freezing). Such deposits characterized by alternating coarse and fine materials in response to seasonal fluctuations is called "varved clays" or "varves".

"Eskers" - Eskers are 5 to 50 meters high, usually less than 2 km long sinuous, steepsided rather smooth surfaced narrow rounded ridge parallel to the direction of the movement of glacier, composed of a mix of assorted and somewhat stratified sand and gravel which are believed to represent fillings of superglacial, englacial and subglacial stream channels (Figure 10). It is far more common in areas of continental glaciation than in glacial troughs for eskers are features formed when ice stagnates and stagnant ice is rare in valley glacier. Three theories are proposed in an attempt to explain the origin of eskers which is still not clearly understood.

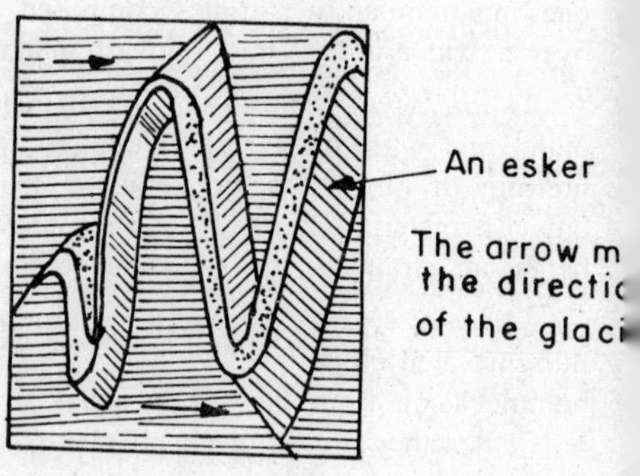

Figure 10 : Esker geometry

(i) Delta theory - It links the formation of eskers with the emergence of concentrated and powerful streams from the subglacial tunnel. The ratio of a subglacial tunnel width with its height primarily determines the slope of an esker and its internal structure. A flat topped esker with fairly undisturbed sediment is related to tunnel which is wider than its height. Narrow but high subglacial tunnel creates a sharp-crested esker with high proportion of slumped material. Each subglacial stream is under great head of pressure in the tunnel, and possessing great velocity, rewashes the moraines and transports the detrital material over large distances. When the stream emerges into the frontal region of the glacier, its flow rate sharply diminishes and so does its kinetic energy depositing all the sand and gravel brought by it at the edge of the glacier, forming a relatively narrow delta. When the glacier and, therefore, its edge too retreat, the ice water stream retreats. Then, the delta is formed at a new place corresponding to the new position of the mouth of the stream. As such more and more deltas are produced which on coalescing may form a continuous or intermittent ridge or esker.

(ii) Channel theory - In this theory the esker formation is related with surface flow water and englacial streams. The streams flowing through channels rewash the morainic material and deposit in their channels in the same way as alluvium is laid down in common river channels. These superglacial and englacial streams are distinguished by a winding course. When the glacier melts away, these fluvio-glacial stream deposits are imposed on the basal moraines or directly on the glacier floor forming sinuous ridge.

The channel theory explains well the origin of eskers on different elements of the relief. The delta theory cannot explain the rise of an esker onto an elevation.

(iii) Proglacial origin - At least some eskers are believed to have developed in ponded water or a proglacial environment formed by annual accretion of sediments in standing water and eskers display coarse and fine sediments in a clearly defined stratum. The coarse sediments form esker centres and represent a single phase of summer ablation. The fine sediments are deposited in winter season. Such successive esker centres give rise to a long continuous ridge.

CAUSES OF GLACIATIONS

An explanation to the causes of the repeated glaciations is rather complicated and knotty problem. Several, often conflicting hypothesis have been advanced.

a. Dubois (1893) proposed to explain the glaciations of the Anthropogene and their intermittent nature due to the evolution of the sun and assumed two stages

(i) yellow star stage (corresponding to interglacials) and

(ii) red star stage (corresponding to glacial periods).

He assumed that before the Eocene and the Pleistocene passed into the former stage and subsequently into the variable stage (alternation of the yellow and red stages). The hypothesis suffers from the fact that

a. the evolution of stars is rather a much more complex and slow in terms of modern concepts and
b. the history of the earth, evidences of repeated climatic changes, are recorded from rocks of pre-Eocene times.

b. F. Nolke's (1937) postulation is based on the passage of the sun through a nebula resulting in the absorption of a part of the sun-rays and the weakening of the solar radiation and, consequently, resulting in glaciation. He further believed that the alternation of glaciation and interglaciation was caused by the variations in the density of the nebula.

c. G. Simpson (1934) opined that glaciation is a function of fluctuations in solar radiation caused by the sun's inner state presumably over long span of time.

d. W. Kopen and M. Milankovith (*cited* in G. Gorshkov, and A. Yakushova, 1967) advanced yet another hypothesis-

(i) some relationship between climatic changes and periodic and combined changes in the earth's rotational elements (*i.e.* the angle between the equatorial plane and the plane of the earth's orbit, the period of variation of which is 40,000 years).

(ii) the magnitude of the eccentricity of the earth's orbit (the distance between the centre of the ellipse and its focus at which the sun is situated varies with a part of 90,000 years).

(iii) the position of the perihelion (at the perihelion the earth is closest to the sun).

A combination of

(i) low obliquities of the ecliptic with great eccentricity and

(ii) the position of given hemisphere at aphelion (farthest from the sun)

in their opinion should cause a temperature drop and provide favourable conditions for the development of glaciations.

e. Swedish scientist S. Arrhenius (*cited* in G. Gorshkov and A. Yakushova, 1967) postulated that the proportion of carbon-di-oxide and the atmospheric variations as the cause of temperature fall and, therefore, of glaciations. It is observed that the carbon-di-oxide of the atmosphere freely passed the sun-rays to the earth's surface but absorb the reflected heat (infra red) rays. He believed that increase in carbon-di-oxide content in the atmosphere causes a temperature rise and consequent climate-warming, while a fall in its content would lower the temperature. Carbon-di-oxide content in the atmosphere is estimated to be about 0.03 percent and if this amount diminishes by half, the earth's temperature would go down by 4-5^0 C, conditions favourable for another glaciation.

Volcanoes may be one of the sources of atmospheric carbon-di-oxide. According to Arrhenius (*cited* in G. Gorshkov and A. Yakushova, 1967), a mild subtropical climate with luxuriant vegetation should correspond to periods of intense volcanic activity, and, therefore, to periods of maximum carbon-di-oxide content. More vegetation would help to slowly deplete the carbon-di-oxide from the atmosphere and lower the temperature. From this it follows that decline in volcanic activity continually over considerable time and consequently lowering of carbon-di-oxide content in the atmosphere is conducive to bring about a glaciation. This hypothesis is corroborated by the relation of coal accumulations with the intense volcanic activity at a number of places followed by a temperature decline, resulting in glaciations in the Tertiary period.

Although the hypothesis sounds convincing, it fails to match with palaeogeographic data related to volcanic activity at different periods of the earth's history and its relationship to ancient glaciations. Role of volcanic activities in changing the climate has been viewed differently by different workers; for instance, W Humphreys (1929) believes that intensified volcanic activity could bring about a cold spell instead a rise in temperature due to great screening effect to the enormous quantity of dust ejected into the atmosphere during volcanic eruptions and estimated a lowering of the earth's mean annual temperature of about 0.5^0 C over the past 160 years due to volcanic activity.

f. Yet another postulation is based on relief changes in response to mountain building movements and attendant climatic variation. Orogenic movement occurred number of times through earth's history. At the end of Neogene globally erected enormous mountain systems - the Alps, the Carpathean mountains, the Caucasus, the Pamir, the Himalaya, the Cordillera, the Andes *etc.* sharply modifying the land and sea relationship and thus

(i) a redistribution of moisture laden winds and

(ii) possibly oceanic currents too,

which had a corresponding effect towards changing the climate. Conditions favourable for ice accumulations were generated in the mountain ridges situated above the snowline. The glaciation centres that developed in the mountain, in turn, tended to lower the temperature of other regions.

Very interesting from this point is the theory of C. Brooks (*cited* in G. Gorshkov and A. Yakushova, 1967) concerning critical climatic changes. In his opinion, a small upsetting of the equilibrium of climatic conditions may trigger major climatic changes. The newly created mountains with their glaciers might have been such an upsetting factor.

The relationship between

(i) glaciations,

(ii) tectonic movements, and

(iii) relief changes

may be established for ancient periods also. For instance, the Devonian glaciation developed after Caledonian folding; the Great Carboniferous-Permian glaciation after the Hercynian folding. Such coincidences could not be purely accidental.

Accordingly the temperature lowering during the Quaternary period is attributable to considerable relief change caused by tectonic movements of the earth's crust. However, it fails to explain either the intermittent nature of glaciations, or their recurrence. In this respect, K Markov and I Gerasimov (*cited* in G. Gorshkov and A. Yakushova, 1967) added great significance to the periodic and combined changes of elements of the earths rotation alongwith the general cooling of the surface of the earth, caused by mountain building.

GLACIATIONS THROUGH GEOLOGIC TIMES IN INDIA

In Indian geological history, prevalence of glacial climate, though not very frequently are recorded.

(a) The earliest evidence of glacial action is evidenced in the scratched pebbles of the Kaldrug conglomerate of Dharwar (Huronian) age in South India.

(b) Tillite composed of angular to subangular rock boulders and fragments, sand-size to a meter across, occurring at the top of the Proterozoic- Bijawar group, are exposed along the banks of the Ken river below the Gangan Dam in Chhatarpur district of Madhya Pradesh. They lay randomly and haphazardly set in a structureless, ferruginous clayey mass. Boulders are generally facetted and a few have striations on the surface. Further more in Sidhi area glaciation is observed in Bijawar (Cuddapah) group of rocks.

(c) Yet another Pre-Cambrian tillite consisting of striated and fractured boulders set in a clayey matrix is reported from the Mirzapur-Sidhi area of Son valley at the border of Uttar Pradesh and Madhya Pradesh. Associated beds in some cases look like varves.

(d) In Bap and Pokran areas in Rajasthan, evidences of two quick and successive glaciations in the Vindhyan period are suspected.

(e) Gondwana Glaciation - However, the first definite glaciogene bed from the Indian subcontinent comes from the Talchir pavement exposed in the Penganga river a little (about a kilometer or so) before its confluence with the Wardha river in the Chandpur district of Maharashtra. The Talchir rocks of the Lower Gondwana (Uralian or Upper Carboniferous) embodies green laminated shales and fine sandstone over the boulder bed. The evidence of glacial environment is now well documented in

  (i) the presence of undecomposed feldspars in the sandstone,
  (ii) green colour of the shales,
  (iii) the polished, scratched, grooved and facetted boulders.

The boulder beds are wide-spread and reported from Hazara, Simla, Salt Range (in Pakistan), Rajasthan, Madhya Pradesh and Orissa, are suggestive of the prevalence of ice sheet of continental type in India at that time. They are identified in

(i) Tannaki Boulder bed in Kashmir-Hazara
(ii) the Mandhali Formation in Garhwal, and
(iii) the Blaini Formation in the Simla area of Himachal Pradesh referrable to the Talchirs.
(iv) The basal part of Agglomeratic Slate Formation in Kashmir also referred to be glacial in origin.

Rocks of similar origin are also known from all over the Himalaya right from Kishtwar through Ladakh and Nepal to Bhutan.

Yet another is suspected based on the presence of undecomposed feldspars in the sandstone of the Panchet beds of the upper part of the Lower Gondwana (Triassic) Supergroup of rocks. However, they may as well be due to arid climate.

(f) *Plio-Pleistocene glaciation* - Widespread glaciation is documented from the rocks of Plio-Pleistocene period during which time four glacial advances took place-

  (i) the first phase is related to the laying down of the Tatrot formation (base of the Upper Siwalik,)

(ii) followed by the Sorna, Potwar glaciations and the last - a feeble phase in Potwar area (Terrace 4) and represented by the boulder beds.

Evidences of glacial advances are also forthcoming from

(i) three well defined terminal moraines of longitudinal glaciers in the western Himalaya,

(ii) two cirque levels,

(iii) three valley floors,

(iv) successive valley-width decrease above each terminal moraines, and

(v) descent of the glaciers from the upper level cirque groups.

Three glacial advances are recognized :

Glacial Advances	*Stage*	*Elevation* (above mean sea-level)
3rd advance	Sheshnag-Millan	3,650 meter
2nd advance	Aru-Martoli	3,050 meter
1st advance	Pahalgam	2,750 meter

Large blocks clearly arranged along well defined lines are indicative of the movement of these glaciers. At times the "erratics' occur conspicuously in a bed of gravel overlying the silt. The blocks and boulders were transported from the Hazara and Kashmir hills by glacial ice and deposited into believed to be once a great lake in Potwar with an exit seemingly close to Kalabagh.

Several warm interglacial period climates intervened the cold climates and it is generally believed that three interglacial periods, separated four periods of glaciations in the extra-Peninsular region. It may be interesting to note that Peninsular India also experienced relatively cold, mild climate, as inferred from migration of plants and animals from higher latitude down to southern side.

CHAPTER VIII

STREAMS/RIVERS

The character of a river is determined by three principal features :

(a) the quantity of water carried by it,

(b) the depth of the water, and

(c) flow velocity.

These do not remain constant but change from season to season and year to year. The pattern of these changes is called the *regime* of a river.

STREAM EROSION

Erosion is defined as the wearing away of the stream channel wetted perimeter in contrast to picking up of loose material dropped temporarily by the transporting stream. The erosional work of a stream increases with discharge. Stream erosion includes in its wide variety of processes :

a) *Corrasion* : It is a process by which a stream uses its load to scrap away its bed. In steep confined sections of stream channels this may lead to fluting and pot-holing the solid rock as also the lowering of the channel.

b) *Attrition* : It is the mechanical wearing of the transported fragments which because of their differential rate of movement by reason of size variations, as the specific gravity for most of the sediments it is composed of is the same, collide with each other,

(i) both, among themselves and

(ii) as also with the bed and the side rocks,

leading to rounding of corners easily perceptible in coarser fractions like pebbles, boulders *etc.* which may also be reduced down in size to sands and silts.

c) *Corrosion* : This refers to the dissolution action by the stream water, more conspicuous in limestone country adding to the solution carried out in weathering.

d) *Hydraulic action* : Occurs where the flow of water is responsible for undercutting banks of unconsolidated alluvium. It is largely dependent on flow turbulence. Greater turbulence induces higher erosional capacity. Hydraulic action of a river is a function of

(i) stream gradient,
(ii) width, depth and shape of the channel,
(iii) stream discharge.

e) *Cavitation* : is a pressure effect at high velocities in which bubbles are developed when form in narrow sections collapse in wider stream sections, leading to the generation of shock waves that deliver hammer blows to the channel sides. Other effects include vortex-like turbulence developed due to the sucking effect also contribute to stream erosion.

f) Water flowing under pressure as in confined conditions may erode more powerfully. This is important (i) in limestone areas, and (ii) in the case of meltwater being channelled through and beneath glacier ice. Under pressure water is effectively squeezed through the ice or rock producing steep-sided, smoothed or fluted rock surfaces.

The erosion accomplished by these methods often work together, is related closely to the

(i) discharge
(ii) load and
(iii) gradient

of the river : the load is involved in corrasion and is itself produced by erosional processes.

STREAM TRANSPORTATION

Stream efficacy as a geomorphic agent is effective only when it is able to move sediment load which depends on one of the four types of the mechanisms involved:

a) *Solution* - The chemically *dissolved load* including carbonates, sulphates, chlorides and oxides are transported by the streams to lakes or seas and precipitate

(i) when oversaturated, and
(ii) on evaporation

b) *Suspension* - Agitated waters remove fine-grained particles with limited mass in mechanical suspension. With the turbulence and/or bouyancy decrease, the suspended load is settled down according to their weight, first the coarsest particles, and the finest last. Clay or siltsized materials are generally suspension- transported.

c) *Saltation* - Particles rather heavy for suspension transport bounce down on the bed. Energy to leave the bed is provided by the hydro-dynamic lift that facilitates their downstream travel with a flat trajectory of a bullet by raising the particles above the ground may be as much as 30 to 120 cm They are energised by

(i) hydrodynamic lift of the system, and /or

(ii) the impact of and subsequent push given by the other jumping sandgrains.

Thus transported load, mainly comprising sand-size particles are restricted to the bottom of the stream.

d) *Traction* - Still coarser sand and certainly pebbles are usually heavy enough to be lifted up for jumping as in saltation. They slide or roll over the bed itself by the mechanical drag and turbulent push of stream waters.

The materials moved by saltation and traction are called *bed load* into which are also included the saltation-transport materials suspended and dissolved loads. The bed load is hard to measure because of difficulties in defining the boundary between mobile and immobile bed load during bankful stage, let alone the limits between bed and suspended load also become arbitrary.

The transport ability of a stream is best identified by its :

a) *Capacity* : It expresses the total potential weight of stream sediment load which varies with particle size; and

b) *Competence* : This refers to the weight or size of the largest particles that the stream can possibly move along its bed.

These two concepts of stream transport are function of

a) *Size and to some extent, shape of the material* - The specific gravity of surface rocks and minerals generally varies within narrow limits (between 2.5 and 3.5) and therefore, particle size is practically considered for particle weight. With identical stream velocities in an area of uniform bedrock, the distribution and proportions of the suspended and bed loads are determined by

   (i) particle size and

   (ii) to a lesser extent, shape.

b) *Turbulent Velocity* - Shear within a stream is closely related to the caliber of particles picked up and is demonstratable; thus stream competence is proportional to turbulent velocity.

   (i) Turbulent eddies support the upward force necessary to keep sediment in suspension,

   (ii) whereas general turbulence imparts the energy to drag, push, and flip particles along the bed.

For the load dynamics much of the energy would be consumed

in overcoming skin resistance. As such, stream competence would be relatively great in shallow, turbulent streams with a small hydraulic radius.

c) *Current velocity* - Capacity in contrast to competence is proportional to the third power of mean current velocity. That is, the total possible amount of sediment-of all sizes-in transport is largely determined by the discharge volume, being greater is larger than in smaller streams.

On overviewing, two principles of stream behaviour most relevant for erosion and deposition are :

(i) geomorphic change is restricted to those periods of time when discharge and velocity are high,

(ii) turbulent velocity is closely linked to stream competence and current velocity is distinctly to the stream capacity.

CAUSES OF STREAM DEPOSITION

The two most fundamental causes of stream deposition are

(i) loss of transporting power and

(ii) inability of a stream to transport all the material brought into it by its tributaries.

The following outline attempts to show the more common and immediate causes:

A. Loss of transporting power resulting from :

1. Decrease in the velocity produced by

a) Decrease in gradient because of

(i) passage from upland to lowland areas

(ii) diastrophic warping or tilting

(iii) delta extension of a trunk stream

(iv) increase in valley crookedness.

b) Increase in the spread of stream waters where the streams

(i) debouch from mountain valleys on lowlands

(ii) overflow banks during flood stage

c) Natural obstruction caused by

(i) more rapid aggradation of main stream than by tributaries

(ii) boulder fans built by steep tributaries into main stream

(iii) landslide dams

(iv) lava flow dams

(v) sand dune dams
(vi) timber rafts
(vii) artificial dams
(viii) minor obstructions in stream channel.

d) Decrease in volume as a result of
(i) decreased run off because of climatic changes
(ii) evaporation losses
(iii) seepage losses
(iv) diversion of waters by stream piracy
(v) diversion of waters by man

B. Excessive load produced by :

1. Great quantities of glacio-fluviatile outwash

2. Increased erosion in drainage basin because of reduction of vegetal cover caused by

a) anthropogenic activities
b) climatic change

3. Excessive load contribution from :

a) tributary valley being cut in easily erodable materials
b) anthropogenic activities.

VALLEY DEVELOPMENT

It includes valley carving of rivers by deepening, widening and lengthening.

I. Valley deepening is effected in the following five ways:

a) hydraulic action
b) corrasion or abrasion on the floor of the valley
c) pot-hole drilling along the valley floor and at the base of waterfalls
d) corrosion and
e) weathering of stream bed.

II. Valley widening too may be accomplished in the following ways :

a) lateral erosion or planation by the stream in a valley may remove material from the base of the valley side through (i) hydraulic and (ii) corrasive action.
b) rainwash or sheetwash on valley sides
c) gullying on sides
d) incoming tributaries contribute to valley widening even though they may be no more than overgrown gullies.

III. Valley lengthening may take place in the following three ways/locales

a) by the processes of headward erosion

b) through increase in size of their crooks or meanders and also

c) at their termini.

DRAINAGE TYPES

Fundamentally stream courses are determimed by

(a) the initial slope of the land

(b) adjustment to diffrences in rock resistance and

(c) random headward erosion.

A stream following a newly developed land surface is a *consequent stream.* Its' direction is detemined by the initial land surface slopes Among common examples of this kind of streams are those which form on

(i) newly developed volcanic cones or lava felds

(ii) recently deglaciated regions

(iii) emergent marine areas, and

(iv) newly uplifed domes, anticlines, or fault blocks

Streams which have adjusted their courses to weak rock belts by differential erosion are known as *subsequent streams.* Most subsequent streams are seen over folded or tilted sedmentary terrains. Faulted/jointed terrains are also prone to this type of river developments, are developed by selective headward erosion, although other types of adjustment are possible (Figure 11).

Insequent streams are those whose courses are developed by random headward erosion or are not related to determinable factors. They are developed due to random headward erosion and do not follow regional consequent slopes or guided by weak rock belts; so tributary streams develop random directions. Insequent streams are common on homogeneos rocks or on horizontal bedded sedimentary sequence without apparent discordance.

An *obsequent stream* flow is in a direction opposite to that of the orignal consequent drainage whereas a *resquent stream* is one which flows in the same direction as the consequent drainage but developed at a lower level than the initial slope after other drainages intervened. It has essentially "retaken" a consequent direction.

Pre-surface relief streams impressed upon the area as for example, the Sindhu, the Sutlaj and the Brahmaputra are of *antecedent type.* They are characteristic of gradually rising terrain wherin rate of river erosion exceeds rate of terrain uplift.

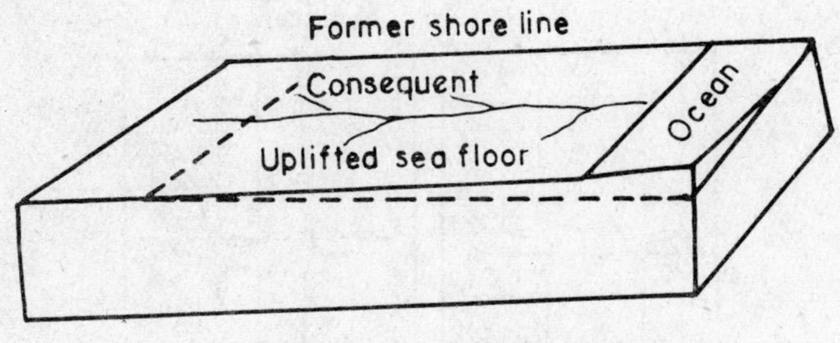

(a) Drainage on newly uplifted seafloor

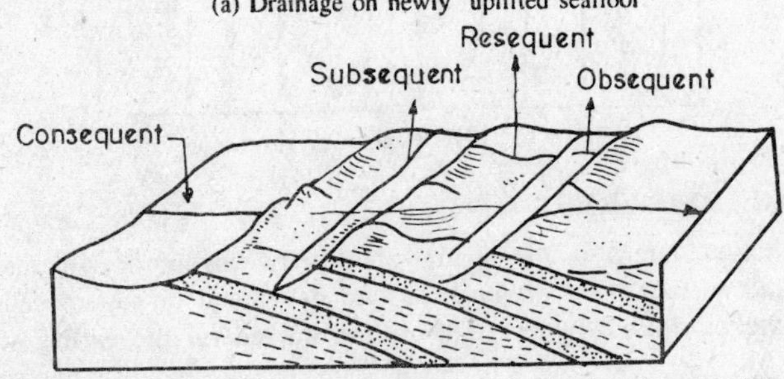

(b) Drainage types developed after tilting and selective headward erosion following week rock belt

Figure 11 : Drainage Types

A stream flowing on a younger formation may drop down on continued erosion onto the basement rocks beneath. This drainage type is referred to as *superimposed* or *inherited* or *epigenetic*

STREAM CAPTURE :

Streams of different drainage systems are separated by ridges of upland areas known as *divides.* On different rates of erosion, as is common, the erosion at the head water regions also vary alongwith the divide. Streams which exhibit the following qualities have an advantage over others :

(a) steeper gradient

(b) shorter distance to base level

(c) erosion in less resistant rock

Encroachment of a stream into another drainage system causes drainage diversion by *capture.* Such types of capture are common in areas of tilted sedimentary rocks of differing resistance to erosion. The former diversion channel is wind gap *in* the abandoned channel downvalley. Local drainage can support only a small stream which is *underfit* or *misfit* with regard in the size of the channel. The bend in the channel at the point of capture in known as the *elbow* of *capture*(Figure 12)

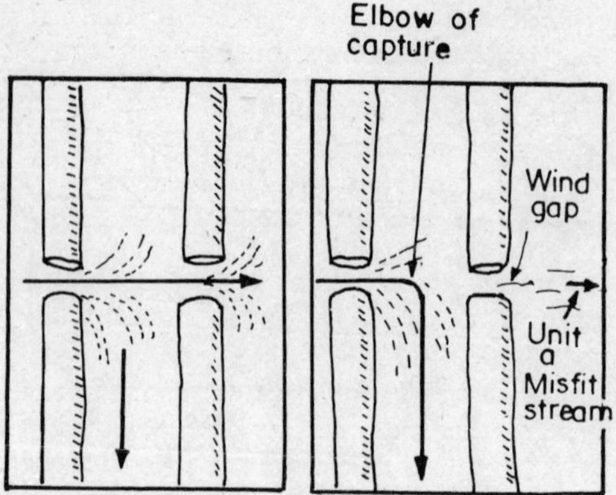

Figure 12 : River Capture by abstraction

DRAINAGE PATTERNS

A drainage pattern is the spatial-planimetric layout of streams, etched into the land surface by a drainage system developed or superposed thereon by aggregate of drainage-ways. The design carved on the earth's surface is governed by

i) topography,
ii) structure,
iii) inequalities in rock grains,
iv) geologic history including time factor in drainage evolution, and
v) climatic conditions.

The drainage patterns may reflect

a) original slope,
b) original structure, or
c) facts about the latter modifications of the surface, including
   - uplift,
   - depression,
   - diastrophism-tilting, and
   - other structural elements like faulting, folding, warping and jointing.

Numerous drainage patterns are evolved, variations therein may be function of one or more of the following elements (Lender 1959).

a) *Degree of Integration :* It refers to the (i) degree of unity, (ii) degree of uniformity of various characters and (iii) presence of modifying factors in an area. Integration is generally observed over particular areas and makes it possible to recognise the boundary of common characteristics.

b) *Density* : The drainage density is a measure of the

(i) degree of erodibility,

(ii) slope on which erosion occurs, and

(iii) special climatic conditions.

Low densities are computed in areas comprising coarse grained intrusive rocks and high drainage density in a terrain of fine grained clastic sediments. The drainage density may be zero km per sq km on a domal landscape while at other extremity it may be as high as 1,000 km per sq km for the highly dissected badlands of impermeable argillaceous country.

c) *Degree of Uniformity* : It is a relative homogeneity of pattern over an area and indicates

(i) the boundary between the areas of different characteristics,

(ii) uniformity of erosional history, and

(iii) the behaviour of the material to erosion.

d) *Orientation :* It refer to direction and indicates

(i) the control, and

(ii) effects due to geologic factors, like structural features, tectonic movement *etc.*

e) *Degree of Control*: This refers to the relative dominance of orientation and acts as most powerful and important tool in drawing geological informatics such as tectonic impulse *etc.* The tectonism produces

(i) anomalies,

(ii) directional changes, and

(iii) adjustments which are useful in tectonic analysis of a terrain.

f) *Angularity* : Directional variation is furnished by the component streams of a drainage pattern which serve as guides for

a) revealing the hidden features like fractures, faults, sheer zones,

b) other features including underground structures.

g) *Angle of Junction*: It is a function of the channel slopes of two streams. Ordinarily an acute angle between streams and its tributary is formed in the upstream direction. Some rocks and structures have distinct and typical angle of junction giving a clue for their identification.

VARIETIES

Several authors (Zernitz 1932; Lobeck, 1939; Von Eugelu, 1942; Parvis, 1950; Honard, 1967) have described certain types-some or other varieties of drainage patterns which are considered as basic patterns. Many modifications of these basic patterns too have been recognized. Zernitz (1932) summarized seven major types. Howard (1967) added

a) modification to the basic patterns, and

b) varieties of modified patterns

to the basic patterns. Parvis (1950) has described 6 basic and 24 modified basic patterns. More common types of drainage patterns are tabulated (Table VIII. 34)

DRAINAGE ANOMALY

In practice, instances of departure, at times marked departure from the regional drainage may be observed. Such local deviation which may include

a) planimetric shape,

b) pattern,

c) depth and width of the valley,

d) various flood plain characteristic, or

e) presence or absence of other features

from the regional drainage is referred to as *drainage anomaly*. It may be (i) present with some casual features and structures or (ii) associated with such features which could explain its interpretation and precision with higher order of drainage and provides clues to the

a) region more correctly and also to the

b) structural features in topographic conditions.

The important anomalies include -

a) *Linear streams* due to alingment of linear features such as faults, joints, fractures, non-resistant rocks or due to such features like sink-holes, springs *etc.*

b) *Domal upwarps*-sometimes basins may cause development of arcuate streams.

c) *Radial drainage* related to domes

d) *Abrupt drainage* may be in response to lithological variations

e) *Bedrock difference, structural variations* and *epeirogenic activity* may imprint sudden differences in the stage of stream development

f) *Stream piracy* is due to tectonic or non-tectonic causes

TABLE VIII. 34 : VARIETIES OF DRAINAGE PATTERNS

(Figures 13 to 15)

Basic types	Form	Causes/Significance	Modified Basic	Form	Causes/Added Significance Locale
Dendritic	Coined by IC Russel (1898) after Greek word 'Dendron' meaning tree; tree-like irregular branching of stream valleys at any (usually acute) angle in many directions. The main stream is consequent type and tributaries insequent (primarily following the local slope) type	Low-level erosion surfaces support integration of channel multiplication to form 'dendritic pattern' *i.e.* irregular branching tree appearance. Commonly i)occur upon horizontally bedded sedimentary rocks or massive igneous or metamorphic rocks. ii) found on homogenous bed rocks without systematic differences . in resistance iii) developed haphazardly by random headward streams on rocks which lack structural controls on gentle regional slope at present or at time of drainage inception.	Sub-dendritic	The acute angle at the junction of the main and tirbutary streams is so small as to give an appearance of psuedoparallelism	Fine textured bed rock with moderate tilt support on which dendritic disposition assumes elongation. This type of drainage is common on gently tilted i) coastal plains or ii) glacial plains *etc.* indicative of minor/subordinate structural controls.
			Pinnate(or featherlike)	Closely spaced more or less parallel tributaries meeting the mainstream at right angle	Developed on fine textured easily erodible materials like loess or the deep soil Close spacing is due to fine textured material and parallelism caused by deep valley side.
			Anastomatic	Examples are characterized by interlocking channel net work of braided type streams separated	Upper Son and Narmada valleys. It may be i) due to excessive load to the main stream ii) because of sudden decrease

Example - Dominant pattern in Sindhu (Indo) Ganga Plain is dendritc,

		It prevails in Chhota Nagpur Plateau over igneous and metamorphic rocks.		by channel bars and oxbow lakes	in stream gradient with resulting loss of transporting power or iii) due to loss of volume through seepage evaporation or diversion. Occur in flood plains, delta region, and on tidal marshes. It's varieties are termed as *reticular* or *reticulate* which are formed on tidal marshes and coastal plains. At high tides sea water flows inward and at low tides flows outwards.
			Distributary (or Dichotomatic)	In this streams diverge from rather than towards the main stream	Commonly found on alluvial fans and deltas where the mouth is choked and do not allow direct flow and therefore diverge the course through smaller multichannels.
				Example - Ganga Delta near Bay of Bengal	
Parallel	Consists of regulary spaced parallel or near parallel master/main and tributary streams	Related to marked gradient in the youth of landscape, developed due to i) pronounced regional slope ii) parallel elongate topographic features like lateral moraines, drumlins *etc*. iii) structural controls (hard and soft rocks alternati-	Subparallel	Streams oriented in similar direction lacking in regularity of parallel pattern	Controlled by a) intermediate or subparallel slope control or b) subparallel alignment of some topographic features, usually met in glacial region. This pattern is a transition between parallel and dendritic patterns.

		on, joints and monoclinal and isoclinal folds *etc.*) All varieties between this and dendritic and trellis patterns possible - a. on slope declining parallel pattern turn into dendritic patterns. b. it differs from trellis pattern in which parallelism is only among the tributaries whereas in parallel drainage parallelism may be between (i) unconnected streams or (ii) tributaries	Collinear	Streams are parallel and flow alternating over and under the surface recognized in a region of loess and sand (as in west of Budapest, Hungary) where wind formed furrows direct the streams to flow in linear and parallel manner. The streams disappear where water table is low and emerge out or reappear where the water table again reaches the surface.

Examples : Seen on the western coastal region where streams originating from Western Ghats flow towards Arabian sea through coastal plain. Also found in the Eastern coastal plains.
Rejuvenated areas of marked slopes, such as northern parts of Chhota Nagpur plateau and adjoining Ganga plains also show parallel drainage pattern.

Trellis	Named by Bailey Willis (1895) as trellis because of its resemblance to a garden trellis. One dominant direction, the main	Dipping or folded sedimentary volcanic or low degree metasedimentary rock areas of parallel features exposed lakes or seaflows	Subtrellis	Parallel elongate lanforms
			Directional trellis	Gentle homoclines Gentle slope with beach ridges
			Recurved	Plunging folds

	streams follow the slope and flow in the direction of dip(generally of subsequent type) but may be under certain circumstances formed by consequent streams with a subsidiary direction of short tributary drainage, usually obsequent or resequent in origin at right angles to it linking up the main streams and flow in direction of strike. Through the right angled bends the major streams cross or pass between aligned ridges and the primay tributary streams are usually at right angles to the main streams and are themselves joined by secondary tributaries whose courses are commonly parallel the main stream.	ribbed by beach ridges. Development of elongate ridges or drumlins by continental glaciers may result in a roughly parallel orientation of streams on glaciated terrain. Parallel orientation of ridges in streams may also be developed in some areas by wind erosion and deposition. All transition to parallel pattern. This type of pattern is regarded as one in which small tributaries are essentially same size on oposite sides of long parallel subsequent streams.		trellis	
				Fault trellis	Branching, converging diverging, roughly parallel faults
				Joint trellis	Straight parallel faults and/or joints
Rectangular	Constituted by streams in two directions at right angles both equally developed along i) intersecting fault lines,	Common in areas underlain by i) jointed, otherwise homogenous crystalline igneous rocks	Angulate (Rhombohedral)	Drainage equally developed in the directions meeting at acute and obtuse angles.	Joints and/ or fault other than right angles in otherwise homogenous strata. A compound rectangular angulate pattern is common.

	fracture lines and/or joints in ii) otherwise homogenous strata. The streams are adjusted to the underlying structure, and thus are subsequent in origin. Streams and divides lack regional continuity.	ii) flat lying resistant sedimentry beds with pronounced joint systems or faulted regions (commonly developed over highly jointed sandstone). Distinguished from trellis drainage by the fact that rectangular patterns a) have right angle bends in their channels b) are more irregular, the side streams have less parallelism c) absence of elongated secondary tributaries parallel to the main stream (characteristics of trellis drainage type)	Barbed	The stream bends abruptly back to flow in the reverse direction	Characteristic of a) river capture which has effected or reversal of drainage of separate river system which point upstream at acute angles b) diversions glacial or tectonic. Often joints, fracutres and faults interest each other at angles other than right angles and so the term *angular* or *angulate* is introduced. In both, rectaangular and angulate patterns the adjustment is more pronounced along one set of joints. The adjusted streams are all subsequent streams.
Radial	Streams radiating in all directions from a common or focal point, largely formed by various consequent streams which flow down the slope; also called *centrifugal*	Developed on structural or geomorphic qua quaversal basement such as volcanic cones, pluton, laccolithic dome, salt dome or circular monadonock, in relatively homogenous strata.	Centripetal	Streams flow inward in an attempt to converge	Craters, calderas, sink holes and other depressions toward which the streams (obsequent in nature) from outer periphery flow in an attempt to converge. If an structurally topographic basin is drained by centripetal drainage and is an original landform the stream would be consequent or if it is developd at lower stratigraphic level than the original, the stream would be resequent.

Example : A good example is provided by the Central Highlands of Sri Lanka, where the rivers rise and radiate in all directions towards the coasts.

		Example : Exellent example in Kathmandu valley, Nepal where river rising in the surrounding mountains of the middle Himalaya from the north and Maharashtra range in the south, flow in the valley centre to meet Baghmati river.
Annular	It is ring like pattern on plain found around maturely dissected structural domes and basins which have alternating belts of strong or resistant and weak rocks encircling the dome. Longer tributaries to annular subsequent stream generally indicate direction of dip and permit distinction between dome and basin.	In such an area with maturely dissected domes and basins characaterized by concentric ridges and lowlands with the advancment of dissection, the circular lowlands are occupied by subsequent streams which tend to assume a circular or annular pattern and some radial streams are developed which are of consequent origin. In fact, the circular subsequent streams originate as tirbutaries of radial consequent streams and become voluminous by capturing the upper portion of successive streams. The two successive *annular* valleys carved out by *annular* streams on domes and basins will be flanking a hogback, an outcrop of harder strata. This pattern is
Deranged (also) called erratic haphazard or disordered)	Drainage not coordinated-numerous drainage basins and lakes.	Blocking or deflecting older preglacial drainage by till moraines *etc*. Insufficient time for coordinations since the present drainage system came into existence. Streams wander aimlessly and their runoff water collects in the lakes swamps and marshes which are intermingled.
Karst	Numerous sink holes, few streams'cryptoheic drainage i. e. underground drainage.	Solution of underlying limestone cause drainage to move underground.
Swallow Hole	In limestone terrain streams are 'swallowed' in sinkholes to continue as *subterranean streams*	

		developed. where erosion of a dome exposes strata of varying hardness and rimining in nature.			
	Example - Parasnath Hill in Chhota Nagpur Plateau area which is an igneous intrusion shows such a pattern				
Multi-basinal	Essentially a descriptive term to include multiple depression patterns whose exact origins are unknown, hence the name multi-basinal. These depressions, may be of similar shapes, size and integration, may have similar appearance but of dissimilar geological genesis.	They are found by i) hummocky surfacial deposits ii) differentially scoured or deflated bed rocks iii) in areas of recent volcanism iv) by limestone solution v) by permafrost.	Kettle Hole	Characterized close depressions with occasionally filled basins.	Such patterns develop in granular moraines or outwash plains
			Thermokarst	Associated with evidence of thawing of permafrost Such a polygonal ground and beaded drainage is called the *thermokarst*	
			Elongate Bay	Found on coastal plains and deltas and is constituted of large elliptical and parallel depressions.	
			Lacunate	The lakes are spaced at random and the individual tributaries connecting them may be dendritic.	It develops over impervious strata.

Contorted	In this tributaries flow in a direction opposite to the main stream but more or less, are parallel to the main stream. Contortion may be a result of stream piracy. Reversal may also occur when the streams meet resistant rocks. This pattern differs from recurved trellis in a. lack of regional orderliness. b. discontinuity of ridges and valleys, and c. generally smaller scale.	Contorted, coarsely layered metamorphic rocks dykes, veins and migmatized bands provide the resistant layers in some areas. The longer tributaries to curved subsequent streams generally indicate dip of metamorphic layers and permit distinction between plunging anticlines and synclines.
Complex	Drainage obviously structurally controlled, but not falling into one of the other group.	Very complicated geological or geomorphological history.
Compound	Designates a drainage in an area consisting of two or more contemporaneous patterns-*e.g.* annular and radial drainage patterns associated with a dome.	

Special types

Palimpset	A drainage pattern that includes traces of older unlike patterns. The older pattern forms the background for the development of present pattern.	Phantom	It is a drainage pattern of seepage ways and looks like cobweb . This drainage is an indicative of shallow surface drainage with no channel development.
Asymmetric	Streams have more tributaries on the upslope side than on the downslope side. More common in mountain region.	Illusory	It is very much similar to phantom drainage except that this is due to artificial drainage.
Yazoo	Tributaries of master streams are not able to break the natural levees of the major stream, and flow for a shorter distance parallel to the trunk stream before joining it.		

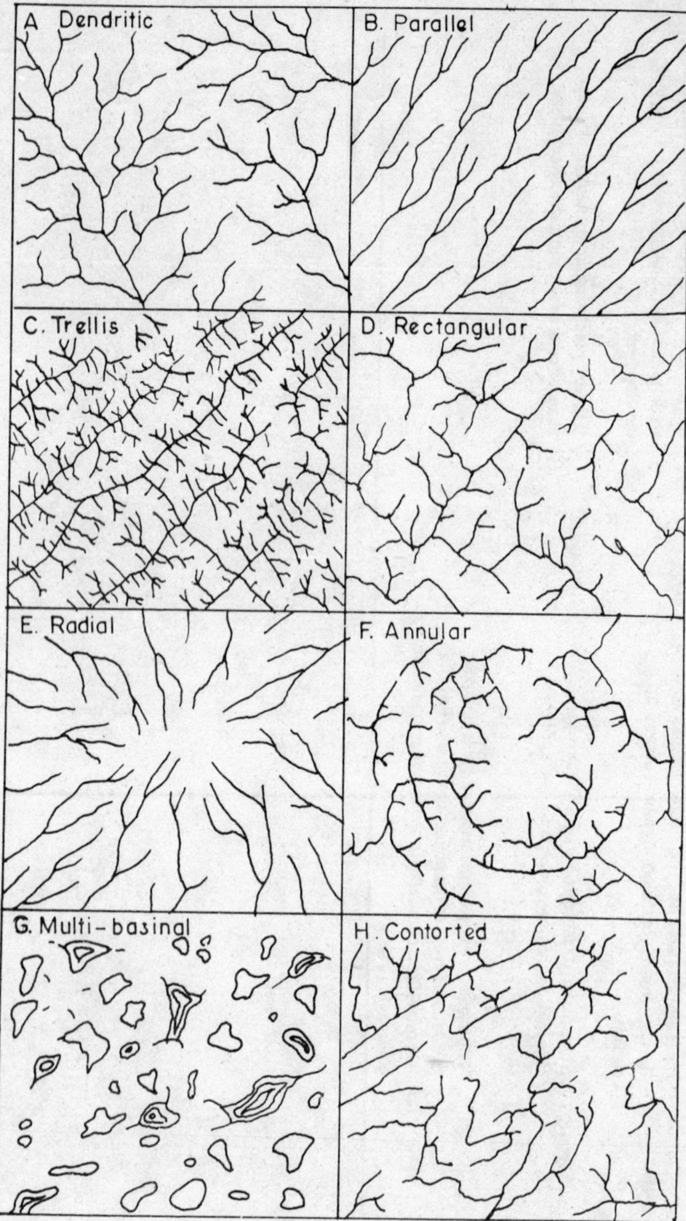

Figure 13 : Drainage patterns

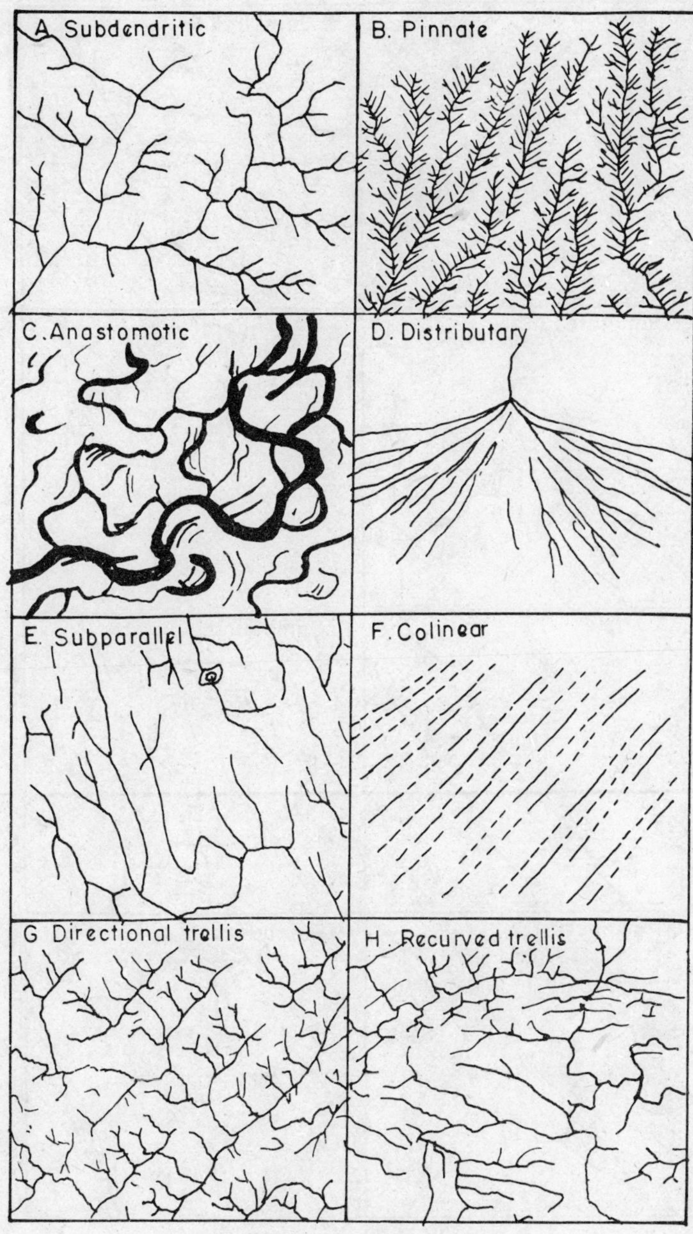

Figure 14 : Drainage patterns

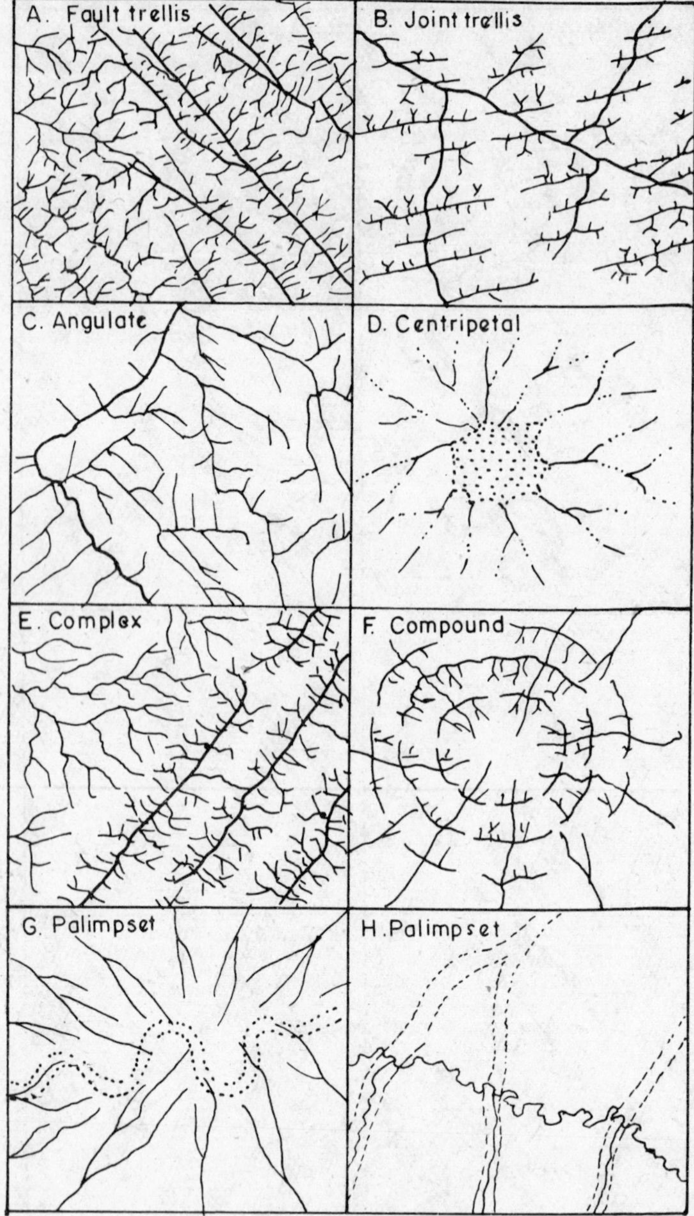

Figure 15 : Drainage patterns

g) Changes in flood plain characteristics may bring about the underlying structures or lithologic changes, for instance, stream may start meandering at a dome. Local widening or narrowing of valleys may as well reflect a structure. Sometimes local braiding is also seen.

h) *Rapids* and *waterfalls* occurrence reveal presence of (i) an outcrop of a resistant rock or (ii) fault/joint

i) Occurrence of an *isolated pond, marsh* or *alluvial fill* along the path of a mature stream may indicate *damming* by

   (a) subsidence or

   (b) uplift directly downstream.

j) At times former channels may subside below marsh level leaving behind only small fragments preserved presumably because they are buried structural features. Since these levee fragments are in total isolation, they are termed *flying levees*.

k) A stream may suddenly (i) disappear or (ii) reappear

STREAM CHANNELS AND CHANNEL CHARACTERISTICS

Water which does not percolate down into the soil or evaporate, flows over the surface may be in sheet-like mass, but generally channelled into the pre-existing lowest depression of a slope and thence to the valley bottom. These channels may range in size from the tiniest rills and gullies on slopes to major rivers. They carry practically most of the water supplied by the pecipitation but for that lost by evapotranspiration or locked in deep storage in rocks together with the debris (as their load) resulting from weathering and slope processes.

Stream configuration in a planar view is described as channel adjustment to the prevailing characteristics of

(i) discharge,

(ii) erodibility of bed,

(iii) bank materials, and

(iv) sediment transport through a river reach at any time. The channel patterns and flow patterns are interdependent in as much as flow which etches the channel configuration in turn is itself dependent on the channel shape.

Channel characteristics include

(i) the form of the channel in map view, and

(ii) cross-section.

A cross-section may show a host of variations between

(i) narrow deep channel, and

(ii) wide shallow channels.

The cross sectional geometry of shape of a stream channel is largely a function of

a) nature of the material making up the floor

b) sides of the channel relative to the erosive power of stream which is determined by

 (i) discharge,

 (ii) quantity, and

 (iii) calibre of the bed load.

Channels with cohesive silty banks develop narrower and deeper channels than less cohesive sandy banks. Vegetation may also be an important factor in bank erodibility.

Streams channels may be grouped into three types : all may be found along the course of one stream -

- straight channel
- meandering stream channel
- braided stream channel.

There appears to be a direct relationship between

a) sinuousity,

b) incidence of channel islands, and

c) the width-depth ratio of the channel.

In general, meandering stream have

(i) high sinuousity,

(ii) few islands, and

(iii) deeper and narrower channels

than braided streams having

(i) low sinuousity, and

(ii) many channel islands.

Straight Channel Pattern

It is rather not very common in nature, and Leopold and Wolman (1957) observed that in length the stretches may be of 100 meters or so generally less then ten times the channel width. However, instances of the higher channel lengths are on record as in case of the channels of uniform material size (Chitale, 1970). In channels of heterogenous bed material a sequence of alternate shallow and deeper section-named *riffles* and *pools* develop. Deeper section may be due to the concentration of main flow of the water along the thalweg, preventing deposition. Experience shows that pools or riffles are spaced approximately at an interval of five to seven times the channel width, pools are the deeper parts of the thalweg and riffles or bars occur on the opposite bank. It suggests that water tends to adopt a sinuous path of flow even when its initial linear path is straight. It is identical to the cut bank and slip off slope sequence in meandering channels. It is for this, the flow and depositional patterns in straight and meandering channels are considered similar.

It is commonly observed that relatively wide and shallow channels tend to be straight over steep gradients, wherein, the bedload movement perhaps dominates over the wash load transport. Studies indicate channel patterns may change in response to the proportion of bedload to that of the wash load. (Figure 16).

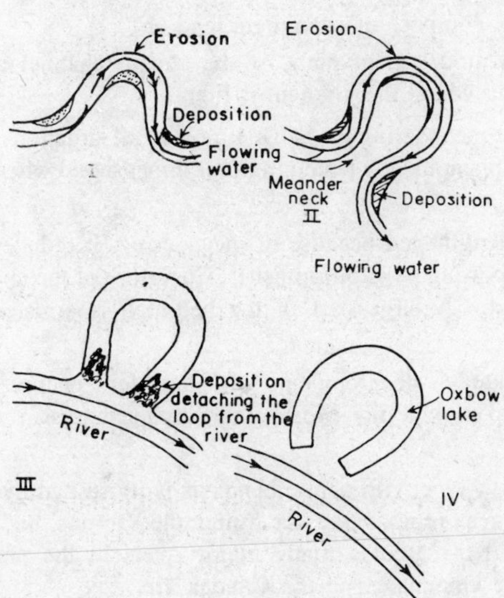

Figure 16 : Strainght stream channel features : broken lines connect the deeper parts (pools) in the thalweg

Meandering Channel Pattern

Streams follow irregularly distributed depressions downslope and consequently in their natural course, they, whatever their size or topographic or climatic situation, have bends. When these courses assume sufficient geometric regularity they are termed *meanders.* The streams meander

- in high mountains, in low hills
- in dissected or peneplained uplands or
- low lands and plains of deposition.

They meander, too,

- near the source,
- in the middle of the course
- as well as near their mouths.

But certainly meandering is more pronounced on an even surface where even minor obstructions lithologic or topographic, may deviate the stream path from a straight course. It is

(i) more pronounced in mellow alluvial fills and

(ii) less so in hard resistant structures.

Meandering sections are particularly common in areas where

(i) silt or clay compose the dominant load or

(ii) there is a moderate gradient of the stream channel and thus less pronounced where the gradient is high.

It thus appears that meandering is visible almost in all situations. Meandering is a three dimensional phenomenon becoming four dimensional when time is also considered.

Stream course is lengthened because of meandering. It reduces its gradient and thus its load-transporting power diminishes. Curvature of the channel hinders the movement of load. A large part of the bed-load is arrested as riverine deposits.

The most remarkable development of meander occurs in mellow extensive plains of deep alluvial fills. Morphologically meanders are classified into 3 types:

a) *Simple wavy type* with gentle bends as is illustrated by the rivers in the Himalayan regions, *e.g.* the Sutlej, the Yamuna, the Brahmaputra and Arun Kosi as also many major rivers in the plains, *e.g.* the Ganga, the Ghaghara and the Gandak *etc.*

b) *Horse-shoe type,* with bends at the curve resembling a horseshoe-shape. They are marked by great curvature in the meanders, and appears to be the most common type specially in low gradient terrains in plains, *e.g.* the Yamuna, the Gomti, the Son, the Punpun, the Burhi Gandak, the Hoogly, the Ganga between Allahabad and Ghazipur.

a. Wavy type b. Horse-shoe types c. Oxbow lake

Figure 17 : Meander Types

c) *Ox-bow or bracelet type (*Figure 17*)* with high curvature practically circular bends of which two hooked ends lie in line with the main streams, giving an appearance of a *kara* or bracelet with open end. It also resembles the face and horns of an ox. In fact this is an advanced stage of horse-shoe type where the shoe ends are brought close together. The numerous ox-bow lakes of middle or lower Ganga plain record the former existence of such meanders. The present examples of the bracelet type meanders are found in the course of the smaller streams *e.g.* the Betwa, Gomti, Sai, Rapti, Punpun, the Tirhi, a tributary of the Ghaghara in plains. When the neck of a bend has joined, conditions in the bend become different from those of the main stream. As the bend has a longer course the velocity of the water is slower than in the main stream which follows a straighter course.

Because of the force of inertia water tend to continue to flow straight around the stream bend whence it impinges on its outer curvature with stronger current than on the inside of the bend. The outside of the channel is thus deeper and the velocity of the water greater. There is a tendency of water to 'pile up' on the outside of the bend near the surface, and a cross-channel component of flow developed. Water at the surface has a cross channel component directed toward the outside of the bend, whereas water near the floor of the channel is directed toward the inside of the bend. The flow pattern thus developed has the form of a *spiral* or *helix*

The higher outer velocity on the outside of a bank causes bank erosion, and cross-channel component of the helical flow pattern contributes to the building of *point-bar* deposits on the inside of the bend. Contrary to the general belief it is observed that material eroded from outside of meanders is deposited on point bars on the same side of the channel from which it was eroded and thus much of the eroded material is not transported across the channel floor, before it reaches the next bend downstream, where the helical flow pattern moves it toward the point-bar deposits on the inside of that bend. Continuation of this process leads to further development of meandering pattern. As curvature of the bend increases, the process becomes self accelerating until two bends come together to form *meander cutoff.* The abandoned channel is then left as a horse-shoe shaped, remnant of the former bend, often filled with water to form ox-bow lake (Figure 18). With time the ox-bow lake passes eventually to swamp, then disappears altogether.

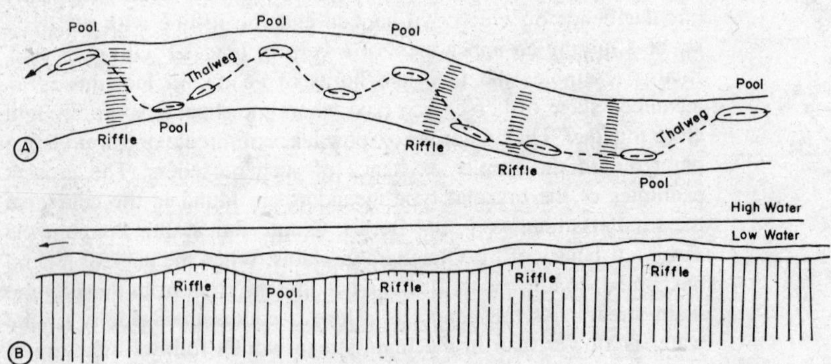

Figure 18 : Meandering river morphology

It is observed that if stream has a relatively narrow deep channel

(i) the inertia effect and

(ii) helical flow

localize erosion on outside of the bends and deposition of bars on the inside of the bends, so that meanders develop. If a stream has a relatively wide shallow channel, helical flow is too weak to maintain order in the whole channel to localize the deposition of point-bar deposits.

Meandering may be insignificant nearer to the source and may be more marked in relatively stable clay stretches of the basin than in sandy or loamy sections. As a result of

(i) hydrological processes and

(ii) the behaviour and movement of sediments,

the meanders migrate downstream. The rate of migration is believed to be 10-100 years. The

a) non uniformity of alluvium,

b) occasional collapse of banks,

c) occurrence of rocky projections of unequal hardness in bed of banks,

d) influence of tributaries, and

e) effect of vegetation

make meanders asymmetrical.

Meandering channels of various dimensions show definite mathematical relationship among various geometrical parameters.

Meander Belt is a zone containing the meanders and is defined by the lines joining the outer ends of the meanders. A relationship exists between the width of the channel and the width of the meander belt and is expressed as

$$W = \frac{W_{mb}}{W_{ch}}$$

where W is ratio

W_{mb} = width of the meander belt and

W_{ch} = width of the channel.

Generally, the value of W is 11.55, but it may be as much as 95 in case of entrenched meanders. The meander belt of the Yamuna is 5-10 km wide between the confluence with the Chambal and with Ganga. The meander belt of the Ganga is 10-15 km wide below Allahabad.

Entrenched Meanders - An entrenched meander is developed in alluvial fills of valleys which has subsequently dug itself deeper into a gorge-like features on land uplift or sealevel fall so that the bed is now seated on bed rock. Examples of this type of meanders are commonly seen in the Himalaya. It may as well be seen in many of the peninsular streams pointing to their development on ancient erosion surface (*e.g.* Central Indian Foreland and Chhota Nagpur) and other parts of peninsular upland which have been recently uplifted during Pliocene-Pleistocene times.

Braiding Channel Pattern

Braiding channels (Figure 19) are interlaced network of channels derived from common trunk. They are best expressed in regions where

(i) the discharge is sporadic and copious,

(ii) the rate of erosion high, and

(iii) the vegetation almost negligible.

and are encountered in two distinct situations

a) high energy streams flowing in noncohesive sediments and carrying heavy sediment load, and also

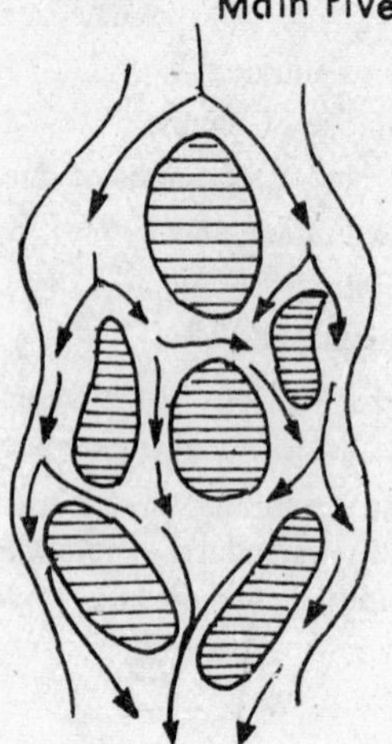

Figure 19 : Braiding river

b) in alluvial fan streams, where the energy available to the stream is rapidly diminishing.

They are confined

(i) to broad, shallow streams

(ii) with great quantities of coarser bed load,

(iii) to high erodible banks, and

(iv) to steep gradients

A braided stream divides and recombines around large *mid channel bars,* that grow

(i) downstream and

(ii) increase in height

on piling of boulders of pebbles, or banks of sand and changing it but never occupying the whole floor of the stream channels. So developed mid-channel bars often become an inter- channel island. It may be later stabilized by vegetation growth on its surface. Subsequently new channels develop rapidly which may be divided again around new multichannel bars. Not all bifurcations

(i) result from bar accretion, but

(ii) may also be developed on cutting into flood plain during periods of overbank discharge.

Braiding is common where there is a marked seasonal variation in stream discharge. As such

a) this occurs below a glacier,

b) around the margins of an ice sheet, or

c) in semi-arid areas.

The load is rather too coarse, entailing a bed-load high proportion, transportable/removable only during a period of maximum stream-discharge.

The horizontal patterns of a braided stream are far more unstable than those of meander belt. It is because of

a) high erodibility of the bed and the banks, as well as

b) limited relief of the channel

which is prone to major changes after each bankful episode. It is self perpetuating process as the obstructing channel bars may be composed of gravels that offer more resistance to erosion than that the more sandy material composing the channel bank. Consequently the effects of channel widening are counteracted by decrease in channel depth. This increased channel roughness and turbulence, provides the necessary competence to transport a large bed load.

The braided channel pattern has been variously explained in terms of

(i) bank erodibility,

(ii) rapid and large variations in discharge,

(iii) channel slope, and

(iv) sediment load of streams.

Most explanations seem to stress the movement of bed load which initiates deposition with a wide channel.

Channel roughness has been suggested as one of the many causes of braiding. Chitale (1970) suggests that an increase of bed resistance due to :

a) a large range of particle size in the bed material,

b) numerous channel bars, and

c) alluvial islands in stream channels,

disrupt the homogeneity of flow. Consequently curvilinear flow in braided channel reaches do not develop. This, however, does not explain the mechanism, that develop the channel bed irregularly in the first place, and the reason for the existence of a large range in the bed material size.

A braided pattern has often been associated with a tendency for aggradation. However, it is also observed that braided channels are not necessarily always associated with an excess of load as seen over years that the braided Kosi River has scoured than aggraded its bed between Balka Hills and Hanuman Nagar in Nepal (Gole *et al.*, 1974)

Bank erodibility is yet another factor cited as cause of braiding. It is shown that flume channels in highly erodible bank materials tend to show wide and braided. In such channels, shoaling on incipient channel bars initiates bifurcation of flow, leading to the development of multiple channel ways. Such development accounts for the unusual width (as Kosi below Chatra is 6 km wide) of braided channels which are originally broadened by the multiplicity of streams in the river. Wide (10-12 kms) braided sections are noticeable in the east of Patna city, west of Bhagalpur city, west of Sahibganj and north of Rajmahal town. The Brahmaputra is a remarkable example of mighty braided river throughout its journey through Assam. The anostomising (dividing and reuniting) channels is frequently 12-13 kms. It has a relatively straight course.

As deposition is an essential precondition for the development of a characteristic braided pattern, the transport of sediment must necessarily take place. It is contended that

a) sediment transport, and

b) a low threshold of bank erosion

provide essential conditions of braiding. The braided channel pattern develops initially by the deposition of a coarser bedload material as a central bar at high flow in a single channel. The bar grows vertically and laterally by the selective deposition of lag deposits and finer sediments. As the bar grows in height, the depth of water over the emerging bar is gradually reduced, but the velocity of water changes only slightly. Some particles get deposited at the downstream end of the bar where the depth of water is relatively large and the velocity relatively low. Lateral growth of the bar narrows the channel and makes it unstable. The channel then widens to attain stability. In this process of channel development, the central bar is dissected into smaller volume of flow and intermediate stages than is conducted by the larger undivided channel downstream on essentially similar bed material. Thus a braided channel is as efficient as any other channel pattern. The pattern of bars once established, remains more or less relatively permanently fixed, changing in detail only slightly with time.

A discriminate function involving the bankful discharge and channel slope to separate braided from meandering channels is derived (Leopold and Wolman, 1957). Braided channels possess a relatively steep channel gradient per unit discharge than meandering channel reaches.

The above theories of the development of braid are discounted by Hong and Davies (1979). In a flume study augmented by time-lapse stratigraphy, it was found that an initial meandering channel splits after it attained a certain critical width. A large volume of flow passed through the cut off and in time the meandering channel was completely abandoned. Such changes in flow pattern caused channel adjustment downstream of the point of cutoff until a braided pattern was fully established.

To a considerable extent braiding and meandering are related to the nature of the country traversed by streams.

a) Meandering as illustrated by the Gomti and the Sai occurs in binding materials like clay which do not collapse easily while in sandy material braiding is common, for instance in Ghaghara and the Ganga particularly the former river tend more to braid than meander.

b) It also appears that larger streams with large discharge and load like the Brahmaputra, Ganga, Ghaghara, Sutlaj tend to braid while smaller streams like the Yamuna, Sarda, Chambal, Betwa, Ken, Tons, Gomti, Sai, Rapti and Burhi Gandak tend to meander.

c) It is observed that

(i) if the stream has a relatively deep channel the inertia effect and helical flow localize erosion on the outside of the bends and deposition of point bars on the inside of bends, so that meanders develop, and

(ii) if a stream has a relatively wide, shallow channel helical flow is too week to maintain order in the whole channel to localize the deposition of point bar (Mackin 1956).

Deposition of randomly placed bars and islands occurs, tending to eliminate the inertia factor by breaking up the current into self-baffling channels. Since the width-depth ratio depends on bank erodibiliy whether a stream develops meandering or braided pattern depends essentially on bank erodibility.

d) If two stream segments, one braided, the other meandering have equal discharge and flow on similar bed material, the braided, segment will have steeper wetted perimeter slope.

(e) In braided segment, the sum of wetted perimeter of two or more channels is greater than the wetted perimeter of a single channel carrying the equivalent discharge. Measurements made on streams which have braided segments in otherwise meandering courses indicate that in a braided channel, slope increases relative to that in a meandering channel

The relationship between discharge and velocity is given (Figure 20) by

$Q = wvd$

where Q = discharge w = width and v = velocity

The cross-sectional area of stream channel is

$w \times d$, and

the wetted perimeter is $w + 2d$

As the discharge increases, both

(i) cross-sectional area and

(ii) wetted perimeter

increase, but w x d increases more rapidly than w + 2d, so the cross-sectional area increases more rapidly than wetted perimeter, result in relative decrease in frictional retardation by the floor and banks of the channel. For this reason velocity increases with discharge.

In the figure 20 (a) the sectional area is 2 x 4 = 8 and the wetted perimeter is 4 x 2 (2) =8 In the figure 20 (b) the area is 4 x 8 = 32 and the wetted perimeter is 8+2 (4) = 16. By doubling the width and depth the cross-sectional area has increased 4 times, whereas the wetted perimeter has increased 2 times, resulting in relatively less frictional retardation of water per unit of cross-sectional area.

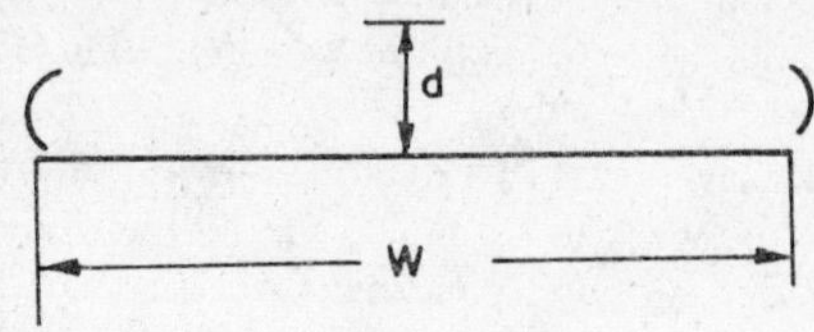

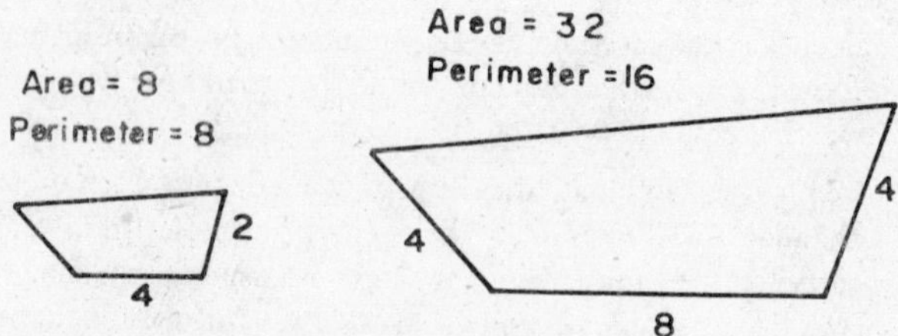

Figure 20 : Wetted primeter and Cross-sectional area relationship

CONCEPT OF THE GRADED STREAM

Grade is regarded as a natural state of balance between

(i) the capacity to do work (with special reference to transport of load) and

(ii) the quantity of the work to be done (quantity of load supplied)

By definition, a graded slope is adjusted for transport of the load given to it. A *graded stream* may, therefore, be taken as the one which is neither eroding nor depositing material at any point along its course. It is, however, difficult to conceive of a stream which is neither eroding nor depositing sediments in various reaches of the channel. This condition can hold only for short stretches in the middle section of the course. As for example, the Ganga is eroding in the Himalayan section. It sheds a good deal of its load when it leaves the hills. In its course in plains it carries its own load as well as that of its tributaries. This may be spread in the surrounding plains during flood overspill. The overspill is a function of flood causing factors like-

(a) short duration storm rain and

(b) the river profile may be irrelevent.

Thus flooding will be associated with deposition. Grade concept cannot be confined merely to the channel of the river. The channel has at any rate to be kept clear for the stream to flow. Therefore, the process of grade must involve the areas, 'along' or 'adjoining' the course of the stream.

Short term discharge also affect the competence and capacity of stream.

Holmes (1952) defined a graded stream as the one in which the size of the bed material is such that it can be transported at high rates. The material is subject to deposition at flows. Kesseli (1941) asserted that graded conditions cannot be attained along the entire length of a stream channel or part thereof. He, therefore, defines a graded stream as the one which possesses a smooth longitudinal profile. It may be pointed out that channels with rapids and waterfalls are efficient in transport load as channel reaches out. Thus a smooth longitudinal profile does not distinguish a graded from a non graded stream.

Mackin (1948) applied the concept of equilibrium to streams and proposed comprehensive definition of a graded stream as follows :

> 'A graded stream is one in which over a period of years, slope is delicately adjusted to provide, with available discharge and with channel characteristics, just the velocity required for transportation of load supplied from the drainage basin. A graded stream is a stream in equilibrium, its diagnostic characteristic is that any change in any of the controlling factors will cause a displacement of equilibrium in a direction that will tend to absorb the effect of the change". Four variables play important roles in profile grading -
>
> (i) slope
>
> (ii) discharge
>
> (iii) load and
>
> (iv) downvalley changes

Slope

Concave-upward long profile of a graded stream decreases in slope downstream is a direct consequence of equilibrium condition in the stream. It may have minor irregularities below junctions of major tributaries could be altributed to changes in discharge or calibre of load, but generally are graded throughout.

Slope, an important main equilibrium factor, is automatically adjustable by the stream itself to accommodate external changes that induces change in velocity and in graded stream it (the slope) is maintained to provide just the velocity required to transport all the load supplied. Load and discharge independent of the stream origin itself are important as factors. The graded long profile of a stream is a profile of equilibrium.

Discharge

Added velocity and consequently increased carrying capacity of graded stream causes discharge increase, settling the otherwise stationary larger particles in motion which facilitates erosion and slope reduction. It in turn reduces erodibility again and thus re-establish equilibrium afresh. The converse of this entire process is also true on velocity decrease.

Short-term discharge changes, such as the seasonal or daily, fluctuations may cause local channel scouring during high-water stages or deposition during low-water stages. The concept of graded stream however, is with regard to long-range changes (Mackin 1948).

Load

The load of a graded stream over a period of years dependent largely on the rock type being weathered and eroded from

- (i) the valley side,
- (ii) topographic relief,
- (iii) weathering and erosional processes operating in the drainage basin, and
- (iv) vegetation.

With stream load increase, a graded stream deposits the larger particles, hard to move, thus increase the channel slope. It goes on until the velocity increase produced by the slope increase is just sufficient to move the coarser load, and a new eqilibrium is established. If the calibre of the stream is decreased, the valley is then in excess of the amount required to move the smaller particles, and hence the stream begins to cut down or *degrade* and the processes is called as *degradation.*

Downvalley changes

The shape of a graded profile is concave upward, which constantly decreases downstream. The slope of the curve represents an equilibrium system in which two of the governing factors

- (i) discharge and
- (ii) load,

change progressively downstream.

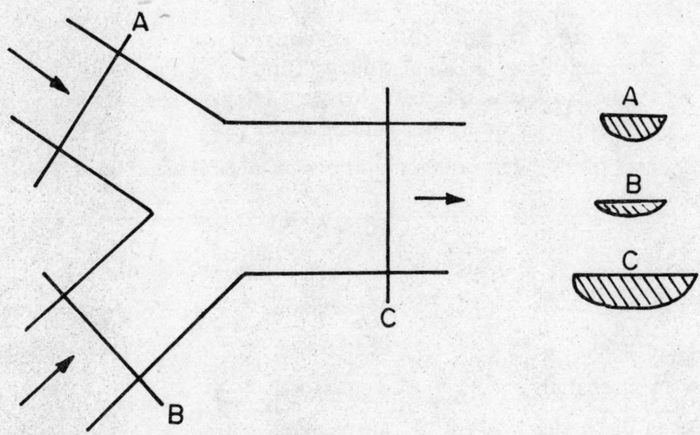

Fig. 21 Wetted perimeter and cross section area on the union of two streams

(where Discharge A+ B streams = C stream but Wetted Perimeter of A and B streams ≠ C stream)

With tributaries emptying into the mainstream the discharge increases downvalley ending up in the cross-sectional area increase related to the wetted perimeter, and a downvalley segment with larger discharge maintaining a given velocity on a slope lower than an upvalley segment with a smaller discharge. Thus from a tributary junction of two graded stream the slope becomes less than that of either confluent as a result of an increase in cross-sectional area relative to wetted perimeter. In Figure 21, the sum of the discharges of the two streams A and B is equal to that of the mainstream C, but the wetted perimeter is less than their wetted perimeters together above the junction causing the slope decline necessary to produce the velocity required to transport the load.

Another factor which changes progressively downstream is the down valley decrease in load calibre. Slope requirements are reduced with reduction in the size of particles transported, load increase relative to discharge, as by influx of coarse debris from heavily laden tributaries or from loss of water, if slope locally increases.

The decrease in calibre of load downstream is due partly to

a) abrasion of particles during transport and

b) selective transportation.

Because smaller particles

(i) can remain in motion longer and

(ii) are more frequently advanced by saltation or traction they move downstream at a higher rate than larger particles. It is rather a difficult task to *distinguish between the effects of abrasion and selective transportation in causing downstream decrease in particle size.* In some streams abrasion is the dominant cause, in others, selective transportation, depending on conditions in a particular stream system are the guiding factors.

Responses of a graded stream to various changes (Easterbrook, 1969) are as below :

	RESPONSE (until a new equilibrium is reached)	SLOPE
increase in load	aggradation	increases
decrease in load	degradation	decreases
increase in discharge	degradation	decreases
decrease in discharge	aggradation	increases

Aggrading and Degrading Streams

When rock, particularly the coarse material carried as bed load, is supplied to stream by its tributaries and by runoff from adjacent slopes in greater quantity than the stream is capable of transporting, the excess load is spread along the channel floor raising the level of the entire channel. Such upbuilding is termed *aggradation,* in contrast to *degradation* which is the process of downcutting carried on by a stream capable of transporting more load than it is supplied with.

FUNDAMENTAL CONCEPT : FLUVIAL CYCLE

The following implications seem to be implied of a fluvial cycle :

1. Terrain uplift provides initial units of geologic material having their own lithologic, structural and geomorphic aspects for sculpturing and reduction and may include forms
   a) recently uplifted sections of a continental shelf,
   b) domes and basins,
   c) alternating anticlines and synclines of folded mountain structures,
   d) complex mountain structures,
   e) volcanic mountains or plateaus,
   f) ice-scoured plains,
   g) glacial till plains and associated features,
   h) lacustrine plains,
   i) areas of sand and loess accumulations,
   j) areas previously reduced to low relief by fluvial erosion,
   k) other types.

2. Primarily through the combined action of weathering, mass wasting and erosion by both concentrated and unconcentrated runoff waters, the sculpturing and reduction of the initial landmass takes place.

3. Reduction may also take place under varying diastrophic conditions after or during uplift. Uplift may be continuous or intermittent.

4. Completed cycles are rather exception for sculpture and reduction may often be only partially completed during given or still stand period before the uplift superimposes a new cycle, and thus partial and multiple cycles result.

5. The various stages of progress in land sculpture and reduction are recognized and designated by the metaphorical stages of youth, maturity and old age. If necessary, substages are indicated by prefixes early, middle and late.

6. A systematic sequence of landforms ensues with a progress of a cycle in harmony with its place of cycle, each stage bearing a particular group of forms.

7. The landforms produced during a cycle are function of structure, process and stage.

8. It is at times possible for and uplifted landmass to be reduced to the ultimate limit called *base level.*

9. Practically, total reduction of an uplifted land unit by the combined processes of weathering, mass-wasting and erosion by run off waters produce a low relief topographic feature called a *peneplain.*

Complication of

(i) geologic structure,

(ii) diastrophic history, and

(iii) climatic conditions

may induce notable departures from the expected sequence of events. The suggestion below are based on the assumption that

(i) the region has recently been uplifted from beneath the sea,

(ii) the geologic structure is simple,

(iii) there are sedimentary rocks of varying hardness having moderate seaward dips uncomplicated by any pronounced geologic structures such as folds and faults, and

(iv) this newly uplifted landmass remains stationary with respect to sea level long enough for a geomorphic cycle to run its course.

Under these situations, Table VIII. 35 indicate characteristic features and stages in the evolution of landform of fluvial cycle.

TABLE VIII. 35 : FLUVIAL CYCLE CHARACTERISTICS

(modified and enlarged after Easterbrook, 1969)

	YOUTH	MATURITY	OLD AGE
River-tributaries	a) few i) consequent trunk streams ii) large tributaries b) numerous short tributaries and gullies extending themselves by deepening and headward lengthening of valley	well integrated drainage system evolved	tributaries to trunk streams fewer than in maturity
Terrain	mountainous-upland tracts and valley bottom	generally comprises slopes of hillsides and valley sides.	
Stream course and lithology	adjustment inconspicuous, if any usually governed by the available relief and slope	adjustment evident as reflected in the existence of some longitudinal tributtaries along belts of particularly weak rock	
Stream gradient steepness	generally rather high	moderate	low
downcutting to laterial erosion proportion/ ratio	downcutting dominant	lateral erosion more dominant	no vertical erosion, lateral erosion too is inconspicuous if any, deposition dominant.
Base level of erosion reference to flood and meander belts	much above the base level of erosion general lack of flood plain meanders confined to valleys incised below the upland surface meander belt > flood plain width	tending to approach base of erosion flood plain tracts constitute considerable portion of the valley floors, meanders conspicuous and are free to shift their position with flood plains generally meander belts ≃ flood plain width	extensive areas are at or near base level of erosion marked development of flood plains : meander belt < flood plain width
Cross-valley profile*	i) have V-shaped cross profile ii) have vigour and energy to remove any obstacle on the way	i) reduced river energy ii) lateral cutting dominant over virtually negligible down/ vertical cutting	i) no energy, or practically negligible erosion ii) deposition dominant-natural levees in flood time, alluvi-

	iii) are shallow or deep depending up-on the height of the region above sea-level	iii) fanglomerate, fl-uvial cones/ fans at the foothhill iv) vertical meanders on rejuvenation	ium in peneplain uneroded islands left monadnocks iii) delta with a) i) topest ii) foreset iii) bottomset beds b) lobes and dist-ributaries in plains.

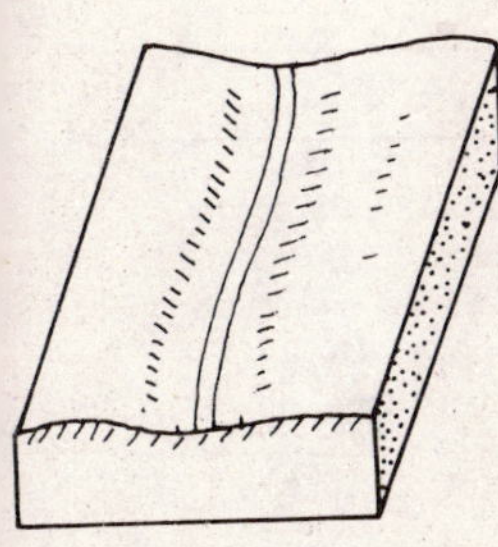

Youth stage

Mature stage

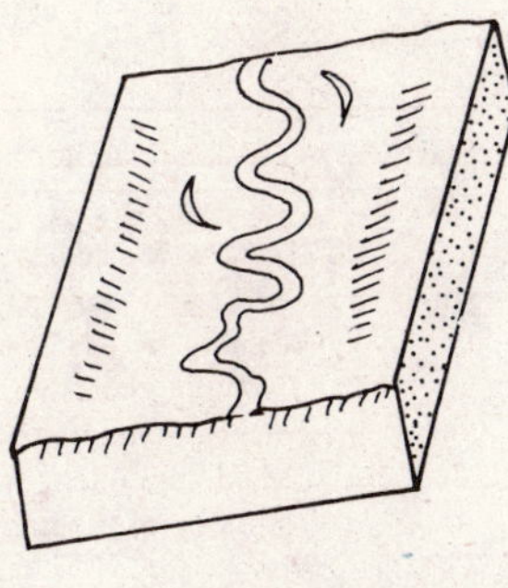

Old stage

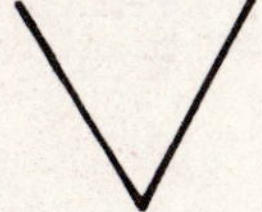

Figure 22 : Cross-valley profile

graded streams	few streams graded mostly degrading	profile of equilibrium attained, most strea-ms graded tributari-es may be still ungraded	all streams aggrada-tional, and longit-udinal profile prac-tically graded : ne-gligible slope down flow.
rapids and waterfalls	present, may be numerous	largely absent	absent
lakes and swamps	may be present, both within valleys and on divides	absent	ox-bow lakes and swamps present on flood-plains but not in the

			interstream areas
relief	moderate	maximum	minimum
divides	broad, irregular poorly defined	sharp crested ridge like resulting in minimum of interstream uplands	broad rolling

* (Gradually the projections of the valley are eroded away and the valley floor is widened. This finally produces a continuous valley with more or less flat floor containing the meandering river in valley much greater than the width of the river. Such a valley is called *strath)*

TABLE VIII 36 : SHOWING LANDFORMS ASSOCIATED WITH FLUVIAL AGENCY

(after S. N. Pandey 1987)

Stage	Erosional landform features	Depositional landform features
Initial and youth	Narrow V- shaped deep valleys, gorge and canyon, Valley lengthening by headward erosion Slumping landsliding; Development of waterfalls and rapids, Maximum inter-stream uplands, Number of tributaries maximum	Paractically no deposition in the initial stage. Lack of flood-plain and no meandering Deposition of alluvial fans, alluvial cones, piedmont. Alluvial plains (or bajada)
Mature	Development of well integrated system General lack of water falls; Conspicuous meandering and formation of cut off lakes; Broad V-shape symmetrical or asymmetrical valleys. Number of tributaries reduced.	Flood plains and natural levees; Formation of river teraces Meander scars, meander belt deposits Channel filling and back-swamp deposits
Old	Very few tributaries Extremely broad and very gentle sloping valleys. Lakes, swamps and marshes developed Base level of erosion practically reached.	Deltaic plain deposits with current beddings.

FLOLD PLAIN MORPHOLOGY

Flood plains and their morphologic subdivisions are indeed depositionsal landforms and thus are genetically related to specific depositional processes (Figure 23). Not all deposits (Table VIII. 37) develop landforms.

Flood plains are built in two fundamental ways:

(i) Lateral accretion and by

(ii) Vertical accretion.

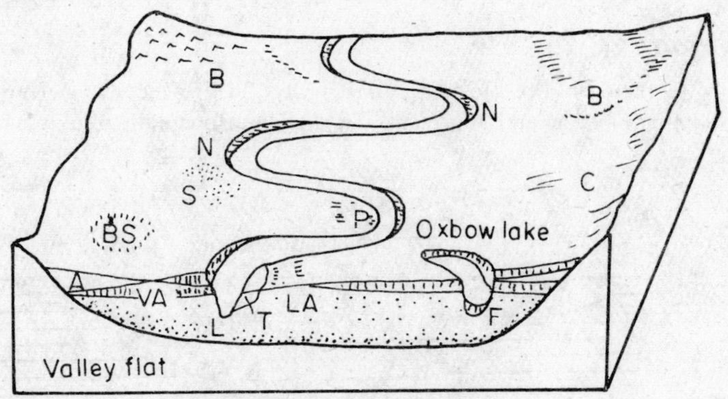

Figure 23 : Valley accretions associations

TABLE VIII 37 : GENETIC CLASSIFICATIONS OF VALLEY SEDIMENTS

(Source: Vanon 1971)

Place of deposition	*Name*	*Characteristics*
	Transitory channel deposits	Primary bedload temporarily at rest; part may be preserved in more durable fills of lateral accretions
Channel	Lag deposits	Segregations of larger or heavier particles, more peristent than transitory channel deposits, and including heavy mineral placers
	Channel fills	Accumulations in abondoned or aggrading channel segments ranging from relatively coarse bed load to fine grained oxbow lake deposits
Channel margin	Laterel accretion deposits	Point and marginal bars that may be deposits preserved by channel shifting and added to overbank floodplain by vertical accretion deposits at top
	Vertical accretion deposits	Fine grained sediment deposited from suspended bed of overbank floodwater; including natural levees and backland (back swamp) deposits.
Overbank floodpain	Splays	Local accumulations of bed-load materials, spread from channels onto adjacent flood plains
Valley margin	Colluvium	Deposits derived chiefly from unconcentrated slope wash and soil creep in adjacent valley sides
	Mass movement deposits	Earth flow, debris avalanche and landslide deposits commonly intermix with marginal colluvium, mudflows usually follow channels but also spill overbank

RIVER TERRACES

A terrace on the flood plain (Figure 24) reflects a definite river development phase and represents an ancient floodplain alluvium.

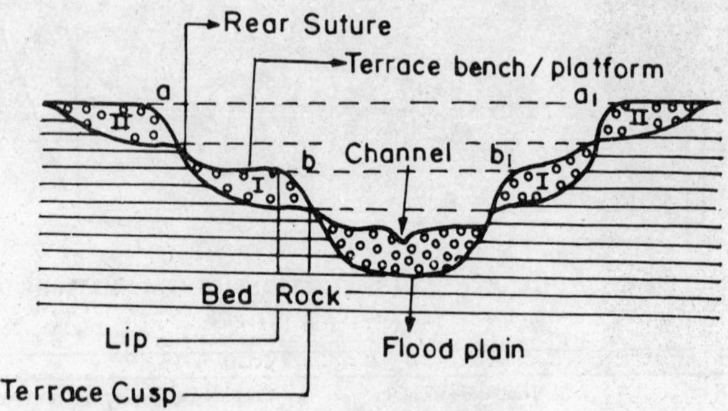

Figure 24 : Stream/River Terraces

[B = Backland, BS = Back swamp, C = Colluvium L = Lag deposit, A = Lateral Accretion, N = Natural levees, P = Point bar, S = Splay, T = Transitory bar, VA = Vertical Accretion]

River terraces may markedly differ in structmre.

Three types of river terraces are recognised :

(i) *Denudation* terraces

(ii) *Aggradation* terraces

(iii) *Compound* terraces.

Denudation terracess are essentially those whose bench and cusp are composed of bedrocks with practically no alluvium veneer. Aggradation terraces are entirely composed of alluvial deposits, a characteristic of the river of long developmental history. Lastly compound terraces are distinguished by the fact that (i) bedrocks lie at the base of their cusp, and (ii) a part of the cusp and the terrace bench, are composed of alluvium.

Factors causing the formation of terraces are :

(i) recent movements of the earths crust that occurred in the Anthropogene and repeatedly disturbed the equilibrium profile;

(ii) increased discharge due to changes of climate, and

(iii) change in the qnantity of detrital material entering the river

By all indications, it is the recent movements of the earths crust that is most important.

WATERFALLS

Drop of water of considerable magnitude both in

(i) volume and

(ii) height

in a stream which plunges over a vertical or nearly vertical rockface is referred to as a *waterfall*, synonymously as the *cataract.*

In some instances water may descend downstream in series of step-like small falls called *cascade*: These may be formed over step-like structure in contrast to *rapids* which are formed where a stream rushes over a steep slope in its bed developed particularly where

(i) hard rock layers cross a stream such that local increase in gradient develops downstream or

(ii) any other obstruction such as boulders from a tributary stream produces the same effects.

The drop of water is less in rapids than in waterfalls but is sufficient to accelerate noticeably the

(i) rate of flow and

(ii) maintain clear water at all stages of discharge.

These conditions approach the waterfall except for the degree of slope. Locally uniform water flow is produced and through the rapids the thickness of water is reduced as the water velocity increases.

a) the normal turbulent flow of water above the rapids become smooth as the water reaches the chute of the rapids. Consequently turbulence flow is transformed into shooting flow condition.

b) down below at the lower end of the rapids, where the gradient becomes gentle and abrupt increase in water volume generates a standing zone of great turbulence referred to as *hydraulic* jump where the water builds up to the height required by the downstream gradient, much of the kinetic energy developed by the fall is dissipated.

An uncommon situation is developed in the plunging waters of the falls which scours the river floor near the ledge, and being reflected from it, forms complex vortices.

(a) A part of the stream flows down channel, and

(b) another backwards towards the waterfall face.

This results in the undercutting of the latter and the formation of *niche* or *alcove.* With continued undercutting the overhang of the ledge crumbles down, permitting the scouring action of the waterfall again to undermine the newly developed precipice afresh causing the waterfall recession. This rate of the ledge recess, and therefore, of the waterfall is controlled by the volume of water that flows over it.

The river scouring beneath a waterfall is further accelerated by the rock fragments of the river bed, which are sometimes quite large. The swift moving plunging stream water in areas of rapids and falls swirls the rock fragments up and when caught in the vortices tend to produce eddies acquiring a rotary motion, boring out round vertical cylindrical hollow or enormous cauldrons in the rocky stream bed or sometimes along the walls of the channel producing an erosion feature, called *giants kettles* or *pot-holes.* Thus, waterfalls help to deepen the river channel.

Waterfalls and rapids are undoubtedly sites of break in the longitudinal river profile which is in broad equilibrium with discharge elsewhere. These breaks are transitory phenomena of the youthful stage which becomes locus of accelerated erosional attack.

Origin

The origin of waterfalls is attributable to several causes

a) *Differential erosion* : In a sequence of hard and soft rocks, erosion is differential, wherein the softer bed is eroded faster than the hard rocks steepening the natural gradients in the river bed. Thus, a rapid or waterfall is developed at the site.

b) *Hanging valleys* (i) A tributary glacial valley meets the main glacial valley at a height. The placement of such tributary valley is termed *hanging valley.* If subsequently, a stream flows over this terrain, it drops down this valley to form waterfall, (ii) Sometimes differential erosion of a tributary and main stream valleys may be such as to the main valley is at lower level in relation to the side valley and if a resistant rock occurs where the side valley comes into the main valley, drop of stream develops into a rapid or waterfall. Such terrain is said to be in *topographic unconformity* and the tributary valley a *hanging valley.* It is not a common occurrence.

c) *Damming of a river* : A river course may be dammed by

(a) landslides

(b) dykes and sills cutting across the valleys

and check the waterflow in the stream whence temporarily a lake is formed and water at its flanks spills out as waterfalls.

d) *Diastrophic movement* : Some minor disturbances in late geological age as most of them in Peninsular India are often, the cause of origin of waterfalls :

i) Wearing of the rock strata along the planes of joints, in many places of Siwalik hills may produce waterfalls. Hundru Fall, Ranchi district, is an example of this type.

ii) Waterfalls may also be produced due to vertical movement of parts of the crust in response to diastrophic movement of fault scarp such as the Ahrabal Falls, Shopiyan, Kashmir. The water of Visho river drops 23 meters vertically down onto the country rocks of Panjal Volcanics in a faulted valley.

Classification : Using genetic criteria three categories of waterfalls have been recognized :

A. Erosional waterfalls : Included in this group are those attributable to the differential erosion potential of the country rock as are developed during the entrenchment of a river channel, sub-type of this category are :

a) Caprock falls- Basal sapling or rapid erosion of weaker rock undermines the overlying resistant capping formation, horizontal or gently dipping cap, the fall recesses or *retreats* upstream, maintaining a vertical or undercut face. Most illustrious citation of this type is the Horse-shoe falls at Niagra.

Step falls - Such falls are formed in strata of relatively uniform erodibility where the undercutting is preferential to the individual bedding planes, for instance Desam falls on the Kanchi river, the Hundru falls on the Subernarekha and the Johnna falls on the Garga in the Chhota Nagpur plateau.

b) Vertical Barrier Falls-Exemplified by the 'Great Falls of the Yellowstone River', in this type water falls over vertical rock bars *e.g.* a dyke. It becomes a fall at the juxtaposition of rocks of greatly differing erodibility along a vertical contact.

B. Erosional waterfalls developed due to the causes not directly attributable to the differing erodibility such as

a) Plateau Falls - wherein the river descends down suddenly from the uplifted plateau to the lowlying and softer rocks of coastal plains, *e.g.* Gersoppa or Jog Falls in Western Ghats on the Sharvati river in Karnataka is a good Indian example.

b) Fault-plane - falls Attributable to a fault-scarp face in the Subernarekha river near Ranchi and Sivasamudram Falls on the Cauvery river in Karnataka.

c) Knick-point Falls - in the rejuvenation in the lower section of rivers either

(i) as a result of a fall in sea level or

(ii) uplift of the land

the down cutting ability is increased developing a sharp slope break, referred to as the *knick point*. It is imminent if the rejuvenation heads happen to be situated in an area of resistant rocks. The rivers produce waterfalls at such knick-points in response to the sudden change in slope.

Several waterfalls developed in the Peninsular Plateau in India are -

Hundru Falls (110 meters) on the Subernarekha

Desam Falls (40 meters) on the Kanchi

Sadni Falls (60 meters) on the Sankh

Gersoppa (260 meters) on Sharvati

Sivasamudram (90 meters) on the Cauvery

Gokak Falls on Gokak river (54 meters) in Belgaum

Dhuandhar Falls at Bheraghat (9 meters) on the Narmada produced in tune with the earth movement in the later geological period.

d) Hanging valley - Water drops down from a tributary valley to the main valleys on *discordant junction*

(i) overdeepened by glacial erosion or

(ii) the junction of two rivers from dis-similar elevations under specific conditions or

(iii) falls over rapidly receding sea cliffs, for example - Rajrappa Fall in Chhota Nagpur on the Bhera nadi, a tributary of the Damodar river hanging over nearly 15 meters above the main stream. Such valleys and therefore hanging valley falls are most commonly found in the glaciated mountainous terrain.

Emplacement of lava tongue or like barrier across stream course are also within the fold of this category of waterfalls.

C. Constructional or auto-consequent waterfalls are rather of specialized type quite different in nature. Developed by the calcite deposition precipitated from solution in underground water as a continuous rim across the tip of a pool in limestone cave of spring. Given a sustained supply of saturated groundwater, growth in height of the rim is self-perpetuating and forming a waterfall on the downstream side. Such falls are generally small and occur in succession of cascades. Individual limestone falls as great as 12 meters are reported.

Recession of Waterfalls

The rate of recession of waterfalls is rather variable :

a) the cataracts on the Nile having inscriptions as old as 4000 years show no recession whereas

b) the Niagra Falls illustrates having migrated a distance of approximately 0.4 km in 260 years, retreating at the rate of 1.2 meters per year between 1842 and 1911 on the Canadian side of the Fall.

Recent C^{14} datings show that the modern Falls originated 12000± 500 years ago. The mean rate of recession since that time has been a little less than 1 meter/year.

Significance of Waterfalls

A. *Geologic*

(a) Waterfalls are distinctive characteristic of the youthful stage in river-history. With the passage of time the waterfalls receds backwards slowly reducing its falling height and ultimately grading in the course of the river. The more rapid the recession of waterfalls faster is the establishment of the gradation.

(b) The waterfall top serves as a local base level for the river above the fall. Consequently, if the recession rate of the waterfall is slow on account of the rock hardness, the part of the river channel above the fall may become graded, the valley widened and the river reflect indications of mature or early old age above the fall, but down below the fall the youthful appearance is preserved because of the presence of gorge and constricted valley.

B. *Economic* or *Applied*

(i) waterfalls are important, being sites of scenic beauty and attraction

(ii) their environmental setting particularly dominate and accessibility are significant in their importance for human being

(iii) they are also storehouses of power supply. Hydro-electricity generated from waterfalls is being used largely for lighting as well as for industry.

Distribution of Waterfalls in India

1. Gersoppa or Jog Falls- It is one of the most impressive 260 meter waterfall etched on the river Sharavati in North Canara, consisting of significant

falls, Rocket, Roarer and Dame Blancha in the Western Ghats. During the monsoon floods when an exceptional volume of water passing over their crest enormously increases, the destructive power is at its maximum and the volume sharply dwindles during the drier part of the year.

2. Sivasamundram Falls - It includes a series of cascades over 90 meters drop of the Cauvery water in Karnataka. It is better known, for this was the first fall tapped for harnessing hydropower.
3. Gokak Falls - The falls is 54 meters on the Gokak river area near Belgaum.
4. Yenna Falls near Mahabaleshwar in Maharashtra have a drop of about 200 meters in one leap.
5. Paikara Falls in the Nilgiri hills descend less steeply in a series of fine cataracts- cascades over the gneissic precipice which have also been utilized for hydro-electric power.
6. Behar Falls - The southern Tons (Tamasa) river emerges out of the Vindhyan plateau in a series of beautiful waterfalls, of which the best is the Behar Fall, which when in flood, majestically presents a water sheet of 200 meters wide and 120 meters high. The course of the Chambal above Kota is punctuated with several cascades, the largest being 20 meters high. Several cascades and waterfalls are recorded in the discordant course of the Sone and Betwa, specially on passing over the Vindhyan sandstone.
7. Dhurandhar or Dhuandhar Falls - Entrenched in the marble rocks near Jabalpur, this picturesque falls on the Narmada river drops from 9 meters height.
8. Mandhar and Punasa Falls - Each 13 meters high. In floods Krishna river becomes a rushing torrent descending some 140 meters within a short horizontal distance of 5 kms near Echampet in Raichur district, Andhra Pradesh. The waters form a series of cascades over granite ledge and are churned into huge clouds of spray.
9. Waterfalls of Chhota Nagpur area - The most important waterfalls of this region, the Hundru Falls, 40 km from Ranchi (110 meters, of which sheer drop is 67 meters and width nearly 7 meters) is on the Subernarekha river flowing over granite-gneiss rock. The maindrop is either
   - (i) locally faulted on a nearly vertical 'scarp-face' like ledge or
   - (ii) could be due now to exposure of large joint blocks on removal of the downstream segment.

The Johna falls, rechristened as Gautamdhan is about 15 km from Ranchi and the Usri Falls is near Giridih.

10. Beadon, Bishop and Elephant Falls are near Shillong
11. There are numerous Falls in the Himalayan region.

The Largest and Highest Waterfalls

Waterfalls may attract attention either by (i) height or (ii) discharge. The table below lists (A) some of the world's highest waterfalls, and (B) some that have greatest volume.

A. Great Weterfalls

Name	*Height in meters*	*Country or State*
Yosenite	2555 ft 850 meters (drops in three stages)	California, USA
Sutherland	1904 ft/670 meters (drops in three stages)	New Zealand
Roraima	1500 ft/500 meters	British Guiana
Kalambo	1400 ft/ 470 meters (drops in two sections)	South Africa
Takkaku	1346 ft/450 meters	British Columbia
Multonomah	823 ft/ 274 meters	Oregon
Bredalnail	620 ft/ 366 meters	Yosenite, California
Gersoppa	747 ft/260 meters	North Canara, Karnataka, India

B. Greatest Volume

Niagara	167 ft/ 62 meters	USA and Canada
Victoria	400 ft/123 meters	South Africa
Iguazu	230 ft/70 meters	Brazil
Kaietem	800 ft/247 meters	British Guiana
Lower Yellowstone	308 ft/102 meters	Wyoming, USA
Grand	316 ft/105 meters	Labrador, USA

ALLUVIAL FAN AND DELTA

Of the most common depositional alluvium landforms two develop out of the same reason *viz*. an abrupt loss of competence in a stream but differently.

(a) Alluvial fans are *subareial forms*, produced on sudden deterring of the stream velocity as it descends down from the mountain terrain to lower level and widens freely

(b) Much of the deltas are *subaqueous forms* the upper surfaces of which usually rises a little above water level.

The latter are developed when stream velocity is suddenly checked on its entering into relatively standing body of water, lake or ocean. Instances may be though not common, where low gradient alluvial fan may supperimpose to cap over delta.

ALLUVIAL FAN / CONE

An alluvial fan is a low cone shaped body of coarse debris stream deposits radiating downslope from the apex situated at the mouth of a gorge or canyon. Contours are arcs of a circle concentric on the head of the fan. In longitudinal or radial profiles, fans are slightly concave upwards. They are excellent examples of water spreading wash slopes point where the stream leaves a mountainous area. Cloudbursts in the mountain upstream from the fan cause surges in discharge of water, and mudflows may transport boulders and much coarse material on to the fan. These fans are diverse of sizes, slopes and types of deposits and source-area characteristics. They are specially widespread in the drier parts of the world and are considered significant landforms in arid intermountain region but are also reported from humid regions. The feature is developed on lateral spreading of dropped debris as aprons in relatively flat-floored valley areas by confined streams loaded heavily with large volume of eroded materials gathered into it through its journey over highlands or mountains.

Bhissenbach (1954) identified three zones (Figure 25) on alluvial fans:

(a) The proximal surface near its apex, the highest point on the alluvial fan is the fan head or upper fan segment. The apex is formed at the stream debouching site from the highland onto low relief area. Closer to the apex, the streams are generally deeply entrenched, forming a fanhead trench. The fanhead area extending far into the highland is known as fan-bay.

(b) The segment lying between the fanhead and the lower margin of the fan is designated as the midfan or middle fan.

(c) The distal lower zone is outermost zone, where alluvial fan of adjacent storms may coalesce, constituting a compound alluvial fan surface.

Many fans in tectonically unstable regions are built of segmented surfaces. The alluvial fans in the Siwalik Hills near Dehradun, Uttar Pradesh are typically segmented along their radial profiles. Relatively the fans are gentler in slopes than cones. The larger ones with radial width of 15-150 km have slopes of less than 1^0 against 5^0 or so in the case of most smaller fans. The cones are usually composed of coarser materials set at divergent inclinations may be as much as 15^0. Hundreds of fans composed of gravels, shingles and coarse sands are observed in the sub-Himalayan belt of *bhabhar.*

Usually alluvial fans are developed where a high gradient tributary stream drops over directly onto the flood plain of a trunk stream or on the flat floor of an intermountain basin. A stream dissecting a fault scarp typically builds a fan at the foot of the scarp. Streams flowing down glacially over steepened valley walls build fans on valley trains of outwash.

Fans are usually composed of bed-load alluvium, especially gravel, may be stratified sheet like parallel to the fan source or may contain many cut-and-fill structures of buried channels. The permeable nature of fan alluvium permits infiltration considerably so that desert fans may contain mudflow deposits recording the progressive loss of water until the suspended load becomes too thick to flow as a muddy stream.

Fans are depositional or aggradational forms. A radial channel on it may remove most of the discharge until its gradient is oversteepend by deposition, when the stream suddenly shifts to a next lower-gradient radius. The process ensures the uniform slope development along all radii and a resulting smooth conical form. Fans may block trunk streams whence a lake is formed. Many underfit trunk streams of glaciated plateaus have low gradients that cannot transport the coarse alluvium brought in by valley side tributaries, and the master stream is pushed from one valley wall to the other by fans. Geometrically similar to talus cone are alluvial cones which are indeed very steep alluvial fans, quite dis-similar in internal structure to talus cones.

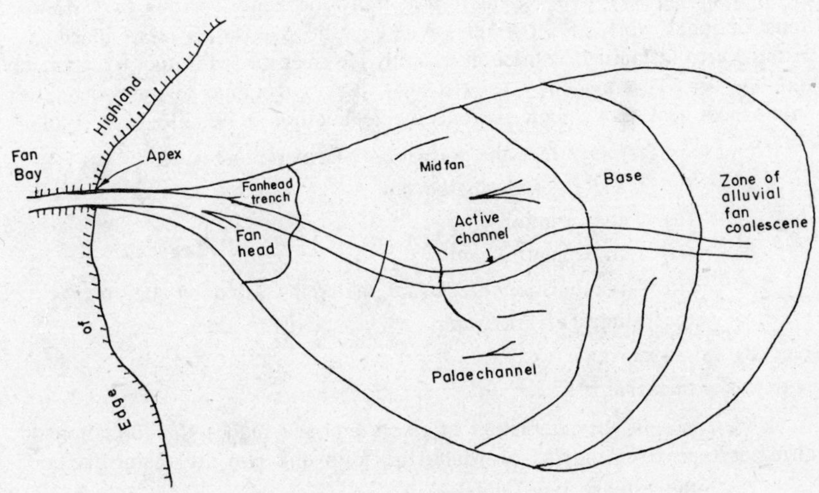

Figure 25 : Morphological features and segments-Alluvial Fan

The fan assumes the form of a sector of a low cone, which is concave upwards so as to steepen in gradient apexward situated at the canyon mouth and grades imperceptibly into the flatter plain at its outer edge. The particulate size constituting the fan is coarsest apexwards comprising bouldery material and progressively grades down to finer size toward the periphery. Large fans of mountainous deserts may be several kilometers in radius from the apex to outer edge, but much smaller fans are more common.

Alluvial fans are of great significance from the standpoint of groundwater reservoirs. Influent seepage water enters near the fan apex and descends downwards and outward along the layers of sorted gravels and sands, filling the interstices. Wells driven into lower slopes of the fan alluvium are often found to be under artesian flow because sloping impermeable mudflow layers interbedded with the permeable gravels form aquicludes which confine the water under pressure. The water resources of alluvial fans are thus great but not for the long term supply as it is effectively limited by the slow recharge rate, so much slow that extensive pumping can deplete the water supply in comparatively short time.

MOUTHS OF RIVERS, ESTUARIES AND DELTAS

River mouths are the sites of complex interaction between

(i) the fresh river and

(ii) the sea

waters where the transition from river to marine conditions takes place. Geomorphologically it is an interesting transition zone, as it is the area of constructional work of the rivers, whereby the deposition takes place, as also the area of marine destruction whereby the river load is further transported into the sea. This evidently testifies that, it is a dynamic region where the subtle sediment movement, is conceivable but not so readily perceivable.

Numerous factors affect the formation of river mouths are :

a) (i) the river discharge and
   (ii) its variation

b) (i) the quantity and
   (ii) composition of detrital material carried by the river

c) the salinity of sea water

d) sea currents

e) tides *etc.*

Differences in the interaction of rivers and sea induce the formation of different type river mouths of which the following two are main types :

a) the estuary type and

b) the delta type.

Again, depending on the mutual relationship of river and sea influences in the river estuaries the following sections are recognized :

(i) The near-estuary section where the influence of the sea is least felt

(ii) The estuary part (the estuary or delta)

(iii) The near estuary foreset beds extending to the boundaries of the shoaling of the near-shore zone of the sea evident in the sea-floor topography

(iv) The near estuary part of the sea including the zone in which the salinity of water is considerably diminished when the river is in flood.

ALLUVIAL DELTA

A delta is a tract spreading outward into the body of water into which the stream empties its detritus. The incoming freshwater spreads out as a sheet on more dense saltwater but may sink beneath the warmer freshwater of a reservoir. In either case the stream velocity is sharply diminished and competence dropped with increase in 'Trap efficiency'. It is simply an extension of flood plains with minimal gradient. Towards upstream the deposits gradually merge into the normal flood plain segment making the point of delta beginning more or less subjective decision and therefore speculative. Similar, but again speculative boundary exists towards downstreams. The subaerial part of the delta merges into the subaqueous part which extends well beyond the visible shoreline with gradual thinning of deltaic sediment and the near sea or lake sediments begin. The term delta was first applied for the triangular seaward part of the Nile abut 2500 years ago by Herodotus, a Greek philosopher, whose penchant for geometry led to find a resemblance between the fourth letter of the Greek alphabet Δ, and the Nile Delta, with base seaward and the apex indicating the stream inlet.

Physical conditions necessary for the delta formations include -

a) large amount of sediment supply by the river

b) relatively calm or sheltered sea at the mouth so that ocean currents or strong waves or high tidal range are absent or feeble and do not effect the depositional process and outgrowth of deltas

c) relatively shallow sea, as deep sea is unfavourable for delta growth because sediments settle down below the sea surface on the marine floor and do not form as a 'landform'. Besides the relatively finer clay particles brought by the river, tends to floculate into larger aggregate by the action of saline water of the sea and tend to settle down and if the sea is deep these sediments too, settle down below the sea surface.

If there are

(i) strong tides or shore currents in the body of water which the stream enters or

(ii) the amount of sediment discharged by the stream is relatively small,

the tendency is for the sediment to be swept so far away from the mouth that either

(i) no delta will form or

(ii) only a small/imperfect one

will develop. Recent sinking of the land, leading to the river mouth submergence limiting time for deposition afresh to develop real delta deposits again around the newly located mouth is yet another reason for the absence of delta from the mouths of many existing rivers (even some large ones).

Certain external features characterize what is generally conceded to be delta. These are :

1. It's surface is wide, nearly flat, usually fan-shaped plain mostly a little above and partly a little below the level of the body of water into which it grows. The great bulk of delta material is however, always under water, and thus it differs from an alluvial cone or fan whose material is wholly or largely on land.

2. Protection into the body of standing water beyond the general shoreline.

3. Outward spreading of the alluvial fill into the body of standing water.

4. A series of low islands bordering the emergent part of the delta.

5. At the point where the flood plain changes to a delta, there is break of gradient and subdivision of the major water channel into several smaller branches called *distributaries.*

These generally point upstream and join the main channel at large angles (*i.e.* as much as 90^0). It is almost universal feature of a delta surface. In fact channel splitting is one of the hallmarks of delta. These distributaries wander over the delta plain in an ever widening network, and so a delta forming river always has several or many mouths. The basis of the bifurcation or splitting at the head of the dela is

(i) the large amount of sediments which fill and choke the channel,

(ii) make it too small to carry the normal or flood discharge.

Each of the distributary forms meander belts that are accomplished by natural levees and associated with flood basins that in turn are studded with abandoned distributary channels and natural levees. But corresponding to the reduced gradients, levee-basin relief is considerably less than on the flood plain. Similarly the relative area of back swamps is greatly increased, and flood-basin lakes may be linked to the open seas by inter-distributary bays.

Many channels build themselves up above the adjoining lowlying areas. The same factor of channel filling or choking causes further 'crevassing' or breaching of the banks whence the distrubutaries keep multiplying. Obviously the size of the delta and the number of distributaries are related to the

(i) size of the basin,

(ii) its structure, and lithology

(iii) its - relief, climate and vegetation *etc.*

Present day deltas display an enormous variety in size, shape, structure, composition and genesis. These differences exist because the same kind of event takes place under wide range of setting. Some of the major factors that influence the characters of a delta include

a) geologic setting and sediment sources in the drainage basin,

b) climatic conditions in both the drainage and depositional basins,

c) tectonic stability of the basin,

d) river gradient and river regime,

e) depositional and erosional processes and their intensities within the delta proper, and

f) tidal range, eustasy and offshore hydrologic conditions.

The numerous combinations between these factors and the element of time result in a dynamically changing complex of environments within delta regions. This is expected since deltas result from interaction of

i) construction and

ii) destruction.

The relations between the effects of these

i) mechanisms of supply and

ii) removal

depend on the interaction and intensities of

i) physical

ii) biological

iii) chemical

processes operative within a delta.

The major part of a delta is due to fluvial processes, deposited by running waters, and known as the *deltaic plain*. Contrarily the *deltaic fringe*, is a coastal environment where standing water and wave action assume considerable or even primary importance. Shallow bars of sand, silt or clay are deposited rapidly in standing waters within the distributary mouths. The levees extend out in the lake or sea as tongues of land, with interdistributary bays fingering their way up the adjacent basins. Lagoons and mudflats, flooded by sea water storms or periods of high tide form a further transitional belt between land and sea, finally a line of sandy beach ridges or barrier islands may form a discontinuous shoreline that connects the various distributary mouths.

Morphology of Delta : The overall configuration of delta expresses different geomorphic environment to which parts of the delta surface responses to. Three major areas of geomorphic influence can be recognized in deltas:

a), The upper delta zone bears testimony to a strong fluvial environment

b) the intermediate zone to both fluvial and marine influences, and

c) the lower zone to marine environment

The three zones have been recognized in the Cauvery delta (Meijerink, 1971). A delta is an ensemble of several geomorphic features :

i) the river regime,

ii) coastal processes,

iii) structural behaviour, and

iv) climatic regime

are the four dominant geomorphic variable which in numerous combinations produce a great variety of deltaic morphology. Active and inactive distributary channels, natural levees, crevasse splays, point bars, swamps, strandlines and dunes provide sufficiently large variations in the morphology of the delta in the intermediate and upper-zones. Subaqueous levees, mouthbars, spits, interdistributary bays and swamps are characteristic forms in the lower zone.

Delta Structure - The delta reflects a characteristic structure because of the special conditions under which deposition of sediments takes place. Thus its internal morphology includes

a. the steep front, so characteristic of a delta, results from rapid deposition of the coarse sediment layer upon layer where the on-rushing sediment-laden stream (or each mouth of the stream) meets the relatively deeper standing water into which the stream flows. These steeply inclined layers are termed *foreset beds*. They make the greater bulk of the delta pile.

b. the fine sediments spread out in layers over the flow of the lake or sea to a greater or lesser distance out from the base of the steep front of the delta. These layers are called the *bottom-set beds.*

c. the *topset beds* are deposited by the stream on top of the foreset beds as the latter advance into sea or lake and shoal the water. They build up, for the most part, a little above the level of the sea or lake in layers which slope very gently seaward or lakeward.

***Deltaic deposits*-** sand, silt and clay-form the greater part in the delta deposits of flat-country rivers ; in contrast to the mountain rivers a large part of the deposit is composed of gravels. Soluble materials mostly of colloidal nature, brought by stream, tend to coagulate by saline sea water salts of Ca, Fe, Si, Al *etc.* and precipitate in the delta. Thick beds of iron oxides were deposited in geological past in this way which are now worked as deposits of brown iron ore. Organic milieu trapped in silt particles and subsequently buried may give rise to oil and gas deposits.

Numerous lakes (oxbow type) are present on delta surface, into which lacustrine, mostly clayey, rich in organic, matter are also laid down alongwith the above deposits. Many old channels are occupied by marhses, followed by peat bogs which are later transformed into coal seams.

Marine sediments also play a part in the entire complex of deltaic deposits.

Accordingly, the composition of river deltaic material is most diverse : alluvial deposits of the channel streams, clayey sediments of deltaic lakes, often with much organic matter, peat formations of bogs, and shallow sea deposits, varying greatly, both horizontally and vertically due to frequent channel shifts which indeed control the transport and accumulation of alluvial deposits, the lake formation and various depressions, reflected in the appearance of clays and peats in the stratigraphic sections. They may also be associated with various economic minerals, such as oil, coal, peat and iron ore

Essentially the deltaic deposition are divided into four different kinds of beds:

a) the upper above water beds composed mainly of river, lacustrine and marsh deposits,

b) topset beds - the upper horizontal - bedded section also composed mainly of river sediments, but those deposited however, below the sea level,

c) the middle cross-bedded foreset beds composed of sediments deposited on steeper slope and rewashed by surf and currents, and

d) the lowest bottom-set beds mostly composed of colloidal material, brought by the river and coagulated from sea water.

Departure in this pattern is not uncommon and is primarily due to the disturbance in the equilibrium between the land and sea in response to alternating movements of the earth's crust (subsidence and uplift). A characteristic feature of deltaic deposits is the oblique (diagonal) bedding.

While deposition of sediment has been going on, many of the great deltas of the world are slowly subsiding developing typical deltas where the rate of sediment supply is faster than the rate of sinking, but where subsidence is predominant, even with other conditions favourable, deltas have not been built up above sea level. Thus the Ganga, Nile and Mississippi rivers have built up very extensive deltas inspite of subsidence of several meters as proven by borings through them.

Delta Dynamics and Growth - The delta growth is a function of base level (i) rise and (ii) fall when the base level lowers down the delta growth enlarges tremendously whereas with base level rise the aggradation is pushed back upstream, and consequently much load does not reach the river mouth resulting in feable delta growth. Also much of the emergent part of the delta is then submerged and consumed by current and wave destruction.

Compound delta growth is commonly noticed because of the development of new distributary channels over the earlier ones. The delta growth intiates with the formation of a lobe in which a single channel is limited between its own natural levees, and gradually extend outward into the body of standing water. With increase in channel length the gradient progressively lowers down which facilitates the flooding the natural levees resulting in (i) break through and (ii) the appearance of new and shorter distributary. The new route commands the largest discharge of water and brings most of the load, a new delta lobe is built. Like this the processes distributary lengthening and break through producing yet another distributary and thus a new delta lobe goes on. New lobe debris in part buries some of the old delta, and slowly the inter-distributaries areas are filled and emerge from the shallow water.

Several features are formed in the delta front which is a zone of the most dynamic delta-building activity. Subaqueous levees develop in a low-energy wave environment along distributary channels carrying large volume of load. Sedimentary jetties at the mouth of distributary channels form distributary mouth bars which are constantly modified by marine activity and variations in stream velocity. The distributary mouth bars grow into spits when the longshore drift is strong. Inter- distributary bays commonly occur in the delatic environment. They are open water bodies enclosed by subaqueous levees and swamps, but are connected to the sea by tidal inlets.

Classification of Deltas : Deltas can be classified on the basis of -

a) the associated body of water into lake, bay, inland sea or marine deltas

and b) the shape-geometry.

The geometrical classification is more common. On the basis of the differences in form (shape), deltas are classified into four major types -

a) Arcuate delta
b) Digitate or Bird-foot delta
c) Estuarine delta
d) Cuspate or tooth-shaped delta

developed in response to the balance of forces between -

(i) wave energy,
(ii) tidal amplitude influences, and
(iii) sediment load

a) *Arcuate Delta* - It is the commonest type of delta resembling the Greek letter Δ with arcuate seaward base of which the Nile delta (Egypt) is the best example. Its shape is like an arc of a circle and resembles a bow. Wave and current activities are stronger here, so that the distributary mouths do not develop rapidly. Instead, they are linked by a well-developed series of beach barriers. The city of Cairo is located at the delta apex, the Nile dismembers southwards into several radiating channels. Since the thickest sediment is on the sides of the main channel, the central relatively lower part tend to project into the sea. The sediments which are carried down the Mediterranean sea by the principal channels of the Nile, form curved bars along the coast under the influence of wave generated littoral currents, behind which there are shallow lagoons. Other examples of delta of this type are the deltas of the Ganga, the Rhine and the Hwang Ho.

b) *Digitate or Bird-foot Delta* - It's shape resembles the finger of a human hand or foot of birds and it is called digitate or bird foot delta, of which the best example is the Mississippi Delta (USA). The distributary channels extend into the standing water body. In this type

(i) sediment influx is high, and
(ii) wave energy and tidal amplitude small.

The finer sediments are carried far into the sea by the stream velocity, where on account of the saline water of the sea, the coagulate get deposited on stream channel side giving rise to long finger-shaped deposits. The Mississippi river flows into the

Gulf of Mexico for some distance in a single channel and then divides itself into four distributary channels, resulting in the formation of the finger-shaped delta.

c) *Estuarine Delta* - A long and narrow delta is formed where the river enters the sea through a single mouth or estuary, and the estuary is filled up with sediments. Such deltas are formed where tides are influential. Such a type is seen along the mouth of the Seine river in France and Sasquehaua river in the Eastern USA. Yet another example is the confluence of the Sacramento and the San Joaquin rivers in the Central Valley of California. Ordinarily the estuary is open and the river sediments are carried away by waves and currents, but when the river mouth is submerged below the sea, a narrow linear delta develop on sedimentation.

d) *Cuspate or Tooth-shaped Delta* - Tiber river delta (Italy) is a good example of this type in which the deposition takes place along the main channel and the delta projects into the sea, but the sediments carried by the river are deposited and redistributed by the waves to spread on both sides of its mouth by waves. Thus on either side of the projected mouth of the river curved deposits concave towards the sea are formed.

Indian Deltas : The coast of India practically all on the eastern side is punctuated with several large deltas of Brahmaputra-Ganga, Mahanadi, Godavari, Krishna and Cauvery in the Bay of Bengal, their growth coincide with the structural/ tectonic activities and they attained their main configuration during the Holocene. Narmada and the Tapti rivers, have no delta as :

a) the tides and currents produced by the strong monsoon winds are so powerful along the western coast that no silt is deposited near the mouths of these rivers and

b) these rivers flow through hilly areas and do not bring much sediments with them.

1. *Ganga Delta* - It is a moderately arcuate shaped delta with an area of 400 sq km from Hooghly to Meghna. The delta is highly crenulated with north-south marine inlets aligning with the distributary channels both separated by intervening marshyland which are partly inundated by the diurnal high tide sea waters, favouring a saline superhumid and marshy conditions quite conducive for mangrove vegetation. These youthful general physical features-

a) high degree of delta indentations

b) alignment of marine inlets with distributary channels

and further corroborated by

c) sinking of *sundri* trees, which grow only above the normal high tide level, as of today

d) occurrence of peat, as also

e) pebble beds at depths of 53-56 meters in them again at 91-147 meters below, as disclosed by boring near Fort Williams during 1835-40

are suggestive of contintinued subsidence of the Ganga delta.

Such subsidence in the case of the Mississippi has been 2-3 meters per century. Comparing the size of the Mississippi and Ganga - Brahmaputra basins one could roughly estimate that the annual subsidence in the case of the Ganga delta is 50 to 70 cm.

The Ganga delta is slowly prograding especially in the active eastern section which is building at a faster rate and where the different stages of lands building into the sea can be clearly seen from the air. Old beach ridges in Sunderbans point to such seaward outgrowth.

It's lobate shape is the outcome of large amount of sediments supplied by the Ganga-Brahmaputra system through the world's highest mountains brought by intense gully and sheet erosion which indeed is facilitated in the tropical and sub-tropical climate with some of the heaviest showers and high range of temperature far in excess over the marine processes it is attempting to washaway partly through its tidal arrows.

A number of islands seaward of the delta face is a notable feature of the Ganga delta. Some islands like the Sagar and Bangaduni are presently isolated from the mainland but may merge in due course commencing with the progressive narrowing down of the separating sea bodies into creeks. Although slow, the progradation is geomorphologically apparent. The banks which are visible above low tide level are north-south indicating the pattern of sedimentation in the offshore zone and the seaward prolongation of the interdistributary land areas 'with further sedimentation of banks will rise into islands which will be finally merged with the delta'.

2. *Mahanadi Delta* - Chilka lake is probably a deltaic lake. Considering that the delta starts at the western end of this lake, the deltaic arc about 300 km of the Mahanadi with reference to a straight line joining the western end of the Chilka lake with the shore point on 21^0 N, is projected to the extent of about 60 km. Enormous sediment supply obtained on intense erosion of rugged terrain, the source of supply under hot and humid monsoonal climate facilitates in the development of markedly arcuate delta.This and other deltas of the eastern coast of India would have been more lobate had they been in protected marine waters and had been free from cyclonic storms and ocean currents. "The Mahanad delta is triple delta where deltaic sediments of the

Mahanadi, the Brahmani and Baitarini" are dropped. Within the arcuate bulge the distributaries arc etched by estuarine inlets with interstream headlands. The structural bases of the Mahanadi delta are different from those of the Ganga. The Ganga delta is located at the seaward edge of a sagging depression associated with the Himalayan mountain-building. Instead the Mahanadi basin on the other hand occurs on the rigid peninsular platform, believed to correspond with Permo-Carboniferous rift valley. However, the present delta is probably much broader than the original graben and the deltaic sediments occupy a wider region. 'The khondalite hills and other crystalline covers with laterites near Cuttack' and further north-east indicate the relative shallowness of the sedimentary cover over the rigid shield. North of Puri, beneath other alluvium are Miocene-Pliocene marine sandstones indicating geologically recent elevation of the 'coastal plain', occupied by the delta. The alluvium of the Mahanadi delta is akin to the Older Alluvium and different from the Newer Alluvium which occupied the Ganga delta. Some islands have grown in protected regions, *e.g.* behind the spit that encloses Chilka lake and off the mouths of Brahmani and Baitarni, *e.g.* Shrott's island and Wheeler island. The latter are probably due to deltaic sedimentation. Some deltaic lakes also occur, *e.g.* Sar Lake (*c* 6 km by 4 km) and Samang lake (*c* 3 km by 1.5 km). These freshwater lakes are the old beds of distributaries.

3. *Godavari Delta* - The delta extends for about 35 km into the sea from the adjoining coasts. Geometrically the delta shore is like three sides of a hexagon with straight seaward face. The longer side of delta is some 90 km, the other two sides being about 35 km. Neglecting indentations, the deltai coast is some 150 km long. It evolved by successive subdivision of active channel into progressively increasing number of shorter distributaries. The development of a distributary network is facilitated by

i) structural dislocations along fractures and faults, and

ii) the growth of morphostructures.

C^{14} dating of calcareous oolite sand sample, indicate Holocene age for the delta. The present deltaic load of the sediment appears to be post-Pleistocene. On the basis of the delta size it is inferred that the subsidence is much less than in the case of Ganga or the Mahanadi delta. Godavari delta is much smaller than that of the Mahanadi delta.

4. *Krishna Delta* - The length of the deltaic coast is about 90 km. The protuberance is about 137 km. The marked projection of the delta related to this leeward position near the mainland and relative freedom from the strong longshore drifts or trimming action of the sea currents. "The dominance of the terrestrial over marine action results from the lee effect of the Eastern Ghats with reference to the SW monsoon". The wave section is also limited because of the short fetch in the small bays on eitherside of the delta, both

for the SW winds. All the factors result in an unusual subordination of marine forces in the Krishna Delta and this probably accounts for the unusual seaward extension of the delta of a river whose basin is about as large as that of the Godavari. The prograding delta of the Krishna River has also evolved during the Holocene. They recognize four and possibly five major strandlines and about seven major distributary lobes. The growth of the delta is interpreted to be the result of slow sinking of the southern coast or the continental shelf, growth of morphostructures and changes in the discharge and sediment load of the river.

5. *Cauvery delta* - The delta developed its main configuration during the Holocene. It is characterized by the fact that it

i) has practically no protuberance,

ii) is unusually symmetrical- the north south seashore of this triangle is about as large as the prependicular bisector of this base, and

iii) has distributaries starting diverging some 125 km west of the coast.

The abnormal straightness of the delta face lacking seaward projection is believed to result from the dominance of marine action which produces bars on the mouth of the river. Differential subsidence of the eastern and western parts of the delta is reported in three-stage bifurcation of the active channel. At present, the delta is being pushed southward by beach drifting.

ESTUARIES

The term *estuary* is derived from the Latin *aestus*, the tide and *aesto*, boil indicating the boiling effect of the rising tide at river mouths where the river and ocean waters meet. It is usually defined as that part of the lower river course that is affected by the mixing of saltwater at the mouths of rivers bringing restricted quantities of sediment to the sea. They are enclosed bays receiving freshwater influx and subject to tidal influence. Heavier salt water underlies silt-laden freshwater, a situation that permits higher concentrations of suspended sediment. Best development of estuaries are known where the tidal amplitudes near the mouths of major rivers are great. In general, they are characterized by mudflats developed due to sinking down and settling out of suspension of tiny particles on flocculation as they meet and react with seawater. In microtidal environments low tide channels cut as deeply as low springs tides will allow. At low tides mudflats are exposed on which soon vegetation grows furthering increased accretion leading to increased rates and providing additional land for reclamation.

Estuary formation is favoured by the ground subsidence near the river mouth. If the subsidence is not compensated by the accumulation of sediments, as is the case in subsiding deltas, the sea transgresses inland forming estuaries.

The penetration is deeper inland during the high tide and during low tide powerful seaward currents are generated due to the superimposition of river current on receding tidal water current which transfer the river borne detritus further into the sea where it is caught by the littoral currents. Notable among river with estuaries are the Amazon, the Rivers of St. Lawrence, the Thames, the Siene and the Elbe *etc*. Conditions favouring the formation of estuaries are :

a) tides and currents,

b) small quantity of material brought by the river,

c) subsidence of the earth's crust in the area of the river outlet, and

d) very deep channel due to the retreat of the sea.

A river debouches into sea in several ways or associations including

a) *Direct,* as with a 'youthful' type stream on a steep coast, there being no delta and no salt water encroachment

b) *Delta building,* where the sediment discharge exceeds the net wave energy available for sediment dispersal

c) *Lagoonal*, the river emptying (often with a delta) into an embayment or lagoon that is almost cut off from the ocean by barrier islands and bars,

d) *Estuarine,* where the river bed is 'overdeepened' *i.e.* lies well below mean sea level and at least from time to time, there is encroachment of seawater upstream.

Classification of Estuaries - Basically there are two fundamentally different type of estuary related to dimensions or genesis, *i.e.* applicable to estuaries of all size and histories :

a) *Funnel shaped estuaries*-completely open to ocean characteristic of gradually widening bays. These are related to

(i) strong tidal effects,

(ii) oceanic waves and swell,

(iii) relatively high salinity, and

(iv) relatively great depths.

b) *Barred estuaries*- partially cut off from the ocean by rock promontories, rock threshholds, sand bars or spits, barrier islands, mudbanks or delatic accummulation. There is often a transition into coastal lagoon. These are directed more towards the river, with strong mixing, but with reduced tidal effects; oceanic wave action is often negligible, and sediment tends to move downstream from the river. Periodically, in semi-arid regions the sand bar may be totally closed and the estuary converted temporarily into a barred basin, marked by stagnation, low oxygen, H_2S production at the bottom and the formation of pyrite concentrations in the sediment.

Dionne (1963) proposed the following subdivision of estuaries :

(i) *Marine or lower estuary,* in the major estuaries a width of 1,000 meters would correspond rather generally to the limit of the marine estuary,

(ii) *Middle estuary,* this stretch is affected by strong mixing of salt and fresh waters, and

(iii) *Fluvial* or *upper estuary*, the limit of tidal movement, this stretch is essentially fresh fluvial, but subject to tidal blocking.

Morphogenesis of Estuaries - The reason for the 'overdeepening' of river mouths that transforms them into estuaries usually depends on several factors, which vary in relative importance according to

(i) coastal stability

(ii) tectonics

(iii) latitude (climatic regime)

(iv) Quaternary history

(v) tidal characteristics and so on.

Practically, the most important single factor is the post-glacial rise of sealevel of several meters. Traces of the course of the ice-age river are usually to be obtained across the continental shelf, sometimes part of extensive 'drowned drainage' network. The seaward end of such drowned channels often passes on the continental slope into a *submarine canyon.* Such features may also develop on tectonic subsidence or in glaciated regions.

Apart from structural aspects, there are dynamic aspects that control sedimentaion and may eventually modify the form of the estuary quite fundamentally. *Bay-mouth bars* formed by longshore drift or in warm tropics, growth of *coral reefs* may progressively isolate the lower estuary until converted into a lagoon.

CHAPTER IX

OCEANS AND SEAS

An enormous volume of water, (1,370,323,000 km^3) is locked up in oceans and seas in about 70.8 percent or 361 million km^2 out of 510 million km^2 of the surface of the entire globe. Being in perpetual motion it continually interacts with the surrounding land, destroys the rocks of the coastal tract and the floor of the shallow sea foreshore, transports and abrade the products of destruction and deposit them as sediments. The seas are the habitat of diverse animals and plants and the receivers of all detritals and dissolved materials brought by rivers from land. *Sealevel* is the surface of the sea-the average elevation of watermark between the extremes of mean low and mean high tides. According to its variations in depth, deep sea and shallow sea is referred to. Sea and ocean are practically synonymous in general use and refer to the whole continuous body of saltwater, including its numerous embayments. The oceans, however, are very large bodies of deep seawater occupying basins between continents. Such names as 'Dead Sea' are misnomers because being inland bodies of water, (salt or fresh) not connected with the great general bodies of salt water, are really lakes. The Pacific, the Indian, the Arctic and the Atlantic oceans have about 3.75km depth on an average.

Classification of seas-

Three distinct classes of seas have been recognized. They are -

a. *Epeiric seas* the word *epeiric* derived from the Greek word *epeiros* meaning a continent, are shallow (seldom over 200 meters) sea within continental regions and are poorly connected with ocean is a very good example.

b. *Epicontinental seas* border lands, nearly everywhere, occupy the narrow platforms (continental shelves). In a way, thus, these seas are shallow landwater extensions of the open ocean and as such referred to as a marginal seas or shelf.

c. *Relic seas or mediterranean type* - were formerly parts of oceans, now cut off by the uplift of land mass. It is a special type of sea much like epeiric sea, but has depths of hundreds of meters. The Mediterranean Sea and Caspian Sea are good examples.

MAJOR RELIFF FEATURES OF THE WORLD OCEAN FLOOR

On the basis of available bathymetric data, several morphological regions or altitude stage of the floor of the World Ocean, as below, are recognized:

1. *Littoral zone* - This includes the area below the beach between the levels of high tide and low tide. Deposits here are the coarsest mostly collected from continental land masses and are called *terrigenous* (derived from land) deposits and the sorting action of the waves is most conspicuous. On account of conditions of disturbances this area does not sustain much organic life, but for those animals which can live exposed to the air for sometime. Width is generally 3 kms on an average from the land towards the sea. The total area is about 150,000sq km It is also called the *strand.*

2a. *Neritic zone* - The word neritic has been derived from the Greek word *neritos* meaning a mussel. This zone includes the area between the low tide mark and continental shelf margin. It is a zone of calmness as well as activity and of visibility and plenty food. Hence organisms, known as Benthos meaning *'bottom* dwellers' such as sea weeds, molluscs, echinoids, corals and other creatures that live on the seafloor are most abundant. Wave action and sorting of sediments are conspicuous. Total area is nearly 25,000,000sq km, depth upto 120 meters and distance from land upto about 30 km.

2b. *Continental shelf zone* - Depth varies from 120 meters to 200 meters and the distance from the land varies from 30 to 320 kms, averages 100 meters. It occupies about 76 per cent of the area of the world ocean fringing nearly all shores. (The Malabar coasts has border shelf than that of Coromandal coasts in India).

This, more or less continous platform, separated from abyssal plain portion of the ocean by relatively much steeper gradient of the *continental slope*, slopes seaward relatively gently (of the order of 2 meters per kilometer) in contrast to the latter which is as much as several hundred meter per km.). Much of the shelf receives a constant addition of sediments, accumulated as gravels, sands, muds and limestone rocks. Other areas receive no sediment at all, and still other parts rise above the sea level as marginal islands. Wave action is noticeable and salinity or turbidity too. In the uppert part sunlight can penetrate easily and temperature changes with season and latitudes where algae thrive best. This upper part is known as *Bathyal* or *Laminarian* zone, the latter from the preponderance of the algae *Laminaria.* In the lower part there is an abruptness in the temperature change on account of low penetrability of sunlight which becomes less and consequently sharp decline in aquatic plants. At the lowest limit finer sediments of mud, lime *etc.* are deposited. Coral flourish and coral islands are notable features of this zone. Brachiopods too are abundant. Continental shelves, in general, are characterized by -

(i) their terraced, irregular, and extremely variable from place to place surfaces and are not smoothly sloping ones,

(ii) particle size distributions of the floor sediments may not necessarily show a progressively decreasing average size seaward. The variability is great in some areas coarser material lying near the peripheral margin,

TABLE IX. 38 : A SYNOPSIS OF SUBMARINE TOPOGRAPHY

(after Karl W. Butzer 1976)

Category		Forms and Features		Processes
A. Continental Shelf	a.)	Drowned depositional forms of terrestrial origin (floodplains, delta and coastal plains)	1.	Terrestrial agencies (running water, ice) operating during glacial-eustatic regressions
	b)	Drowned and partly filled erosional forms of terrestrial origin (river valleys, glacial troughs)	2.	Coastal agencies (waves, wind) operating during glacial eustatic regressions
	c)	Drowned coastal landforms (barrier islands, sea cliffs, shore platforms, coastal dunes)	3.	Mass movements (debris slides, slumping, turbidity currents)
	d)	Basins and flats of marine deposition	4.	Glacial drift (iceberg debris)
	e)	Submarine valleys and canyons	5.	Tidal and bottom currents
			6.	Marine deposition or precipitation of suspended or dissolved materials
			7.	Diastrophism and vulcanism
B. Continental Slope	a)	Inclines and platforms of marine deposition	1.	Marine deposition or precipitation of suspended or dissolved materials
	b)	Submarine canyons	2.	Mass movements and turbidity currents
C. Sea-floor Plains	a)	Submarine fans and valleys	3.	Glacial drift
	b)	Flat plains of marine deposition (brown clays and organic oozes)	4.	Diastrophism and vulcanism
	c)	Seamounts and tablemounts		
D. Mid-Ocean Ridges	a)	Fault-block plateaus, ridges, and rises	1.	Diastrophism (including plate tectonics)
	b)	Volcanic seamounts	2.	Vulcanism
	c)	Spreading rift fissures	3.	Mass movements and turbidity currents
	d)	Transform fault scarps	4.	Marine deposition or precipitation of sediments
E. Trenches and Isalnd Arcs	a)	Deep trenches	1.	Diastrophism and vulcanism
	b)	Volcanic seamounts and islands	2.	Mass movements and turbidity currents

TABLE IX. 39 : A CLASSIFICATION OF COASTAL LANDFORMS

terrestrial		Features due to wave action		Features due to tidal or processess
I. Coasts Modelled Primarily By Wave Action				
A. Rocky Coasts	1.	Cliff sectors		
	a)	Notched or plunging cliffs	a)	Chemical corrosion or solution
	b)	Abrasional, depositional platforms	b)	Frost weathering
	c)	Stacks, reefs, arches, caves	c)	Talus and rockfalls (consolidated rock)
	d)	Longshore bars	d)	Slumping and debris slides (unconsolidated rock)
	2.	Sand and shingle sectors		
	a)	Beach ridges		
	b)	Shingle beaches		Coastal dunes
	c)	Nips		
	d)	Longshore bars		
	3.	Attached bars		
	a)	Bay bar	a)	Tidal lagoons or mudflats, including saltmarsh or mangrove swamp, tidal creeks, and inlets
	b)	Spit		
	c)	Cuspate bar	b)	Non-tidal lagoons and marsh
	4.	Coral reefs		
	a)	Fringing reef		Lagoons
	b)	Barrier reef		
	c)	Atoll		
	5.	High Pleistocene shorelines, including cliffs, platforms, ridge and coral platform		Fossil dunes
B. Coastal Plains		Offshore bars, including sectors with cuspate bars, spits, or longshore bars	a)	Lagoons and mudflats, with salt marsh or mangrove swamp, tidal creeks, and inlets
			b)	Estuaries and deltas
			c)	Coastal dunes
II. Coasts modelled Primarily by Terrestial processes (Wave action subordinnate).				
A. Fluvial	1)	Contemporary stream action	a)	Estuaries
			b)	Deltas (including birdfoot, arcuate, and estuarine types)
	2)	Pleistocene stream action : drowned landscapes shaped primarily by fluvial dissection (including *rias*)		
B. Glacial	1)	Contemporary glacial action : valley glaciers or ice caps calving into open sea or feeding ice shelves		
	2)	Pleistocene glacial action: drowned glacial landscapes	a)	Drowned glacial troughs (fiords)
			b)	Drowned ice-scoured plains
			c)	Drowned drifts plains

(iii) the average shelf sediments appear to be coarser than the present day river supply

(iv) of most glacial areas shelves show unique topographic features.

3. *Continental Slope zone* - With depths of 200 to 2,500 meters and an average gradient of 3.5^0 to 7.5^0 it occupies about 15 percent of world ocean area. At places the continental slope show very complex relief with deeply incised canyons. Submarine canyons generally appear where (i) the continental slope is near to the continent and (ii) the shelf is not wide. At times these canyons, 3 to 4.5km wide between 650 to 1000 meters long appear to be counterparts of the terrestrial river canyon. Usually
   (i) the area occupied by this zone is nearly thrice that of the continental shelf zone,
   (ii) except in the upper most part (light does not penetrate beyond a depth of about 220 meters in open oceans) no light can penetrate,
   (iii) no wind action is felt,
   (iv) temperature is low and water pressure is great, and
   (v) the sediments are very fine.

Black, blue and green mud, coral mud, volcanic mud *etc.* are found here comprising primarily clay minerals, quartz and calcite. This, due to non-oxidation of iron sulphides in the presence of organic matter, imparts blue colour to the mud and the green colour is attributed to the presence of glauconite- a hydrated silicate of iron, magnesium and potassium. Mainly calcareous deposits with glauconite are known as green sands. Deep sea deposits like oozes are also present in the lower part.

Submarine Canyons - Unlike the continental shelf, the slope is anything but regular in its configuration. Beyond doubt, the most conspicuous features are *submarine canyons.* By definition, they are rock-walled, V-shaped, winding valleys extending down the continental slope. Tributary valleys, often dendritic, are common. The canyons begin abruptly on the lower shelf. Occasionally they extend to coastal proximity near the mouth of some, but not all, large rivers. Such canyons generally terminate on the lower continental slope, but may as well continue as shallow levee bordered channels across broad sediment fans debouching onto deepsea plain. The submarine canyon floor normally includes alternating sand and clay layers, and similar beds build up the associated *submarine fan.*

Dimensions of submarine canyons vary and some are comparable to those of the Grand Canyon. Basically they differ with deeply incised stream valley in that the submarine canyons are comparatively straight, without entrenched meanders of small radius. The length of the rock-cut segments of the submarine canyon may exceed 120 km and the longitudinal gradients decrease downvalley from initial averages of as much as 3^0 to 8^0. Valley walls commonly have a relief of 300 meters or more.

Contemporary research indicates that submarine canyons have multiple origins as arranged below in decreasing abundance incidences (i) faulting (ii) turbidity currents and mass movements and (iii) stream cutting during period of low sea level. The last type may be modified by marine process or maintained by turbidity currents. Large submarine canyons are often situated against the mouths of large rivers of great age which might have long provided suspended sediments for turbidity currents. Finally, 'built in' patterns of up-and-down canyons currents, only in part related to tidal alternations have been recorded. Available evidences suggests that many, if not most submarine canyons require millions of years to form.

4. *Abyssal zone* - About 76 percent of the areal extent of the floor of the world ocean 2,500 to 6,000 meters deep is occupied by this zone. The gentle gradient is between 0^0 20' and 0^0 40' intersected by long submerged ridges at places.

5. *Deeps* - Occupying about 1.2 percent of the floor with depths of more than 6,000 meters, it is limited to near island arcs and continents.

THE MOVEMENT OF THE SEAWATER

The movements of seawater are various, but may be conveniently classified as under :

a) waves, caused by the action of wind

b) tides, caused by the attraction of the moon and the sun

c) ocean and sea currents

d) slow and circular movements from the high latitudes to the equator and *vice-vesa.*

e) compensatory movements between two basins differing from each other in salinity and density.

CURRENTS

Turbidity current is developed on account of relatively higher density of turbid or muddy water as compared to the clear water and therefore permitting it to sink beneath. Several phenomenon in the present oceans as also of the geological past offer most plausible explanation of certain deposits under deep waters. These basins contain thin layers of sand which could not ordinarily have been transported so far away from shore by the slow water drift along the ocean's floor, but for turbidity currents quite capable of moving the sand down the slopes of the oceanic basins to the deep floors. Thus *graded bedding* is the structure produced by a turbidity current carrying particles of various sizes out of which the larger ones were the first to be dropped on the ocean floor and successively finer followed. They are aptly referred to as *turbidites.*

Motion in these currents may be set (i) by the slumping and sliding of material along the slopes of the ocean basin under the influence of gravity, or (ii) aided by the jarring of an earthquake, or (iii) by the churning up of bottom sediments under the influence of violent storms.

WAVES

Waves are generated when the wind while passing over standing, more or less flat, body of water, transfers its energy, and produce drag by the friction which disturbs the water body, as manifested in rippling the surface and setting water particles in motion. At their limiting boundary of water and land which constitute the shoreline, a relationship is observable, between (i) the wind velocity and (ii) the wave motion, along the interface. A *shoreline* is thus the line of separation of land and water and represents the interaction between the land and marine/lacustrine environments. It keeps fluctuating from moment to moment under the influence of tides and waves and therefore timewise it is not a fixed line. The area of influence is subdivided into :

i) the *offshore* - the shallow bottom seaward of the breaking waves

ii) the *nearshore* - between low tide level and breaker zone

iii) the *foreshore* - the sloping zones between the low tide level and the limit of high tide level; storm wave effects, and

iv) the *backshore* - from the limit of recent or frequent storm-waves landward to the base of a cliff, dune or vegetated beach ridge.

The extent of lakes, ponds as well as oceans are defined by shores, but in the general usage, the word *coast* applies only to the geographic adjacent to an oceanic shore. Although the distinction between shore and coast may not be carried too far but it is real. The zone between high-tide and low-tide (including the spring and neap tide levels) is known as the *shore* or shore zones. The term coast on the other hand applies to the land 'immediately behind cliffs'. Even in the absence of cliff, the surface immediately behind the zone is related to the coastal geomorphology in several ways *e.g.* the distribution and alignment of dune ridges, in progradation processes, in the silting of lagoons *etc.* It includes all those land surfaces modelled conspicuously by waves-now as well as in the recent past.

The size (height) of the waves generated over the water surface by the wind drag is a function of

a) the wind velocity,

b) time duration of consistency in wind direction, and

c) the fetch *i.e.* the outward open water surface distance over which the wind builds up waves without interference.

The wave geometry may be modified by any of these determinants under the influence of three closely related parameters *viz*, (i) the wavelength, (ii) the wave velocity and (iii) wave period.

Waves tend to be somewhat choppy, relatively small and irregular. The small waves reach a maximum height and then destroyed.

When a crest becomes

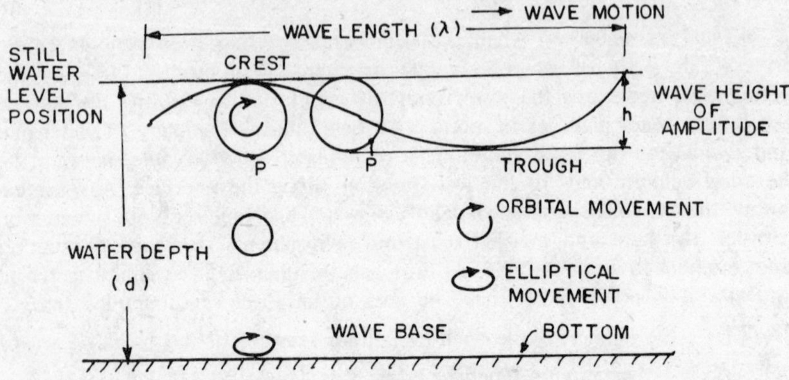

Figure 26 : Wave motion at the surface and down below

i) steeper than about 60^0 and (ii) the wave height 'h' become>1/7 λ (Figure 26), the wave become unstable and is usually blown off by the wind. Constrastingly waves with the longest wavelengths characterised by high λ/h ratio, advance onwards.

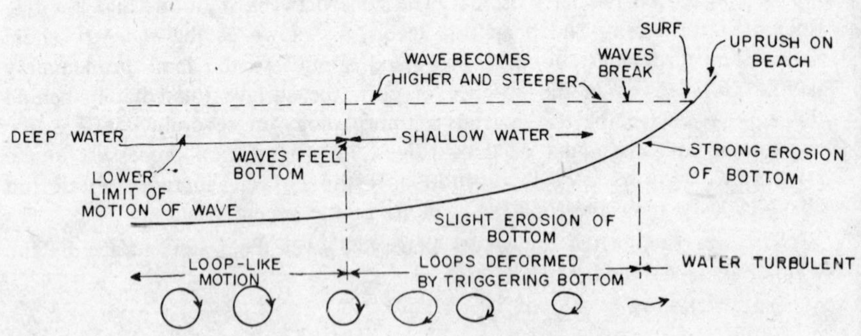

Figure 27 : Wave motion deformation with depth

Wave motion is oscillatory in which a particle describes a nearly but not wholly circular path, returning approximately not precisely to the same point on successive passing of wave. The orbital diameter is equal to the wave height-the vertical distance from the crest of a wave to the trough. As a particle in a trough moves upward and towards the crest when a wave approaches and then, moves downward and back towards its original position in the trough when the crest passes the orbital motion or water particles is much reduced below the water surface. The extent of motion at depth decreases in circle of decreasing diameter (i) wave length and (ii) the ratio of wave height to wave length, h/λ.

Waves developed during a storm assume different dimensions, lengths and heights and some of them may be superimposed upon or interfere with one another producing *compound waves.* The crest of two wave trains on coinciding cause an exceptionally large wave but when the crest of one wave train coincides with the trough of another, the two practically cancel each other. Thus some waves approaching a shore are larger than others.

Waves travelling away from region of origin or source generating them have their crest more smoothly rounded and usually lower than those near the wind source. They are seen during fair weather with characteristically relatively long wavelengths known as *swells*, which may travel long distances, far beyond the stormy area of its origin transmitting energy outward in two forms :

a) potential - the height of a wave which determines the potential energy of position above still-water level that moves forward with the wave form, and

b) kinetic - the motion of individual water particles as a wave passes which is expanded in the nearly circular orbit of the particle.

A particular type of ocean wave often of unusual size is the *tsunami* or seismic sea wave, which is sometimes associated with earthquakes and related landslides. They possess tremendous havoc potential when they bang the shore. They occur only rarely but when they occur, they are capable of great destruction bringing about significant modifications of shorelines in a very short period of time.

The study of wave environments has suggested a world division of coast (Table IX. 40) which appears generalized, but is a valuable basis for comparing the effects of wave action on coasts

TABLE IX. 40. : MAJOR WORLD WAVE ENVIRONMENTS
(after Davis, 1972).

Wave Environment	Potential Wave Engergy	Wave Characteristics	Directions
Storm wave temperate storm belts	High all year from south, less in summer from north Swell waves too	Height often over 5m. Short waves (period 6-12 s) with shallow wave base	Local storms cause variation, Refraction has less effect near shore.More constant swell wave background
West coast swell	Medium to high and diminishing from south, little variation	Low waves, declining in height from centres. Occasionally subject to tropical cyclones. Long waves (period 14-16 s) from south	Very consistant, especially from the south. Refraction important due to deep wave base.
East coast swell	Medium to low and less regular, High variation	Low waves, and long. Subject to tropical cyclones	Fairly consitent except in tropics. Refraction important
Protected	Very low, Little swell penetrates some with monsoons	Very low waves. If subjected to tropical cyclones, extreme contrast occurs	Variable

SHALLOW WATER WAVES, WAVE REFRACTION, BREAKERS AND SURF

As deep-water swell approaches a shoreline the orbital motion of the water particles, encounters interference with seafloor under shallower water, the (i) wave form and (ii) velocity and the method of (a) water and (b) energy, transmission change. The wave form is modified (i) the wave length and (ii) velocity, are decreased, the wave height is increased with the same wave period. Usually these changes are initiated when the water depth is approximately one half that of the wavelength (1/2λ.) i.e. (d= 1/2λ.). Nearly circular orbital motions in deep water, become elliptical in shallow water, with progressively flatter orbits near the bottom until in the back-and-forth motion, seaward in particle movement on the sea floor is set in under passing wave troughs, and landward under passing crests whence, the conversion of wave energy is translated into geomorphic work which begins. The depth at which waves begin to stir bottom sediment is determined by

(i) wave height,

(ii) wave length,

plus(iii) period,

(iv) the grain size. and

(v) sorting of the sediment.

Very long period swell waves may stir water and bottom sediment to great depths, ordinarily, such waves are not significantly effective at depth greater than about 10 meters because of inadequate bottom-sediment transport or erosion. This depth, which is called *wave base* or *surf base* is a convenient reference level. The wave velocity is decreased by bottom friction when a swell wave begins to 'feel' the 'bottom' and become a geomorphic agent. The wavelength is shortened and correspondingly the height begins to increase sharply maintaining its wave period. As the crest of each wave approaches, the water moves rapidly forward as well as upward. Ultimately, the forward motion of the water mass in the wave equals the decreasing forward motion of the wave front. The wave assumes a steep face and the crest begins to move ahead of the lower part of the wave. Eventually, the crest moves too far ahead of the waves, and the unsupported water in the crest curls over, and a *breaker* or *surf* is formed. At this moment the particles of water within the wave are thrown forward against the shoreline. The energy thus developed is then available to (i) erode the shoreline or (ii) set up currents along the shore that are able to transport sediments produced by erosion. Breakers form in the surfzone where the bottom depth is about one-third times the height of the wave (breaker). This produces an uprush of water at the shore- the *swash* followed by a return flow of water- the *backswash.*

a) If the *slope of the bottom is relatively steep* and a large swell abruptly impinges on the sea floor the water necessary to fill the wave form decreases rapidly, and a *plunging breaker* forms. The wave crest curls over in a half-cylinder form and in doing so trap the available air flinging the crest into the preceeding trough with a great roar.

b) If the *seafloor is more gentle,* a *spilling breaker* forms, in which the crest of the wave spills over the advancing front of the wave without completely destroying the wave.

c) *Surging breakers* - where the wave surges up the beach face and the crest collapses and disappears especially associated with steep beach slopes.

Waves approaching a uniformly sloping shoreline at right angle to the shore, would break practically simultaneously, and all the waves would strike the shore in straight parallel lines. The swell/waves are *reflected.* When a breaking wave hits a cliff or sea wall, thousands of tons of water in motion impinges against the structure.

However, when the waves approach the shoreline at an angle, a part of each wave impinge on the shallow sea floor before the rest of the wave, and that part will be slowed down, resulting in a bending, or *refraction* of the wave, similar to the bending of light rays on passing through a lens. The segment of each wave in the shallowest water is slowed so that in plan view the crest-becomes concave forward, until the waves begin to parallel the coastline approximately. Wave crest converge

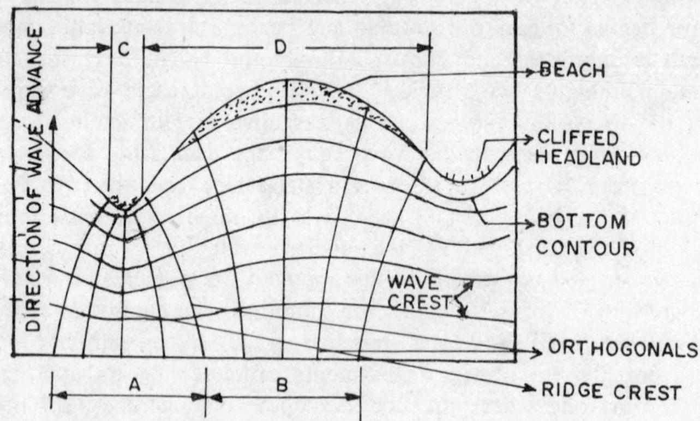

Figure 28 : Irregular coastline : Wave refraction and energy distribution-concentration (C) and dispersion (D)

[Wave refraction and distribution of wave energy over an irregular shoaling bottom. Orthogonals are perpendicular to the wave crests, equally spaced along the swell. The two wave segment A and B, have equal energy, but because of wave refraction, the energy of A is concentrated on the projecting headland C whereas the energy of B is distributed over a much larger area of the bay D]

on the headlands and diverge on the bay (Figure 28) and thus the headlands are characterized by high breaking waves, rocky shores and intense erosion, while bays are typically quiet, with sandy beaches. A shallow submarine ridge thus focuses wave energy against its flanks, whereas a submarine valley or basin causes wave orthogonals to diverge and wave crests to be attenuated.

Wave refraction is the basis for two important generalizations in respect of the evolution of coast -

(i) First, an initial irregular shoreline becomes more smooth or regular as initial seaward protrusions caused by sloping ridges tend to be eroded faster than adjacent shoreline segments in coves or bay at the heads of submarine valleys.

(ii) Second, wave refraction generates currents flowing along the shore from headlands, where the breakers are focussed, to adjacent coves, where the water is lower. These wave-generated *longshore* or *littoral* currents. along with tidal and other currents, transport the sediment eroded from headlands into adjacent coves where beaches are built. Refracted waves thus erode the headlands as well as fill the bays.

Waves are *diffracted* when a segment of wave crest passes between headlands or artificial constrictions into open water such as a river mouth or bay. The littoral currents generated within bays and harbours by diffracted waves erode and transport shore sediment until the bay shoreline 'fits' the diffraction pattern closely.

The mechanics of wave erosion can be outlined as follows :

a. Wave impact against rocks or cliffs at the watermark exerts great pressures, particularly when water or trapped air is suddenly compressed in fissures or joints by shocklike impacts. The wedging effect pries rocks apart, while suction drags out loose material.

b. The swash hurls pebbles and cobbles against shore rocks, with a battering effect. Projecting corners are knocked off; loosened well jointed rock masses are disloged; and the rock is ground down.

c. Sand and rock are swept back and forth over the beach or rock bench by the alternating swash and backswash. This brings about mechanical abrasion beneath the watermark, as far down as there is significant wave agitation.

d. Chemical solution and weathering at the watermark or in the spray zone combine with mechanical attack, particularly in warm environments. Breakdown of most rocks is accelerated, and, in the case of limestone, corrosion may even outpace mechanical abrasion.

e. Materials loosened and removed are carried by traction (rolling and sliding), saltation, suspension and solution. This transports due to both the bacakwash and the *undertow* a diffuse seaward return of the water along the bottom. In some situations, too, longshore currents and *rip currents* - concentrated seaward flow lines- transport sediment away from the watermark.

f. Effective wave cutting is generally restricted to times of stormy and high seas, when wave energy is at a maximum.

TIDES

The periodic rise and fall of the sea due to attraction of the moon and the sun is termed as *tide*. Gradually rising to highest level it is designated as *high tide*. Conversely, in *low tide* the sea falls to the lowest level.

The difference between the height of low and high water levels is referred to as the amplitude or height of the high tide which may be a few meters in the open ocean, but it may be as high as 9-12 meters or even upto 18 meters in some shallow seas, particularly in narrow channel and straits communicating with the ocean.

The effect of the moon is by far more powerful than that of the sun. According to the period of rise and fall, the tides are divided into

a) *semi diurnal* tides recur twice during the lunar day, as seen in the Atlantic Ocean

b) *diurnal* tides occur once a day as noticed in the northern Pacific and Indian Oceans. Tidal movements are affected

(i) by changes in the declination of the moon and the sun,

(ii) in the distance from the earth to the sun and the moon,

and (iii) in position of moon and the sun, relative to each other and to the earth.

These sometimes result in mixed tides. Imagine the earth with a uniformly distributed water envelope. The water envelope looses its spherical shape (a) under the attraction of the moon and assumes the form of an ellipsoid (b) this is because the water centered at A is pulled up towards the moon more than the earth centered at E, while the earth in turn is attracted more than the water centered at A, The water at the farther side is thus left behind, as it were, and the water at B and B_1, is attracted to A and A_1 where high tide results (Figure 29).

The high tide magnitude is affected by the relative position of the earth, the moon and the sun. Twice a month at *syzygy* (new moon and full moon) the earth, the moon and the sun fall along the straight line, and twice a month, at *quadratures* (the first and the last quarters) the earth-moon straight line is at right angles to the earth-sun line. The tidal rise changes somewhat accordingly.

Tidal processes in coastal proximity have geomorphic impact. But few, if any characteristic landforms are developed primarily due to tidal influences. The role of the tides in coastal evolution include a number of processes :

a) Wave erosion is most effective with tidal amplitudes when the water mark moves up and down the shore, in as much attack a broader zone which has become available and can sculpture more prominent forms.

b) Tides may cause and maintain inlets between lagoons and the open sea and subsequently, they may be widened, deepened and preserved by the scour of ebbing tides that remove sediment out to sea.

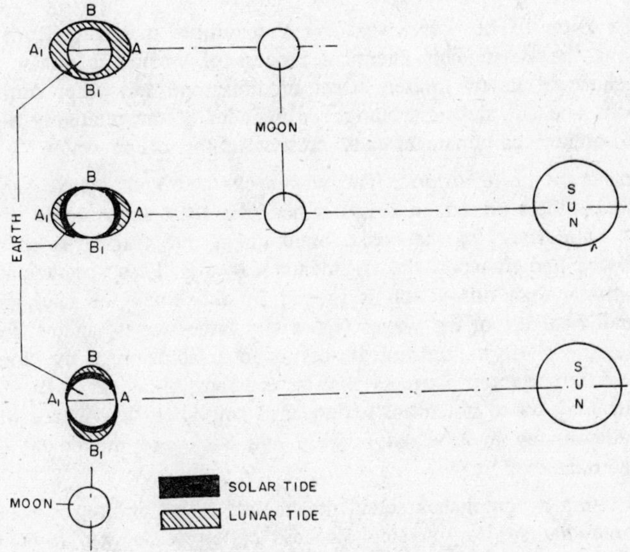

Figure 29 : Illustration depicting tide generation

c) The salt-water environment resulting from periodically flooding and draining coastal lagoons in tidal zones is colonized by low vegetation known as *salt marsh* in mid-latitudes and by swamp forests or *mangroves* in tropical regions. The tides enter through inlets and surges up dendritic channels that traverse low-lying mud flats. Ebbing waters retreat through the same *tidal creeks.* are devoid of sediment even as the lagoon gradually fills up with suspended matter and organic residues.

d) Broad, shallow coastal streams may be conspicuously affected by tidal activity. Such *tidal rivers* behave like an arm of the sea, and occasionally *tidal bores* travel upstream upto 16km/hr and with amplitudes of 4.5 meters or even more. The mouth of the tidal river, an *estuary,* is characterized by intertidal sandbars, mud shoals, often forming a salt marsh.

e) Tidal flows may cause acceleration in longshore currents along some complicated shorelines and in a general way, tidal currents can affect extensive tracts of the continental shelves.

PROCESSES OF MARAINE EROSION

Five distinct processes-corrosion, corrasion or abrasion, attrition, hydraulic action and the shock pressure of breaking waves are identified to operate in marine erosion by waves.

a) *Corrosion* : The waves dissolve the soluble materials from the rocks along the coast. This chemical process of erosion is not so significant because of its (i) limited effect on limestone and other similar soluble rocks and (ii) slow dissolution on account of the relatively high amount of calcium carbonate already present in the ocean water.

b) *Corrasion or abrasion* : The waves charged with coarse sand, pebbles and boulders present in seawater near the coast act as tools strike against the solid rocks of the coast, break them into fragments to accomplish erosion and transport the fragments seawards. Less spectacular but more common than this action is to and fro movement of sand, pebbles and small boulders of the waves (especially large waves) against the cliff and over the platform, causing the corrasion or abrasion of the floor. It is the most important process of marine erosion. Abrasion at the foot of the cliff may cause notching. It may also smoothen the surface of the shore platform due to repeated forward and backward movement of abrasive materials over it.

Abrasion accomplishes actual mechanical disintegration of cohesive rock, but somewhat similar physical process called *quarrying,* it is the pulling away by the waves of particles of material which have been prepared by some other agency specially blocks provided by jointing or cleavage. Closely jointed rocks like some basalts or those with pronounced cleavage (slates and schists) are especially favourable to quarrying process. Quarrying also embraces the sea cutting into unconsolidated material and freeing pebbles and sand particles directly. However, in this process the sea disaggregates rather than disintegrates. Quarrying is most effective in high energy coasts of temperate latitudes where big storm waves are more common. Greater quarrying means more tools for the abrading waves.

c) *Attrition* - In the process of corrasion the rock fragments also collide with one another reducing themselves into smaller pieces. Such mutual breaking up of rocks by impact and their conversion into finer fragments is known as *attrition.* Ultimately the fragments become smaller enough to be easily transported by the undertow and rip currents towards the sea.

d) *Hydraulic action* - The waves erode the coastal rocks by their movement and forceful impact even in the absence of rock debris in them. When the breaking waves strike against the coastal rocks like a big hammer, this impact is sufficiently powerful to break and disintegrate the rocks.

e) *Shock pressure of breaking wave* - Enormous pressure is exerted when the breaking wave strikes against the coastal rock, especially on the rock joints and cracks. In these joints and cracks, the air in the opening is compressed in response to wave pressure and it appears as suddenly wedges have been thrust into the cracks. Then suddenly as the water returns with the waves, the pressure falls sharply and the air expands and the rock starts exploding and breaking along the weaker rock parts. When water enters these cracks, not only there is increase in mechanical stress, but solution activity also increase in soluble rocks.

The marine erosion is a function of the following factors -

a. *Nature of the coast* - The wave action is violent in precipituous coast. Unobstructed seawaves dash against the shore directly eroding a great amount of coastal rock by impact. In contrast in a shallow sea over a lowlying coastal plain the wave movement is much obstructed consequently much impact force is lost before it reaches the shore leading to relatively less violent action in such areas.

b. *Nature of the coastal rock* - (i) Erosion is less rapid on a hard resistant rock and in contrast if it be soft comparatively the erosion is more rapid. Igneous and harder metamorphic rocks have thus possibility of longer survival than the sedimentary rocks. (ii) The nature of the individual constituent minerals of the rocks is also a parameter in as much as rocks consisting of soluble materials are eroded easily. (iii) The disposition in stratified sedimentary and some banded metamorphic rocks too are important factors. Generally, horizontal beds and in the case of inclined beds those which dip away from the sea are eroded more rapidly.

c. *Presence of joints and fissures in the rocks* - The joints and fissures in rocks appreciably facilitate wave erosion. The sudden blow of the waves on striking against the jointed rocks compresses the inside air under tremendous pressure in all directions and acts like a wedge, highly accelerating the erosive action. It also facilitates dissolution of soluble materials by allowing seawater to percolate into the rocks.

d. *Presence of rock particles with waves* - A great deal of abrasion is accomplished by the sea waves, armed with hard rock fragments acting as a force of an artillery on dashing against the coastal rocks. These fragments of rocks are thrown against the shore wearing off much of the coastal regions. In this process of dashing, the particles themselves are also worn out no doubt.

e. *Chemical action of seawater* - Seawater which becomes a much better solvent particularly on its containing some dissolved materials, ordinarly can take into solution many substances of the coastal regions.

f. *Wave strength* - The more the strength of the sea waves, the more is the marine erosion.

The above factors are operating since the formation of continents and oceans and modifying the coast through its gradual retreat as also the development of irregular coast lines bordering the land masses. Where the erosion is less the rock is hard it stands projecting into the oceans. Such projecting landforms are called *headlands* or *promontories.*

FEATURES OF MARINE EROSION

By the above work, number of erosive features are formed along the shore.

Scarplike feature facing the sea shore is evolved at the moderately highland on wave cutting. The basic characteristic of a cliff developing into star-like step topography, its profile comprising of

a) the sandy and pebbly platform showing seaward declivity sloping at an angle of 50^0 or so and varying width of several hundred meters,

b) the attendant cliff itself, either steep or vertical wall-like, may be of the order of 3 to 60 meters in relief, and

c) the knick-point or undercut notch at the intersection of the platform and at or just above the watermark.

To begin with cliff-face is available for wave erosion over its entire surface, but with passage of time only the lower part of the elevated cliff is open for wave attack leading to heavy top and weak bottom. With such undercutting processes, aided by weathering, the weakly supported heavy top of the cliff topples down to the base

(i) steepening or maintaining a constant angle through time as, also

(ii) provide additional tools to the waves to further cut the cliff.

Cliff retreat further widens the abrasion *platform.* As the sea cliff retreats, a shallow-water shelf called *wave-cut terrace* (Figure 30) develops, over which the water increases in depth seaward to the limit of wave action, that is to a depth of tens of meters. The width of such terrace is in fact limited, is not several kilometers wide, as the waves, moving over the shallow-water shelf, loose much of their erosional- energy gradually on account of bottom friction. Thus some broad relationship is established between the height of the cliff and width of the platform, *platform-cliff ratio* of a wave built terrace.

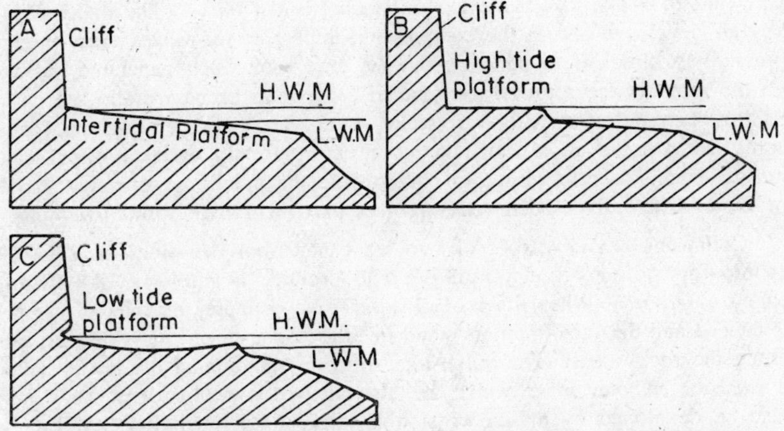

Figure 30 : Classification of wave cut platform (after Bird 1969) : Cliffed coast with
A. an intertidal shore platform
B. a shore platform at about high tide level
C. a shore platform at about low tide level

The ratio of the platform width to the cliff height provides a useful geomorphic index for correlation of uplifted marine platform (terrace). Marine terraces represent former high sea levels and or rise of land surface. Verma (1973) has shown that the platform-cliff ratio can be used for the correlation of marine terraces occurring at different levels in the tectonically active coasts of California.

Much of the material eroded and ground up by the waves is removed seaward usually to develop the *wave-built terrace* a feature something like a submarine delta.

The rate of retreat of sea cliffs is dependent upon various factors, particularly,

(i) the force and persistance of wave action, and

(ii) the nature of the rock attacked, besides

(iii) the climate which exercises indirect influence on the development of sea cliff since weathering progresses, development of regolith and growth of vegetation all are climatic functions.

SEA CAVES, COVES, STACKS AND ARCHES

Many irregularities often develop during the sea cliff retreat and the cutting of the wave-cut terrace. At the bases of cliffs sea caves are produced by wave action, often localized near masses of weaker-lithologically or structurally, rock lie at or near sea level. Sometime such caves proceed landward to a considerable extent producing a *sea tunnel.* The undercutting and air developed due to the compression impact of sea waves exert enough pressure to blow off to puncture the sea cave roof. Such punctured passage on the roof of the sea caves are called *blow holes* or *gloups* and sea caves having such holes are called *spouting holes.* Sea water, cutting the caves, spray out from such holes. The wave cut relatively easily eroded rock-masses back faster to form *sea coves* while the adjoining separated harder or more differently eroded rocks project into the sea to form *headlands.*

Collapsing sea cave roof or two sea caves from two sides of a narrow promontory join by undercutting it and produce a gate-way like feature called a *sea-arch.* When the roof of a sea cave collapses an inlet of the sea is caused and the far off pillars stand in the sea due to unequal wave erosion along the rocky coast. This pillar-like structure standing in the sea is called a *stack* or *chimney rocks* which are also known as *pulpit rocks.* They may also be developed by marine erosion in vertically jointed rocks. When the coastal rock is weaker, erosion is much pronounced and a small inlet of the sea is made which is called *core* or *bay* .

PLAINS OF MARINE EROSION

With sufficient uplift of land relative to sea level a wave cut terrace becomes a *plain* of marine erosion. The surface of such a plain like that of a stream-developed peneplain, cuts across all kinds of rocks irrespective of their composition and structure. On a peneplain however the rock waste is characteristically a residual soil, while that of a plain of marine erosion consists of water-worn transported material rather uniformly spread over the surface. The surface of a newly upraised marine plain is usually smoother than that of a plain of stream erosion, and sides erosion remnants left by th waves are steeper than those formed by stream. A remarkable example of a long plain of marine erosion with steep-sides isolated masses (formed island) rising above its surface occurs on the eastern coasts of India, south of Mahanadi delta. Similarly, the plateau of Tibet is supposed to be an elevated plain of marine denudation of the ancient Tethys sea.

MARINE TRANSPORTATION

The marine transport mechanism for the materials include *e.g.* suspension, rolling, sliding, saltation and solution as in rivers. The maximum transporation is by movement at the bottom. The material towards the beach is carried by shore currents, whereas outwards into the sea is carried by the undertow to be brought back on the shore by advancing waves. If the currents are inclined to the coast, the material swings forward and backward.

Actually the suspended particles as well as those on the bottom of the shore zone are carried further off by means of sea waves. After disintegrating the coastal rocks the waves move them seawards. They are lifted and are carried offshore. Parallelism of sediments, along the shore that builds up *spits* or *bars* across the coastal region enclosing a lagoon (*e.f.* Lake Chilka or Lake Pulicat) are brought up by longshore currents, the sediments parallel to the shore. There is conspicuous sorting action of the sediments carrying away the finest particles and leaving behind coarser particles on the beach to form *beach shingle* by the wave action. The coarse sediments, in course of time, are ground fine by abrasion caused by waves and subsequently carried away. The terriginous deposits are not found in deep sea bottom. They are only to be found in continental shelf and margins of the continental slopes.

The materials in solution are also carried farther off. Sometimes by reacting among themselves *i.e.* other salt solutions that may be precipitated, but more often minute organism, animals and plants, extract some of the dissolved mineral matter from them to build their protecting shells or tissues. $CaCO_3$ and SiO_2 are thus utilized by marine organisms.

MARINE DEPOSITION

The marine sediments may be deposited

(i) in shallow water comparatively near borders of the land, that is on the continental shelf and continental slope as also

(ii) as the abyssal (deep sea) deposits laid down on the floor of the deep ocean.

SHALLOW - WATER DEPOSITS

a. *Wave -Built Terrace* - The sediment removed from the shore are carried offshore close to the farthest end of *wave-cut bench* for deposition seaward margin resulting in and gradually the development of another more or less flat region showing somewhat seaward declivity. Such platform-like features developed by the constructive action of sea waves is referred to as *wave-built terrace*.

A transverse section across shore thus include *shore profile* which changes gradually by the sea waves that perform two functions

(i) erode the coastal region, and

(ii) transport the debris farther off

A delicate balance is established between marine destruction as work done by the sea waves corresponds with the (i) amount of erosion, which produces more (ii) load to be handled, more energy and less erosion. If the shore is steep, it is exposed to the full vehemence of the sea waves and consequently the cliff retreats and tends to become flat. On the other hand if the region, is flattish, deposition over the beach near the shore would have the tendency to steepen the shore with additional load on it. In such situation an equilibrium is maintained between marine destruction and construction at a later stage and so balanced shore profile is called a *profile of equilibrium.*

b. *Spits and Bars* - A spit is a narrow sandy accumulation deposited by wave refraction obtained from cliff erosion and river deltas and is transported in the direction of longshore drift extending from the shore at one end towards the adjacent shallow water body.

A spit grows in the direction of longshore drift as a submerged feature (i) in response to the strong effect of wave refraction, the free end of the spit may curve towards the land, forming a *cuspate spit* which may also develop (ii) from convergence of currents (iii) eddies in longshore currents, (iv) wave approaching from opposite directions, and (v) frequent coastal configuration changes relative to prevailing waves. Long continued deposition of spits in a series results in a cuspate foreland. When such body more or less encloses a water body portion it has encroached, it is transformed into a *bar*. Spits growing from opposite sides of bays may in some case unite into a continuous bar which joins the headlands and extending across the bays to form *baymouth bars* and the enclosed protected waters landward of the bay- mouth bar developed into *lagoons*. The shore drift from an island develop sand bars which connect it with the mainland. The feature is called *tombolo*.

Spits mostly of cuspate type are developed from the prograding delta front of Krishna and Godavari rivers in the Bay Bengal (Roa, and Vaidyanadhan 1979; Rao and Vaidyanadhan, 1979). Due to rapid deposition from above average sediment charge in tidal creeks and currents the enclosing lagoons have become shallower.

c. *Lagoons* - The lagoons are shallow waters cut off by an offshore or attached bar, connected to the open sea by narrow channels or tidal inlets. It remains water filled even at low tides (Figures 31, 32 and 33). Wave activity is at its lowest in it. But some fresh water influx suspended and / or dissolved from the land are introduced in lagoonal complex, the martix of a specialized vegetation, which may be colonized by vegetation adapted to salt or brackish waters, that in turn help to bind more sediments. Finally, the lagoonal area may shrink down and transform into marsh or tidal mudflat with salt tolerant plants. Further, dunes may encroach from the offshore bar, at times sweeping right across the flats. Nearly 13 percent of the coastline is estimated to be punctuated by lagoons, some of these may be several kilometers long.

Traces of relict deep channel traverses may be observed on the lagoonal floor are once traversed by deep channels, representing lower sea levels before coastal valleys drowning in the period of marine transgression. The

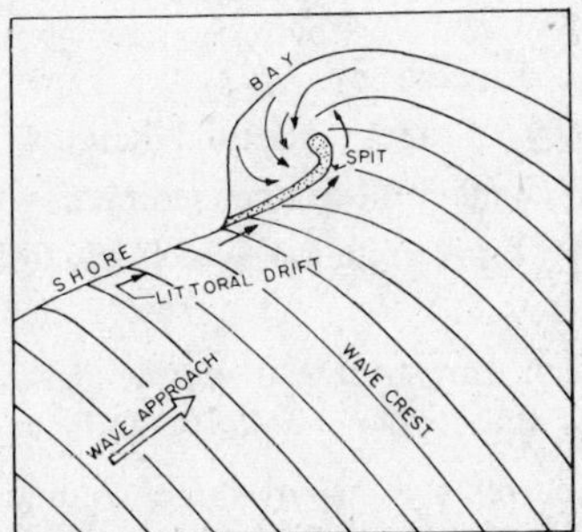

Figure 31: Wave Refraction causing Spit Formation

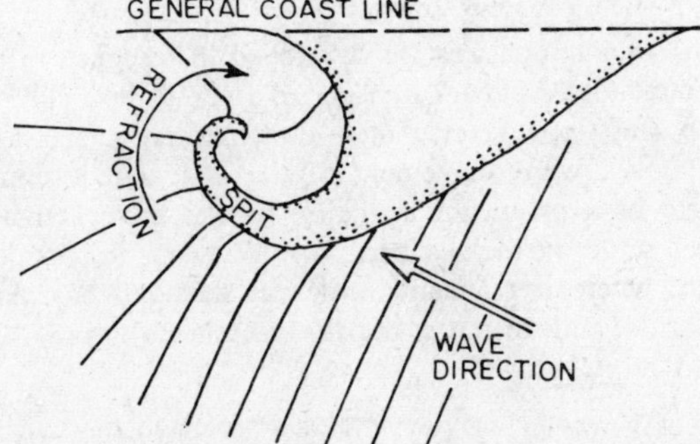

Figure 32 : Spiral (or hooked) shaped Spit formed by Wave Refraction

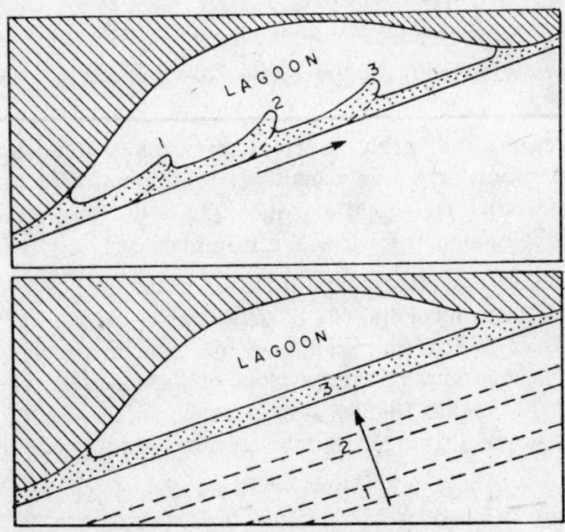

Figure 33 : Shoreward migration of a barrier that originated offshore (after Bird 1969)

lagoon development is affected by

i) the frequency of tides and tidal heights,

ii) the rate of channel flow and sediment yield, and

iii) the amount of precipitation which control the level or water in lagoons.

The manner in which terrestrial and marine processes interact at or near that turn-zone, controls the lagoons' salinity and accounts for their growth.

d. *Mud Flats* - Tidal currents transport large volumes of silt and clay-sized fractions which settle on the floor of bays and estuaries which progressively accumulate over a period of time forming mud flats that are exposed at high tide and submerged at low tide.

A mud flat is an intertidal feature developed in response to tidal rhythm. The saline water facilitates stream transported fine silt and clay to flocculate easily. Tidal ebb and flood reverse their direction every six hours, causing periodic land and sea ward movement of the coagulated sediment (mud). Being in the state of a practically agitation by the wave activity ordinarily, the mud does not settle on the sea bed. The following flood tide returns the material landward where it gradually settles in sheltered bays and estuaries. Burrowing animals and organic matter which colonize the mudflat subsequently add to its growth.

Mudflats are extensively developed along protected bays in the Rann of Kutch which provide a suitable environment, being for tidal flats development. Two dominant characteristic of the Ranns are

A. Salt - encrusted flats are common due primarily to the

a) yearly flooding of the Ranns in the three monsoon months, and

b) year-round high thermal efficiency leading to high evaporation rates causing (i) high salinity in sediment effecting (ii) capillary movement of water and (iii) the development of surface efflorescence

B. Because of lack of perennial drainage in the area, the flat-building material from local sources is scant and is derived only from the uplands of Kutch and Kathiawar by some monsoon torrents. As such the mudflats have grown out of the material transport from across the Sindhu (Indus) system further northwest of the Ranns by longshore drifting and tidal currents throughout the years;

e. *Tidal Marshes* - These are saline wetlands characteristic of low-wave energy environment such as in sheltered bays and lagoons. They evolve from aggraded mud flats exposed to only half the tidal cycle. Halophytes salt-tolerant vegetation and burrowing animals colonize the surface and thus also trap mud. Rapid accretion builds the surface to the level of the mean tide height. Tropical mangrove swamps, such as the Sunderbans in the Ganga delta, are characteristic tidal marshes. A tidal marsh transforms into a salt-marsh plain on growing above high tide mark level.

f. ***Coastal Dunes*** - These are typically observed in virtually on all coastal sectors at or above high-tide mark where sands are abundant. If there is a constant supply of fresh sediment, disturbance by storm waves can impede colonization of plants while the salt splash and spray inhibits most form of vegetation. Consequently, the plant cover and root mat may be inadequate to bind loose sand.

Most common are the lines of nondescript-beach foredunes sometimes developed parallel to beach ridges. These (i) form irregular mounds, (ii) typically 3 to 6 meters high and (iii) locally undercut by wave action or deflated to form blow-outs. Farther there may be more extensive fields or dunes of types common to the continental interiors.

g. ***Beaches*** - Unconsolidated loose material-very fine sand to large boulders under the influence of

- waves
- undertow and
- shore currents.

constitute the *beach*. It may include zone of rock fragments within the reach of waves along the shore as well. It is widerspread coastal feature that extends between the upper most limit of wave action and low tide marks. The lower margin is beneath the water, a little beyond, where the great storm waves break. Its upper margin (on shore) is at the level reached by storm waves, and is usually several centimeters above the water. Sediment sorting is conspicuous, arranging the coarser material upwards, acted upon by the oncoming waves only, or constantly under - water portion of the beach, is made up of finer material. But along very young coasts, beaches are as a rule, not that well developed at the sea cliffs bases. They are well developed generally around the shore of coves and recesses of the coasts. Beaches are often also finely developed where the land slopes down to the sea gently, as on coastal plain. A beach may be classified into backshore, foreshore, and a part of the near shore zone (Figure 34, 35 and 36).

The beach sediment is derived from many sources :

(i) the erosion of the coast and the land margin behind the beach,

(ii) the shoreward drift of material by waves after they break in the near shore zone,

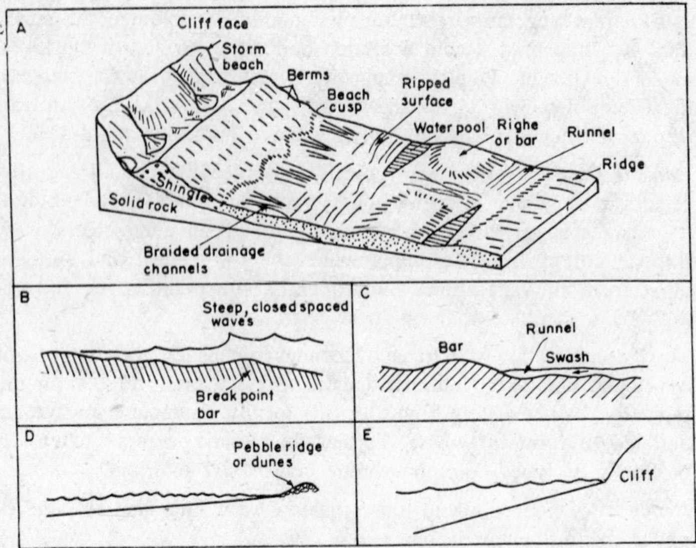

Figure 34 : Beach Profile

A. Idealized beach features B. A break point bar formation
C. A swash bar formation D. A beach profile facing storms waves and swell
E. A beach profile facing more variable wave conditions

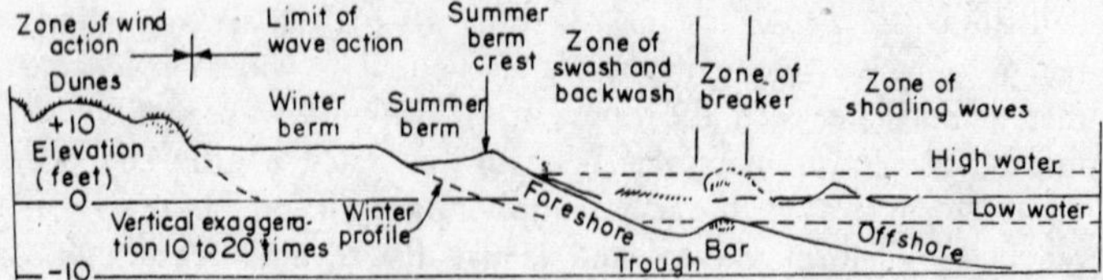

Figure 35 : Characteristic element and zones of the profile of a sand beach

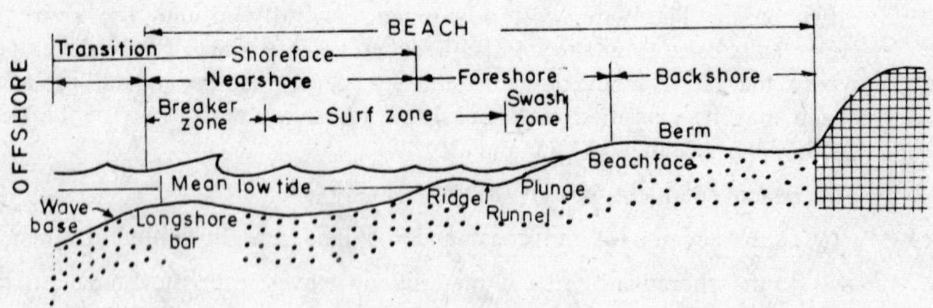

Figure 36 : Beach Features

(iii) the transport of material alongshore in the surf and intertidal zone,

(iv) the sediment transported into the sea by geomorphic agencies like rivers, glaciers, and to some extent wind.

The main bulk of the beach material is provided by longshore transport of sediments acquired from any other above sources.

Beach sediment composition varies widely and may be broadly grouped as (i) sand beaches and (ii) non sand beaches. The former are most common and comprising rounded and well sorted grains of quartz and a wide surf zone.

Transportation of beach sand by swash an ever ongoing process and affected by it, the sediment is moved upslope to the maximum uprush limit. Generally the incoming waves are not always parallel to the coast and consequently the material is transported at an angle to the beach-face slope. The backwash alongwith the contained sediment move back downslope following the steepest beach gradient, when the uprush energy is exhausted. Such up and down slope to and fro sediment, move in a series of motion takes place in parabolic paths known as *beach drifting.*

Features of beach.

Backshore - It encompasses the zone lying seaward from the cliff foot, coastal dunes or vegetated beach to the berm crest limit, which separates it from the foreshore by sharp slope discontinuity beach, one or more beach ridges composed essentially of coarse material deposited several meters high beyond the limit of storm surge assinged almost parallel to the coast may occur in this zone. Somewhat similar beach ridges are *chenier* ridges. They are abandoned wooded beach accumulation of sand and shell materials several kilometers long upto 450 meters wide and 6 meters high. Chenier ridges are distinctly related to the periods of outbuilding activity of coasts.

The development of chenier ridges is related to variable sediment supply from nearby rivers - when a stream's load is reduced, the waves rework the sediment in a ridge form. The shallowing of water in front of the ridge develops a marsh. When surplus sediment arrives in the stream, the ridge grows in height and extends outwards. Over a period of time, a number of sandy ridges alternatively with marshes in front are produced.

In a prograded coastal environment, cheniers and intervening, mudflats with marshes and associated vegetation make a *chenier plain*, when the volume of sediment transported by the streams is fairly large. It is transported to the near shore zone and eventually developed into mudflat. A chenier plain can also develop when a balance exists between longshore sediment transport, deposition and erosion.

Nonsand beaches, are of several types, may include :

i) the uncommon mud beaches

They developed in (a) low wave and current energy environment and (b) in areas where coastal erosion has exposed the former deltaic material; associated with the seaward margin of deltaic sediment

ii) the more common shingle and cobble beaches derived from the erosion of bedrock or of stony glacial drift associated with high energy environment or sand beaches associated with swell environment common in higher latitudes associated with seaward margin of deltaic sediments

iii) biogenic beaches, mainly associated with coral reefs and commonly occur in tropical environment.

Beaches may be narrow and steep, or broad and flat. Field and laboratory investigations indicate that the dependence of beach slope is a function of three variables, namely

- the grain size
- wave length and
- wave steepness.

The relationship between the grain size and the beach surface gradient is explained in terms of the volume of swash and backswash. In coarser beach material such as coarse sand and shingle, greater percolation being permeable results in loss of swash and backswash in a steeper gradient than in the finer sand beach sediment. Similarly longer waves and greater wave steepness cause greater volume of both swash and backswash, resulting in a relatively flat beach gradient. A casual relationship between the velocity in the nearshore zone and beach gradient is shown (Huntley and Bowen, 1975). Accordingly steep beaches result when waves break rapidly or collapse over a narrow surf zone. In the case of broad gentle beaches, the wave energy is dissipated.

Foreshore - It is a zone between the berm crest and the low-water level and include coastal features like

- beach face, plunge or step and
- ridge- and runnel-

The beach face usually slopes at 1-3^0, but at times can be steeper. Seaward of the beach face occurs a plunge or step that sets the lower limit of the beach, corresponding to the point where the waves finally break. The *ridge-and-runnel topography* is formed by temporary accumulation of sand bars in the intertidal foreshore zone. The sand ridge and its associated trough-like topography is the result of waves adjusting a swash gradient in relation to their dimensions.

Nearshore - A broad zone of shallow water forming the seaward extension of a beach is known as the shore face. Submarine sandbars, of which longitudinal bars are the commonest, develop in this zone. The bars mark the position where the waves first break. Observations show that usually two to three longitudinal sand bars develop parallel to the coast. Smaller inner bars are destroyed by steep waves but the sand grains rapidly regroup when normal wave conditions are restored. In case longshore current is present, the bars develop into crescentic type with their ends pointing landward. The nearshore zone is followed by the deeper waters of the offshore zone.

Barrier Beaches and Associated Forms - A *barrier* or offshore bar composed entirely of beach sand is a partly emergent, narrow sandy ridge separated from the coast by a lagoon that is more or less aligned parallel to the shore across the mouths of inlets and embayments.

Barrier beaches are usually absent from

(i) rocky coasts,

(ii) coasts of low-energy wave movement, and

(iii) areas of limited supply of coarser sand.

A succession of barrier ridges and intervening lagoons makes barrier complexes. Such a sequence suggests intermittent coastal emergence due to repeated tectonic activity or eustatic changes.

The origin of barrier beaches is debatable. At one time they were identified with overgrown and uplifted submarine bars. Barriers may evolve from longshore growth of spits and emergent beaches offshore. Some barriers may be of composite origin.

The longshore development of spit may enclose in embayed section of the coasts forming a barrier bar. It is noted that barrier initially develops from longshore bars on prograding coasts. The observations suggest that longshore bars establish both

i) during storm surges, and

ii) in high tide season.

The bars gradually build above the high tide mark by swash. They are equilibrium forms, as even the material of exposed, unvegetated bars is not known to be displaced or destroyed by the high tides. Many of the barriers are interpreted to have been formed by the shoreward drifting of sediment since the Holocene marine transgression.

Attached Bars develop along irregular rocky coasts with abundant sediment and strong longshore currents. *Bay bars* close off large coastal indentations, while *spits,* develop down current of headlands that mark major changes in coastal geometry. Converging longshore currents can build up *cuspate bars*, in front of convex coastal sectors, and a variety of bars may be found attached to islands and stacks, including tombolos tied to the coast. These bars rise from the coastal platform and extend above sealevel, creating zones of shallow water. If not breached periodically by storm waves or kept open by tidal action, attached bars can eventually close off embayments, forming lagoons or mudflats.

Barrier Islands : The detached portion of a broad barrier beach between inlets is known as a barrier island. Like barrier beaches, barrier islands can form in diverse ways. A view is that the feature originates from dune and beach ridges. The ridges develop as partially emergent features above the high-water level, and continue to grow upward with a slowly rising sea level. The growing ridge eventually encloses a lagoon landward of the feature forming a barrier island.

DEEP SEA FEATURES

About 84 percent of ocean area is occupied by deep sea beyond the continental slope. Earlier it was believed that most of the deep seas formed flat, featureless surfaces, referred to as sea floor plains. Now, however, it is too well known that there is a broad range of forms included, limited in sea-floor distribution, of different origins, lending to the ocean floor a complexity and diversity almost comparable to that of land surface.

Three major topographic constellations can be recognized :

a) *Sea floor plain,* locally studded with groups of seamounts
b) *Mid-ocean ridges* and *rises*, a geometrically organised topography of great submarine relief and irregularity
c) *Trenches* and *island areas,* that is parallel chains of volcanic islands accompanied by linear deeps.

Complex tectonic and vulcanism are responsible for most of this topographic relief.

The deep sea sediments mantle varies in thickness on the sea-floor plains, near the foot of the mid ocean ridges and the continental slopes and thinnest to practically absent, at the surface of the ridges and rises. These sediments vary in composition and are of several different sources.

(a) Terrestrial derivations, as, for example

(i) the silts and sands introduced from the continental shelf by turbidity currents and mass movements,

(ii) glacial drift deposited by melting icebergs,

(iii) eolian dust, volcanic ash, as well as meteorite particles further contribute to bottom sediments.

(b) Organic in nature, such as the calcareous and siliceous skeletons of the algae or *plankton* float on the surface waters. The skeletons of these organisms accumulate in a clayey matrix providing *organic oozes*, variously named after the characteristic forms (calcareous foraminifera and pteropods, siliceous radiolaria and diatoms). Furthermore, slumping around coral reefs can introduce coralline sand and white, lime mud.

c) Inorganic precipitation including manganese, iron, nickel, calcium or magnesium in their make-up and alteration of existing deposits which produce clay minerals is yet another type.

Sea-Floor Plains

Sea-floor plains are best developed between the continental slopes and mid-ocean ridges. Typical depths are 30,000 to 6, 000 meters. Many such plains are truly flat and featureless, mantled by organic oozes or brown clays of mixed origin. Other forms shallow submarine fans best developed downslope of deep submarine canyons and adjacent to large rivers with much suspended matter, built up of silt, sand, and fine siliceous plant debris originally derived from river discharge. In fact, the greatest fans are found beyond the Ganga and Sindhu, (Indus), deltas. Typically, the surface of such turbidity deposits is runneled with fan valleys bordered by levee like ridge that ultimately bifurcate in delta fashion.

At some places the ocean floor is several hundred meters higher than the neighbouring abyssal plains (sea-floor plains) and extend over thousands of square kilometers. These are known as *Oceanic Rises.* A good example is the Bermuda Rise (3 to 15 sq km wide and 40 to 100 meters high) in North Atlantic. In some sections of the abyssal plains, deep sea channels, called *mid-ocean canyons* are found. These are either (i) continuation of the fan valleys of the continental rise on the abyssal plain or (ii) some form of erosion has cut the gentle slopes of these plains, developing winding channels with relatively steep walls and flat floor 1 to 8 km. wide. They have a general trend which is more parallel to the distant coasts than at right angles to them. One of the best known is the Mid Atlantic channel, characterized by flat floor, at least 2 km wide. The north east Pacific has a large number of deep sea channels.

The major relief of the sea-floor plains are groups of submarine volcanoes called *seamounts.* They rise abruptly 2,700 to 3, 700 meters above the seafloor, with steep sides and conical forms. If the volcanic activity persists over periods of 10 million years or more, sea mounts may show up above the ocean surface as volcanic islands, such as the Canariese or the Hawaian groups.

Oceanic ridges are apparently not permanent features of the crust. It can be demonstrated, for example, that a major ridge, called the 'Darwin rise' after Charles Darwin, existed in the Pacific about million years ago but has since disappeared. The evidence for this rise consists mainly in the finding of large number of broad, flat-topped submarine mounds grouped in a generally elliptical pattern throughout the Pacific, west of Hawaii. Mounds of this type have been found in all oceanic areas, though they are most concentrated in the western Pacific, and are called *tablemounts* or *guyots* (Figure 37), after Arnold Guyot, the nineteenth century Swiss-American geologist-geographer. They have the general shape of volcanic islands from which the tops have simply been shaved off by wave action before they sank beneath the oceanic surface 1,500 meters or more below sea level. They commonly provide the foundation for coral atolls in tropical waters.

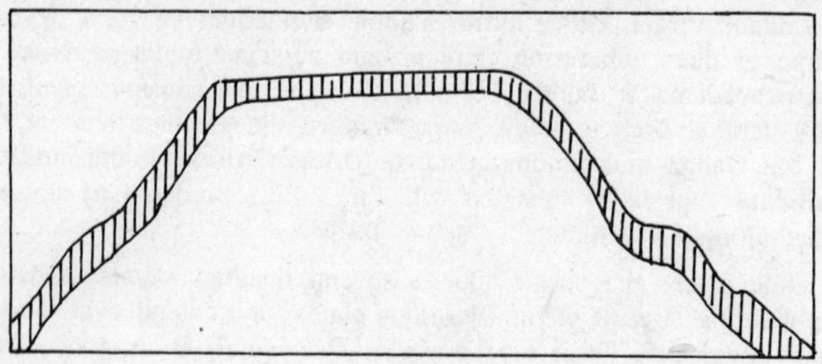

Figure 37 : Guyots

The origin of guyots has been a topic of considerable discussion. It appears obvious that the only agent in the ocean capable of cleaning a flat surface several kilometers across is the wave and waves obviously do not extend to depths of several thousand meters. Guyots, therefore, must have been at or near the surface at some time in their history. This conclusion has recently been reinforced by the finding of shallow water fossils of approximately 75 million years in age on the top of some of the Pacific guyots.

Mid-Oceanic Ridges

One of the most remarkable of the major discoveries coming out of oceanographic explorations of the mid twentieth century, for little was known about them beyond their basic existence, before about 1960, has been of a great submarine mountain chain extending for a total length of 640,000 km. down the middle of North and South Atlantic basins, into the Indian Ocean basin then passing between Australia and Antartica to enter the South Pacific basin, where it branches into several arms. The other mid-oceanic ridges extend across the Arctic basin and a part of North Pacific basin. Today, these mid-oceanic ridges are a major focus of attention, holding a key to the understanding of crustal deformation.

The mid-oceanic ridges extend down the axes of the major oceans, often following a rectangular, zig-zag pattern and branching off into less prominent 'plateaus' or rises. The major ridges are approximately 1,200-3,000 km wide, with an average relief of 1500 to 3000 meters Yet the crests typically lie 1,500 to 2,400 meters below sea level. Only a few peaks with relief of 3,600 to 6,000 meters rise above the sea surface (*e.g.* Azores Islands). Consequently, in their dimensions the mid-oceanic ridges rival or exceed the Rocky or the Andes Mountains.

Of the great mid-oceanic ridge the part-best known from detailed studies is the Mid-Atlantic Ridge. The ridge in its entirey is a belt some 2,000-2,400 km wide in which the surface rises from abyssal plains on both sides by a series of steps toward the central or median line where the ridge assumes mountainous proportions. The higher points lie at depths (1,000-1,500 fathoms). With reference to adjoining abyssal plains, then, Mid-Atlantic Ridge has a height of roughly 2,000 fathoms. A distinctive feature of the principal continuous ridge is that, instead of having a single crest line, as many mountain chains of the continents have, there is a characteristic trench like depression, or *rift valley* running precisely down the mid-line of the highest part of the ridge. Along with other parallel scarps and steplike rises on both sides, the rift strongly suggests that faulting of the crust was a dominant process in forming the mid-oceanic ridge.

These great underwater mountain ranges do not, however, resemble the changes of the continents in anything but size.

a) Seen in profile, a mid-ocean ridge resembles two slabs of lithosphere upraised at their contact. A grabenlike rift runs between them, trapping magma reservoirs of batholith dimensions. The slabs of the earth crust consists largely of successive lava flow, extruded along rift like fissures and cooled at the bottom of the sea with strange pillowlike surfaces. Rising above these great slabs of lava there are endless rows of gigantic volcanic cones, the great majority of which never reach the ocean surface. In the absence of glaciers and streams, these underwater ranges are not subject to erosional forces other than mass movements and turbidity currents. Slopes are therefore steeper and probably smoother than those of most mountains on the continents.

b) Seen in plain, the mid-ocean ridges resemble a jigsaw puzzle of rock rectangles separated by great faults. The rift running down the center is a zone of fissure flows, with lavas being added to either side of the rift as the segments move apart. The fractures that cross transverse to the ridge axis are called *transform faults,* and individual segments move sideways along them while they spread outward along the zigzaging central rift.

The mid-oceanic ridges are centres of continuous crustal deformation and vulcanism on an impressive scale. The frequency of seaquakes along these belts of ocean gives but small inkling of what goes on in the depth. The dynamism of the ridge is fundamental to the principle of continental drift.

Deep Sea Trenches and Island Arcs

At various places the ocean floors are marked by long narrow deep gashes/*trenches* also referred to as *foredeeps* with sea level depths of 75,000 to 10,500 meters or more. These long narrow depressions are steep sided and V-shaped, with rocky bottoms relatively free of sediment. The floors of these deeps lie 3,000 to 6,000 meters beneath the sea floor plains. Almost invariably the trenches run parallel and immediately adjacent to and on the oceanward side of long, narrow submarine ridges or to subcontinuous *island arcs,* mainly volcanic in origin or coastal ranges of the continental margins, known as the *mountain arcs.* Trenches are also found offshore the Circum-Pacific cordilleron belts created during the Alpine orogeny.

An example of a particularly deep trench, and one that is relatively straight throughout its length, is the Mindanao Trench, lying close to the eastern edge of Phillipine Islands. Deepest may be the Merianas Trench, where 5,940 fathoms depth has been measured. Although the Pacific ocean basin contains most of the great trenches and island arcs, the Atlantic Ocean basin is not entirely lacking in these features.

The oceanic trenches are zones along which the earth's crust has been forced downward and that this activity is closely tied in with the elevation of the adjacent island and mountain arcs. Even though partially filled by soft sediment from turbidity currents, so that the floors are locally abyssal plains, the filling process is so slow that the trenches maintain their great depth. Should a similar trench be produced within the limits of a continent, it would be kept filled with sedimentary deposits brought by debris-ladened streams issuing from adjacent mountain ranges, as in the case of Sindhu(Indo)-Ganga plain bordering the Himalaya Range.

As in the case of mid-ocean ridges, trenches and island arcs form belts of intensive earthquake activity. Their origin is intimately linked to continental drift.

DEEP SEA DEPOSITS

Microscopic organisms, animals and plants habitat (in suspension) themselves in millions in ocean waters and their dead bodies drop down as rain drop, forming deposits composed of their skeletons on the deep ocean floor. When dry, they form generally light grey coloured shiny mud like deposit called ooze and clays/mud showing flour-like in texture. Such oozes are named after the organism supplying the materials, of which the principal types are (i) *Globigerina* (ii) Pteropod (iii)Diatom and (iv) Radiolarian, called *pelagic* (ocean derived) deposits. The marine organisms belong to a group called *Plankton* (wanderers), *e.g.* single-celled sea plants and animals.

Pteropod ooze : or sea butterflies, as they are also called, floating gastropods (molluscs), which thrive well in (i) shallow (ii) warm and (iii) open oceans *e.g.* Atlantic ridge and on submarine elevations of the Pacific within the tropics at an average depth of about 2,000-3,000 meters in an area of 10,000,000 sq km. Their shells are of calcareous materials generally aragonite and dissolve in water before reaching the deep sea plains and thus are limited to shallower water only. These oozes constitute an important phosphate source material for use as fertilizer.

Foraminiferal or *Globigerina ooze* : These protozoans are yet another but most abundant type of marine animal, (i) minute in size and (ii) their shells are composed of calcium carbonate which constitute calcareous foraminifera with *Globigerin*a pre-dominant. They occur at 4,000 meters depth in (i) tropical and (ii) warm temperate regions, like Atlantic. Indian and South Polar oceans, covering about 36% of the ocean bottom area in about 11.68 lakh sq km ooze white, pink or yellow when wet and resembling chalk when dry. Other organisms radiolaria, diatoms, and sponges spicules may also be present. Much of the shells of these tiny animals while sinking down when die get dissolved in seawater before landing on to the sea bottom.

Radiolarian ooze : Yet another group of silica-shell radiolaria a minute marine protozoan sometimes mixed with sponge spicules and diatoms live generally in (i) deeper and (ii) warmer oceans and accumulates at an average depth of about 6,000 meters in the Indian and Central Pacific oceans and inter-tropcial seas. The shells sink down and accumulate at the deep sea bottom form ooze. Rocks formed from such deposits are called *radiolarite*.

Diatom ooze : A minute plant, the diatom thrives in quite shallow water upto the sun light penetrability depth of about 220 meters and abounds in cooler oceans. Their shells are formed of silica and together with some sponge spicules are found in a belt surrounding the Antarctic in high latitudes. Generally *diatomacaeous earth*, kieselgular, diatomite, infusorial earth or Tripolite are such deposits.

Red mud/clay : These clay bulk, constitute minute particles of silt or clay with a small admixture of mineral grains (chiefly quartz) and with much organogenic $CaCo_3$ owe their red, brown or yellow colour to the presence of iron oxides reflecting to the conditions under which it had formed. The regular pattern of areal distribution of red muds is widespread near the great river mouths that trap abundant red weathering products rich in iron oxides across which they flow and transport into the sea. Red mud may also develop on smaller areas of sea bordering on shores with widespread red beds where conditions are favourable for their runoff, *e.g.* the red sediments found in the southeastern part of the Baltic Sea.

Red ocean clay occupies vast areas of ocean floor beyond the zone of massive plankton organisms accumulation of remote from the coasts at depths of 3,500 to 8,000 meters. Both in appearance and properties it is a true clay, coloured brown, or brick red by iron and manganese oxides. Until recently, it was held that terrigenous material could not reach remote parts of the ocean, and red ocean clay was considered to be the product of submarine weathering of rocks of the ocean floor, and or decomposition of volcanic eruption products.

However, extensive studies of bottom sediments carried out in different parts of the world ocean have demonstrated that red ocean clay was a complex polygenic sediment. Taking part in its accumulation are :

i) insoluble remains of foraminifera, the calcareous skeletal parts of which are dissolved at high pressures and never reach great depths,

ii) a certain part of the fine clayey and other colloidal particles brought by rivers and carried by currents to remote parts of the ocean,

iii) aeolian dust,

iv) detritus brought by icebergs,

v) products of volcanic eruptions, both submarine and subaerial,

vi) meteoritic dust - tiny gloubles (0.2 mm. in diameter) of nickel, iron (the main constituent of iron meteorites),

vii) insoluble organic remains, such as shark teeth, the oolites of whales *etc.*

The

a) huge accumulation of these organic remains, and

b) also meteoric dust globules (upto 20-30 per sample)

as compared with rare inclusions (2 or 3 at the most) in the same volume of oozes, point to the great length of time over which the abyssal red clay has accumulated. Another evidence is that some of the teeth found in the sediments of small thickness (upto 40 cm) are remains of extinct fishes that lived millions of years ago. The composition of red clay varies with depth and locality.

Green mud and sand - first occur at the depth of 80 to 100 meters the shelf region and godown to a depth of 2,300 meters. However, mainly they develop in the transitional zone from the shelf to the continental slope (near and below the 200 meters isobath). Generally, green mud is coarser in relation to the blue and the red muds, and only sometimes is composed of fine sand. Occasionally it contains more than 30 percent of $CaCO_3$ and then becomes lime mud. It accumulates where steep slopes are composed of igneous rocks, with a

i) meagre supply of terrigenous material, and

ii) where strong cold currents pass.

The green colouration is caused by the presence of glauconite. Mixing of warm and cold currents, *i.e.* a change in the regime of the basin to the most favourable locale for their formation, and the glauconite develop from the decomposition on the sea floor of alumosilicate sand grains, spiculae and volcanic glass or precipitates in sea water as got out of colloidal solution brought from the land. The presence of iron oxide in glauconites indicates that green mud had formed in an oxidizing environment. Phosphorite nodules occur frequently both in green glauconite muds and in ancient sedimentary rocks of similar composition.

Blue mud - Most abundant on the floor of the Atlantic Ocean, consisting most fine silty or clayey particles, it is widespread within the limits of the continental slope and the adjoining part of the ocean floor, and extends to a depth of 5,000 meters, and is usually bluish-black or grey colour with a steely tinge containing 30 percent of calcium carbonate - mostly of organogenic origin- calcareous shells of plankton organism (when the $CaCO_3$ content increases to 40-50 percent, it is called lime mud often gives hydrogen sulphide smell). The typical -bluish colour is suggestive of considerable oxygen deficiency reducing environment condition. It facilitates pyrite formation in the abundance of organic matter and oxygen deficiency. It may also contain siderite, manganese, limonite concretion, and in places, increased concentration of organic matter (organic carbon C). Some oxidation of ferruginous compounds, in the upper reaches, the blue mud may show brown hue.

CORAL REEFS

These are organic structures developed around islands or some volcanic peak islands made of essentially corals and other lime-secreting minute animals and plants which abounds deposits in tropical and sub-tropical oceans. These are abundant in the Pacific Ocean as also in Indian Ocean. The corals assume by secretion protective calcium carbonate shells around their bodies. These shells sink down and accumulate on the sea bottom, when the animal die.

Coral consists of the calcareous skeletons of primitive organisms that accumulate layer by layer on long stalklike projections (solitary coral) or irregular rocklike masses (colonial coral).

Coral polyps in fact are biologically animals of which the fleshy part resembles a jelly fish, with small tentacles protruding from a base fixed to dead coral or rock, although in appearance they like plants assume shape (s). Typical dimensions are about 1 cm. The polyp feeds from the seawater, and lime particles build up in the cells at the base of organism. As the polyp is replaced, the calcareous base remain behind as a layer of bizarrely patterned lime rock.

Corals live in swarms in (i) shallow (ii) clear and (iii) warm water of the oceans. Millions of them supply calcareous material to form the reefs when they die. Coral families build structures in plant-like fashions and the interspaces between the branching coralline structure are filled up in due course of time by the deposition of calcium carbonate either

(i) by other lime secreting organisms like the nullipores (algae) or molluscs, or

(ii) by the debris broken by the seawaves.

Slowly but progressively an island-like structure is developed which is exposed only at low tides. Since the minute animals cannot stand exposure to the atmosphere for a great length of time, coral reefs cannot grow above the low tide mark. The following are favourable conditions for the formation of the coral polyps and the associated reef building organisms and their successful growth met in specific geographical conditions -

i) Corals are exclusively found in salt water, with salinity of 2.7% to 4%.

ii) They also require clear water, free of suspended sediment.

iii) Waters, preferably agitated by wave action to ensure adequate food supplies too, is a factor.

iv) They as well require a solid base of rock to grow out from, oriented to the open sea, but at shallow depth (adequate light and oxygen), coral growth is normally impossible in turbid, silt-laden waters or below a depth about of 60 meters.

v) Some corals, mainly of the solitary kind, can grow in cool or temperate waters. Most of the colonial forms, however, thrive only at temperatures less than 25 to 30^0 C (77 to 86^0 F) and do not tolerate winter temperatures of less than 18^0 C (64^0 F). As a result reef-growing coral is not found outside of the tropical oceans, and it is absent near the mouths of rivers and muddy streams.

vi) Corals cannot survive for more than an hour or two above sea-level at the time of low tide and this is why they are usually found below the average sea-level.

vii) For the growth of corals in open seas, it is necessary to have submarine platforms which may act as foundation for the corals.

These platforms should not be deeper than 50 fathoms (or about 90 meters), because corals cannot start growing and developing their colonies on deeper surfaces.

Some of the more important geomorphic features of a coral reef include:

a) The steep external slope, formed of coral, coralline sand, and talus, dipping steeply at angles of 25 to 45^0 down to considerable depths.

b) The reef crest above the external slope, formed of growing colonies of lime- secreting algae, active coral, ground-up coralline sand, and loose coral rock, the crest surfaces is about mean sea level.

c) The reef platform, to the rear of the crest, often several hundred meters wide, composed of dead and living coral below low-tide level and

d) The lagoon (if present) with a sandy floor, interrupted by sporadic coral growths below the low-tide mark.

These constituent parts are developed in varying degrees on most coral coasts, although they are best represented in fringing reefs.

Several shore forms (Figure 38) are provided by coral reefs:

1) *Fringing reefs,* attached directly to the shore platform

2) *Barrier reefs,* extending up from a rocky base at depths of more than 90 meters and separated from the shoreline by broad, reef lagoons.

3) *Atolls,* coral islands consisting of circular or arcuate reefs, interrupted by inlets or boat passes surrounding a lagoon with water depths of 30 to 90 meters.

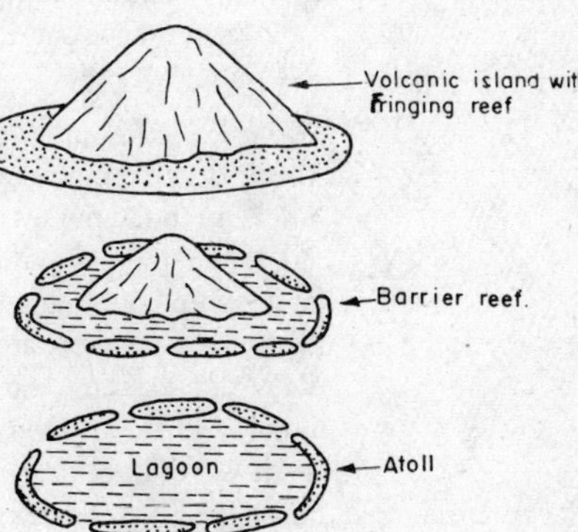

Figure 38 : Forms of Coral Reefs

1) *Fringing reefs* : These are attached to the coasts, continents or islands; however, in some instances, not very common a narrow waterfilled depression resembling a miniature lagoon may intervene the landmass and reef. The reef surfaces rugged and

uneven may rise upto to low tide water level and the outer edge of the platform slopes steeply seaward. These reefs are not generally very wide, and becomes broken and discontinuous where a river enters the sea from the land.

2) *Barrier reefs* : these

(i) rise up from a rocky base at depths of more than 90 meters which is greater than the depth of coral growth and

(ii) enclose conspicuous broad deep lagoons like barriers. 0.5 km to 15 km or more wide and the depth from 20 to 40 fathoms are always built away from the mainland. [(The Great Barrier Reef, world's largest barrier reef, off the coast of Queensland in east Australia is about 1,800 km long, 16 km wide and separated from the shore by a channel of about 135 km wide)]. More than 90 meters depths of the rocky foundations tend to support a case of subsidence after the reef formation.

Barrier reefs are long, wide and extensive, and slope seawards at an angle of 25^0 to 45^0. The lagoons are sometimes isolated *coral knolls* also called *patch reefs or pinnacles* rising from the lagoon floor to the sea- level. The barrier reefs rarely extend in continuity for long distances and is often dismembered by gaps called *passes* or *tidal inlet* which link the lagoon with the open sea providing passage way for the water brought into the lagoon by waves return again to the sea.

3) *Atolls* : Atolls are the most common type of coral, discontinuous roughly circular horse-shoe or arcuate shaped, interrupted by inlets or boat passes reefs enclosing lagoon with water depths of 30 to 90 meters with an average of about 45 meters. Their outer and steep slope leads to deepwater and a lagoon in the center. Most of surveyed atolls indicate that the lagoon floor is irregular with depressions and coral knolls which in turn have surrounding almost circular elongate ellipses and or irregular-shaped masses with several protruding corner reefs which show wide variations in shape. A south-east spur is particularly common and spurs are also usually found on the north-east. A relatively straight reef is found between the spurs facing the trade winds where both are present. The reefs at their outer margins are exposed only at lowest tides, but further inside islands rise above much of the encircling reef. There are more islands in the south-east windward side and the reefs are also wider than the leeward side of the atolls.

Most of the world's atolls lie in the western Pacific and the nieghbouring Indian ocean in the tropical zone. Some of the well known examples are Funafuti atoll in the Elice Islands in the western Pacific, Bikni and other atolls in the Marshall islands. The Laccadive and Maldive islands in the Indian Ocean are composed of atolls. The atolls in the Maldive islands have numerous faros both in the lagoons and on the encircling reefs. They are also rounder than most other atolls.

THEORIES OF CORAL REEF FORMATION/ORIGIN OF CORAL REEFS

Corals generally thrive near an island or a volcanic cone when they get some shelter against the ocean currents where they gradually construct a platform like structure to build up a fringing reef. The barrier reefs or the atolls generally point to depth of formation where no coral can live. They can only live in shallow, generally less than 50-60 meters of depth. Hence their formation can evidently be due to either submergence of the ocean bottom or the rise of the sea level keeping pace with reef building.

a. *Subsidence Hypothesis of Darwin* : It says fringing reefs, barrier reefs and atolls are three successive stages in the growth of corals round a slowly sinking land area so that the reef was able to keep pace with the subsidence. It was subsequently slightly amended by Dana, Davis and others. The inlets of the atolls are supposed to be due to drowned embayments.

   Suppose that

   (i) a fringing reef has been built around the summit of volcanic cone and

   (ii) the volcanic cone is subsiding owing to diastrophic movements of the sea floor

   Now as the subsidence continues, the coral migrates outwards and upwards. The growth of coral reefs is rather rapid nearly 0.3 meters in 10 years. Fringing reef will thus pass on to a barrier reef enclosing a body of water between the reef and the island. Ultimately when the volcanic island totally disappears the barrier reef is transformed as atoll with a circular lagoon inside.

   It explains the phenomenon associated with coral reefs well and submergence of the seafloor is a possibility and actually occurred in geological history. It however, suffers as it fails to answer for the flatness of the lagoon floor inside the barrier reefs and atolls.

b. *Glacial Control Hypothesis of Daly* : It is envisaged that during the Pleistocene the sealevel was lower and the reef building activities of coral became restricted so that benches and submarine plateaus were cut and built by sea waves on old reefs. With the advent of warmer times sea level became higher, corals not completely exterminated began reef building again on the former submarine wave cut, wave built platforms so that lagoons were left within the barrier reefs and atolls. This also shows why most of the larger reefs have almost same depth, 70-80 meters. The reefs are narrow and hence young, which according to Daly, is due to their post-glacial origin. The objection to this idea however is that it cannot explain the growth of coral reefs for more than 110 meters which is the estimated rise of sea level following Pleistocene deglaciation.

It has been criticized on the following grounds :

(i) Some of the platforms are so wide that it does not seem likely that they could have been formed by marine erosion during the Ice age.

(ii) The similarity in the depth of lagoons emphasized by Daly is rather overstated. Further this depth is much greater than one would expect resulting from the Pleistiocene fallen sea level.

(iii) His estimate of the Pleistocene fall in sea-level is much less than the actual.

(iv) In the event of stability of land masses and fall in sea-level, cliffs should be formed along the coasts of continents and islands. It is probable that with the subsequent rise in sea-level, some to the cliffs would be submerged but they will not be completely destroyed. But there is complete absence of cliffs coasts in Coral seas.

Boring at different coral islands like Funnafuti in the Ellice Islands, Bermuda *etc.*, in the Pacific Ocean have not yet yielded any conclusive evidence in favour of any of these two hypothesis. None of them is able to explain all the features of coral building but they together combine to explain all these. Darwin's theory of subsidence by diastrophism and Daly's theory of submergence due to rising sea levels are complementary to each other and both taken together solve many reef building problems.

c. *Solution Hypothesis of Louis J. Agassiz and Sir John Murray* :

It proposes the formation of barrier reef as a result of solution of the inner parts of the result, the fringing reef grow wider but the solution of calcareous material from the inner side removes the connection between the reef and the main island around which the reef grows. The fringing reef thus becomes a barrier reef with a body of water inside. The atolls have grown in similar ways around submarine platforms formed by marine denudation. This idea is unacceptable on account of the fact that the enclosing body of water is saturated with calcareous material, and hence any more solution of calcareous material is not possible by this water.

PERMANENCY OF OCEANS AND CONTINENTS

Earlier it was believed that the landmass (continents) and ocean basins are not stable and they frequently interchanged in respect of their position. However, subsequent revelations on close examination attest that no significant such changes took place and that neither deep ocean floor formed land masses, nor the continental blocks deep sea basin at any time of the history of the earth particularly after Archeaeozoic time but the marginal parts of the continents have been inundated by oceanic transgressions, and shallow parts of the continental shelves of the oceans have formed land masses at intervals.

a) Some surveys which have not revealed anywhere submerged continental blocks at deep oceanic depth.

b) Deep see oozes too have not been reported from the land surface to form sedimentary sequences, but for certain deposits in some oceanic islands like Barbados, Bornes *etc*. The sedimentary rocks like sandstone, mudstone and limestone are formed on the shallow continental shelf of the oceans. The fossils contained in them too reveal shallow water organisms. Such rocks are seldom to be found on distant oceanic islands.

c) Sinking of light sialic continental blocks in heavy oceanic simatic layers is rather a physical impossibility. Small portions may however sink down under special conditions but sinking on large scale more so of vast continental sectors appears to be beyond comprehension. So also of vast continental sectors is the situation with the emergence of ocean floors.

The concept of permanency of oceans and continents does not rule out the possibility of continental drift/plate tectonics which attempts to explain the origin of the Himalaya, as the former deals with the vertical movements of the continental areas and oceans and the latter with the horizontal sliding only.

Marine Transgressions - The sea has invaded the continental borders from time to time transforming these marginal areas into epicontental seas. Such temporary invasions by the sea on the land are called marine transgressions which may be due to either

(i) the decrease in ocean capacity some water will evidently result in spill over of sea water on land or

(ii) increase in sea water volume.

caused by the

a) melting of ice sheets releasing voluminous quantity of water to the ocean and thus permitting of sea level rise,

b) elevation of the sea bottom.

c) sinking of the continental masses

Joly's thermal cycle explains such changes. The transgression deposits present some characteristic on submerged continental border areas and are called coastal system of deposits, *e.g.* the Upper Jurassic rocks of Cutch and the Upper Cretaceous rocks of Trichnopoly.

The invaded water recedes sometime after each marine transgression and this phenomenon is is called *marine regression*. The cause of such recessions are just the opposite to those of marine transgression.

THE CHANGING LEVEL OF THE SEA

Sea level is not constant. Evidences of marine animals in the rocks of the highest mountains, and deep drilling recorded shallow marine sediments hundreds and even thousands of meters below the present sea level indicate its fluctuating nature.

Isostasy accompanying continental glaciation and deglaciation has led to some changes in the shoreline. The weight of continental glaciers is so great as to cause major lowering the land down below it. When the glacier melts and the load of the ice is removed, the land slowly recovers and achieves its original balance.

The water that is now locked up in the ice of glaciers originally came from the oceans. It was transferred landward by evaporation and winds, precipitated snow, and finally converted to firn and ice. If all this ice were suddenly to melt, it would find its way back to the ocean basins and would raise the sea level to an estimated 60 meters elevation. A rise of this magnitude would transform the outline of the earth's landmasses and would submerge coastal towns and cities. For the last several thousand years melting glaciers have been raising the sea level and modern records of sea level indicate that the sea is still rising at about 1 mm/year.

During most of the glacial periods of the past waters impounded on the land in the form of ice was more extensive than at present. Consequently the sea level must have been lower than it is now. It has, in fact, been estimated that during the maximum extent of Pleistocene glaciation the sea level was from 105 to 120 meters lower. Most of the geologists accept this estimate, although some feel it is too conservative. On the other hand, during the interglacial periods, when glaciers were somewhat less widespread than are today, the sea level must have been higher.

The level of oceans and connected seas respond to the following few major mechanisms.

A. *Tectonic Deformation* : On a local or regional scale tectonic deformation may raise or lower coastal strata -

(i) Individual earthquakes may even be sufficiently severe to displace local shorelines

(ii) Volcanism can change the elevation of the land

(iii) Explosive eruptions can also reduce a land mass.

However, tectonic deformation is generally restricted in terms of area and is neither world wide nor synchronous, although entire tectonic provinces may be affected on an appreciable scale. Most of the mountainous coasts have apparently been upraised by 90 to 150 meters during the past 2 or 3 million years and uplift along the California coast has been still greater.

B. *Isostatic Readjustment* : A different effect has been registered in areas once under the Pleistocene glaciers. The weight of the ice masses served to compress the Earth's crust. This temporary isostatic depression at any one time may have amounted to over third of the thickness of the ice sheet. As the ice melted, the crust rebounded with degree of uplift. Accordingly, there are many well developed but abandoned shorelines of post-glacial age along Canadian and Scandinavian coasts at elevations of 180 meters or more.

C. *Capacity of the Oceans* : Through geological time the shape of the great ocean basins has changed very slowly but nonethless appreciably in response to large scale movements of the Earth mantle and crust. In response, it is probable that the capacity of the ocean basins to hold water has also changed as the basins changed their configuration and became deeper or shallower. Further more, the continental shelves are also liable to change and , depending on their slope, can also readjust to other changes by isostatic readjustment. Finally, all segments of the continental margins appear to rise or sink, relatively to the sea, at slow rates but over long periods of time. Whatever the explanation, on some occasions

(i) the oceans appear to have spilled far up over the edges of the continents for long periods of geological time, on other occasions

(ii) they have receded.

There is at least some reason to believe that relative sea level on a world wide scale, was as much as 100 meters higher during the Pliocene period, some 3 to 5 million years ago. Since then the overall trend has been downward.

D. *Amount of Ocean water* : The growth and wastage of large glaciers on continents provide a proved mechanism to lower or raise ocean level. A hydrological balance exists between the water in the oceans, in the atmosphere, in lakes and streams, in the groundwater and in ice sheets or glaciers. If the ice caps grow, water must eventually be withdrawn from the oceans, which provide the only significant reserve. If the present glaciers of Antartica, Greenland and other polar or mountain regions would melt today, the meltwater would all be returned to the oceans, so that the world sea level would theoretically rise by about 66 meters. A sudden thaw of this magnitude is of course quite improbable. Though a slight but widespread recession of glaciers during the twentieth century is slowly adding more water to the oceans, and over the past 70 years sea level has been rising at a rate of 11 cm per century.

Estimates have been made of the volume of ice held in the Pleistocene glaciers. At the maximum of Pleistocene glacial stages world sea level was according to various estimates 85 to 159 meters lower than it is today. During the waning stages of the last glacial stage, meltwaters were returned so rapidly to the oceans that sea level may be at times have risen by as much as 1.5 meters per century.

Such fluctuations in the volume of ocean water are known as *glacial eustatic* changes. They provide a mechanism that can theoretically raise the sea level to about 66 meters, above the present or lower it to perhaps 145 meters below that value. This is a surprising range of 210 meters.

Pleistoeene Sea Levels :

The rate at which coastal landforms evolve today is in most instances slow. If world sea level can and does fluctuate rapidly, it becomes pertinent to know whether sealevel is sufficiently stable to permit full-fledged 'mature' coastal forms to develop. In fact, do the visible coastal features reflect present or past stands of the sea? For reasons, such as these, it is necessary to know about sea level changes of the Pleistocene and more recent past. Only then can coastal landforms be seen in a proper time dimension and evaluated in a meaningful perspective.

Three trends can be recognized during the 2 million year course to the Pleistocene.

a. There has probably been a gradual deepening of the ocean basins or an uprising of the continental margins to simulate a downward trend or regression of world sea level of the order of perhaps 100 meters.

b. Superimposed upon this slow downward trend were a series of rapid glacial eustatic spasms. During each glaciation, sea level dropped perhaps 90 meters or more below 'normal'. Slight positive fluctuations or *transgressions* in the range of 3 to 8 meters may have been caused by slightly greater melting of residual ice masses during several interglacial phases.

c. In addition to these worldwide changes, varying degree of local deformation can be detected in all coastal areas next to mountains affected by Alpine orogeny. Here intermittent uplift has been the rule, and a few areas such as California coast have experienced appreciable change during the past 100,000 years or so.

In more stable tectonic regions, positive or negative movements of the land have resulted from isostatic readjustment to the weight of water added or withdrawn from the continental shelves in response to glacial-eustatic transgressions or regressions.

For the purpose of coastal geomorphology, two significant applications emerge from the understanding of Pleistocene sea-levels. Minor advances and retreats of the existing glaciers are well documented in sea level fluctuations over its present level, even during the historical era, a complex sequence of sea- level oscillations (Lind 1969), are as follows :

Fourth and third millenia B.C.					Near present level
3000	(?)	-	2000	BC	As much as 3 meters *above*
2000	-	-	1200	BC	As much as 3 meters *below*
1200	-	-	200	BC	2 meters *above*
200	BC	-	400	AD	As much as 2 meters *below*
400	AD	-	1200	AD	1 meters above
1200	AD	-	1400	AD	As much as 2 meters *below*
1400	AD	-	1600	AD	50 cm. *above*
1600	AD	-	present		Present level or slightly *below*

The oscillations of sealevel since 4,000 BC have been too brief to produce distinctive landforms in most parts of the world. But rapid oscillations are a fact in areas with narrow continental shelves, and they do help explain why contemporary coastal platforms are poorly developed along sheltered coasts where tidal amplitudes are small. The instability of sea level also helps explain the multiple beach ridges and barriers is seen in many coast. Repeated oscillations of sea lends former's renewed adjustments of the erosional-depositional balance, as the resulting landforms are better developed.

a) World sea level was higher during successive inter-glaciation than today, specially during the Last Interglacial (for about 125,000 to 75,000 years ago), shore features were developed-carved or built that remains conspicuous at anywhere from 1.5 to 30 meters or more above the water marks on many coasts. Going into finer details even during the Last Interglacial, sea level was not always stable and multiple beaches frequently formed at individual locations. As a result, in many places, a series of high or 'raised' shoreline features above the modern beaches can be observed. Although fragmentary, these relicts of the Pleistocene are definitely a part of the constellation of coastal forms.

b) For almost 60,000 years, during the course of Last Glaciation, world sea level was much lower than today as is partly reflected by now submerged coastal features at depths almost 150 meters on the continental shelves. Far more important, however, is what happened on land surfaces now exposed at the coasts. Rivers erode down their profiles to establish a new equilibrium, developing, sometimes fluvial landforms that have been difficult for subsequent coastal process to camouflage.

Holocenr Sea Levels - Except for a few technically unstable areas generally the relative movements of land and sea during the past 10,000 years have been primarily of glacio-eustatic nature. About 25,000 or 20,000 years ago, the earth was coldest in the Last Glacial period, and the ice sheets began to recede rapidly by about 14,000 years ago. In about 10,000 years over 90 percent of the ice sheets volume melted and traversed down to the oceans, an amount that, if not compensated by coastal readjustment, would cause a sea level rise, of the order of 100 meters. This comparatively rapid transgression flooded out broad sectors of the continental shelves, drowning whole constellations of fluvial and glacial landforms. Infact, both fluvial and coastal processes have barely had sufficient time to adjust to a new equilibrium during the past 6,000 years. The result is that many coasts still appear to be 'drowned' as an aftermath of the postglacial rise of sea level.

NORMAL CYCLES OF SHORELINE DEVELOPMENT

A. *Cycle inaugurated by Submergence -*

a. *Initial stage* - Imagine the submergence of a coast of moderate relief. It would be rougher and more irregular because of drowned river and glacial valleys, forming fiords, separating headlands and promontories.

b. *Youth stage* - The relief is accentuated by prolonged erosion in early youth, prominent cliffs, erosion remnants such as stacks, pulpit rocks, bridges as prominent features. Subsequent wear of promontories and headlands results in the formation of high sea cliffs. Bays are filled up by sediments. Still further erosion causes recession of headlands, extending spits, bars into embayments and lagoons. The latter begins to fill with sediments.

c. *Maturity* - Headlands are cut even further back than original embayments, and there is developed an extensive wavecut shelf with sediment deposited beyond it perhaps in part as a wave built terrace. The remaining sediment filled embayment largely or wholly obliterated. Sea cliffs loose height on account of weathering which lowers general relief. The shorelines begin to show irregularities because of differential erosion but the irregularities are gentler than those of the initial stage ; wave cut benches become wave built terraces.

d. *Old stage* - In this stage land surface becomes low, waves become weak, cliffs loose their heights, less of land derived sediment reaches the shore. The wave cut bench is clearer. The shore has a gentle slope.

e. *Final stage* - It is reached when the whole landmass attacked by the sea is reduced to *base level of wave erosion* or in other words, when a rock platform is cut everywhere to the lower (depth) limit of wave erosion and sediments have practically all been swept off the platform. The depth of water over such a platform is usually several hundred feet, the maximum probably being about 200 meters. This final stage may be likened to the peneplain stage in the normal cycle of stream erosion. However, this appears to be a theoretical stage and its examples are not known because sinking and rising of land takes place sooner or later. Partial submergence of a nearly flat land area would, of course, inaugurate a regular or relatively straight shoreline free from indentations. In such case, the shoreline cycle is essentially the same as that inaugurated by uplift of flat sea bottom into land.

B. *Cycle inaugurated by Emergence* -

If part of relatively flat sea bottom emerges into land, the resulting shoreline is, of course, regular or almost straight and free from indentations because the line contact between the two practically flat surfaces makes it so. If the emergence produces a gently sloping coastal plain with very shallow water offshore for some distance, the waves, due to strong frictional drag on bottom have practically no erosive effect upon the new shoreline.

a) Instead, the waves, by their heavy bottom drag, transport loose materials to a zone some distance offshore where a barrier beach (or offshore bar) is built up. A lagoon, usually with marshes, is thus found between the original shore and the barrier beach. This stage is of *youth.*

b) The depth of water on the seaward side of the barrier beach is increased enough so that waves cut into the barrier. Some of this material, especially during storms , is carried out to sea, but some of it is carried over the barrier beach. In this way the barrier migrates towards the lands, and in the meantime the lagoon becomes partly or wholly filled with sediment and marsh-plant remains.

   In time the barrier beach is completely removed, and the water offshore is then deep enough for the waves to attack the whole original shoreline. This shoreline is now *mature.*

c) Indentations may then develop where some (weaker) rocks are cut back faster than others, but such irregularities are small. Bars may be built across such indentations temporarily , but, during the *old stage* to the *final stage,* the history is practically like that of the same stages in the cycle inaugurated by partial submergence of a mature land area.

In case emergence leaves sufficiently deep water offshore, a barrier beach does not, of course, develop, but wave cutting starts right on the shoreline and succeeding stages are much like those of the latter part of the cycle.

Comparison of Coastal and River Cycles

There are several points of similarity between coastal cycles and the erosion cycle of rivers-

a) First a river system roughens the surface of its basin, increasing its relief; finally it reduces it to a smooth plain, near sea level. Likewise, waves and currents normally increase to a maximum the irregularities of a coast, and finally reduce them to a minimum.

b) An essential difference is that

   the irregularities of the river basin are vertical while those of the shoreline are horizontal.

   In each, the cycle of development is introduced by diastrophism

CLASSIFICATION OF SHORELINES

The three basic components of geomorphic explanatory description

(i) structure

(ii) process and

(iii) time

are not applicable as such to coastal studies in quite the same manner as it applies to subaerial landscapes. The coastal developments rather, display wide diversity and compound nature under varying influences of

a) tectonic activity and

b) instability of the Pleistocene sea levels caused due to alternation in glaciation and deglaciation

which renders most shorelines to fit into a unified scheme of classification difficult. However, thus far no single hypothesis has emerged from two centuries of geomorphic research to satisfactorily account for the varietal morphology and complex evolution of all the major type of shorelines. Too much of the coastal landscape is relict, inherited either from

(i) earlier shoreline process now inoperative or

(ii) subaerial or submarine processes.

A. Johsnson's classification (1919) : It is a commonly used classification, based on the sea-level fluctuations in relation to the land, and separates the shorelines into four categories

a) *Shorelines of submergence*; formed by partial submergence of a landmass as a result of

(i) rise in sea level or

(ii) subsidence of the land

Such shorelines are characterized by cliffs, wave-cut platforms, headlands, caves, natural bridges, offshore islands, drowned valleys, deep embayments and a very irregular configuration. They are illustrated by

(i) *Ria shorelines* - dissected landmasses associated with drowning of river valleys, on its partial submergence.

(ii) *Fiord shorelines* - formed by partial submergence of a region of glaciated valleys or troughs.

b) *Shorelines of emergence;* formed by

(i) uplift of land or

(ii) lowering of sea level

and are characterized by relatively straight coasts of low relief, marine terraces, offshore bars and lagoons

c) *Neutral shorelines* : whose essential characteristics are independent of either

(i) submergence or

(ii) emergence.

These include

(i) *Delta or alluvial/outwash plains* shorelines produced, building forward of a delta or broad alluvial slope

(ii) *Volcanic shorelines* - where an active volcano builds up its cone above water surface.

(iii) *Coral reef shorelines* - where coral builds upward from the seafloor or outward from the landmass.

(iv) *Fault shorelines* - where fault blocks are depressed enough to allow sea or lake water to come against the fault scarp.

d) *Compound shorelines* : are those which include more than one category listed above, occur side by side, such as a shoreline which display both marine terraces and drowned valleys as a result of fluctuations in sea level.

Johnson's classification is simple in perception, genetic in nature and stresses on factors likely to exert profound influences on a shoreline. However, it suffers from the following disadvantages:

(i) Many shorelines have undergone both submergence and emergence under the influence of sea-level changes in response to Pleistocene glaciations and interglaciations.

(ii) Isostatic rebound following disappearance of continental glaciers, differential plates and earth movement including neotectonic activityhad a profound influence on the stability of the sea level.

(iii) Shore terraces built by waves and drowned valleys have been found side by side at certain places. This means that all shorelines are compound

Moreover

(iv) the 'neutral' shorelines have, truly speaking, undergone both emergence and submergence as in the ice ages when changes in sea level took place.

B. Shepard (1937, 1948, 1963)'s classification : Realizing limitation of Johnson's classification in terms of sea-level rise following retreat of the last continental glacier, Shepard brought out a two-fold classification based on

(i) shoreline maturity and

(ii) nature of the agency(ies), terrestrial or marine producing the shoreline.

I. Primary or youthful coasts and shorelines sculptured primarily by nonmarine terrestrial agencies and include subtypes resulting by

a) eosion on land and drowned subsequently due to deglaciation or downwarping

(i) drowned river coasts (*ria coasts)*

(ii) drowned glaciated coasts

b) deposition on land by

1. river

(i) deltaic coasts

(ii) drowned alluvial plains (*e.g.* Tamil Nadu coast)

2. glacial coasts partially submerged

(i) moraines

(ii) drumlins

3. wind (prograding sand dunes), where dunes have been built by seawards blowing winds.

4. vegetation (mangrove swamps)

5. landslide coasts where landslide material has prograded coasts -

c) volcanic activity

1. recent lava flows where land surface has been built by seaward blowing winds

2. volcanic collapse or explosion (calderas) as in volcanic oceanic islands.

3. in slopes built by unconsolidated fragmental volcanic material.

d) Diastrophism

1. fault - scarped coasts (*e.g.* Atlantic type of Suess)

2. coasts related to folding (Pacific type, *e.g.* Arakan coasts)

3. salt - dome coasts (scarce, but are present in some areas *e.g.* Persian Gulf)

II. Secondary or mature coasts and shorelines developed chiefly by marine processes and include three subtypes :

a. shaped by marine erosion -sea cliffs.

1. straightened by sea waves

2. made irregular by sea-waves

b. shaped by marine deposition
   1. coasts straightened by deposition of bars across estuaries
   2. coasts prograded by deposits
   3. shorelines with offshore bars and longshore spits

c. shaped by animals and plants
   1. coral coasts (*e.g.*Lachchadive coasts)
   2. mangrove coasts (*e.g.* Ganga delta).

These groupings are not always clear-cut and well marked as some border line coasts are referrable to both or either categories. Erosion and deposition are likely to be occurring on a coast simultaneously. Besides often it is difficult to spot the stage at which a primary coast becomes sufficiently modified by marine process to be given a status of a secondary coast (Bird 1969). While advancing his new classification of coasts, Sheparad argues that the sea level change since the last glacial event of the Pleistocene has been one of the sea level rise during last 20,000 years as indicated by carbon dating of stable coasts. His primary type where the sea is regarded to rest on the land (submergent type) takes special notice of this eustatic change. This classification is clear-cut, comprehensive and useful, but suffers from the same difficulties as Johnson's.

C. Cotton's classification: Cotton (1952) proposed landstability as the main variant in the classification of coasts and identified the following two categories and subsets, therein :

1. *Coasts of stable regions* : These have all been effected by recent submergence-coasts dominated by features
   a. produced by recent submergence
   b. inherited from earlier episodes of emergence
   c. of miscellaneous identity (*e.g* fiord, volcanic *etc.*)

2. *Coasts of mobile regions :* These have been affected by (i) uplift or (ii) depression of the land as well as by recent submergence and include coasts
   a. on which the effects of marine submergence have not been counteracted by uplift of the land
   b. on which recent uplift of the land has led to emergence
   c. folded and faulted
   d. of miscellaneous nature (fiords, volcanic *etc.*)

He further classified mobile coasts by the style and orientation of tectonic movement, as below Table 41 :

TABLE. IX 41 : CLASSIFICATION OF INITIAL COASTS

(source : Cotton, 1942)

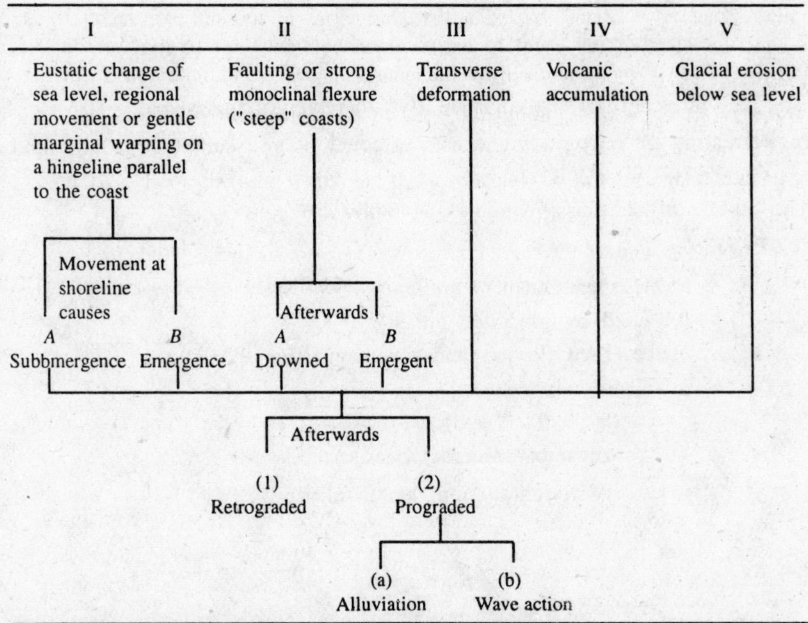

D. Velentin's Classifiation : In his 1952 classifiation of coasts, Valentin onsidered the stability as a function both of

(i) the fluctuations of climate and

(ii) diastrophic instability

in the Quaternary. An apparent stability of the sea level results when the rise of the sea-level is compensated by an equal amount. Valentin compounds the effects of

(i) sea-level fluctuations and

(ii) coastal dynamics

in devising a partly genetic and partly descriptive classification of coasts. The scheme stresses the recency of evolution of coasts by ignoring

(i) the initial conditions

(ii) historical aspects in morphogenesis of coasts.

He proposed two major classes -

a. advancing or prograding coasts

(i) building outward by-
fluvial
marine or
organic activity

as also those that have

(ii) risen from sea- bottoms due to tectonic causes and

b. retreating or retrograding coasts effected by the submergence of land due to rise of sea level.

The outline of his classification is as follows (Figure 39) :

A. *Retreating coasts*

a. Submerged coasts

(i) drowned by glaciated areas

(ii) submerged fluvial landscape consisting of
folded and
flat bedded and
volcanic (calderas) coasts

(b) Coasts of retrogradation due to cliffing .

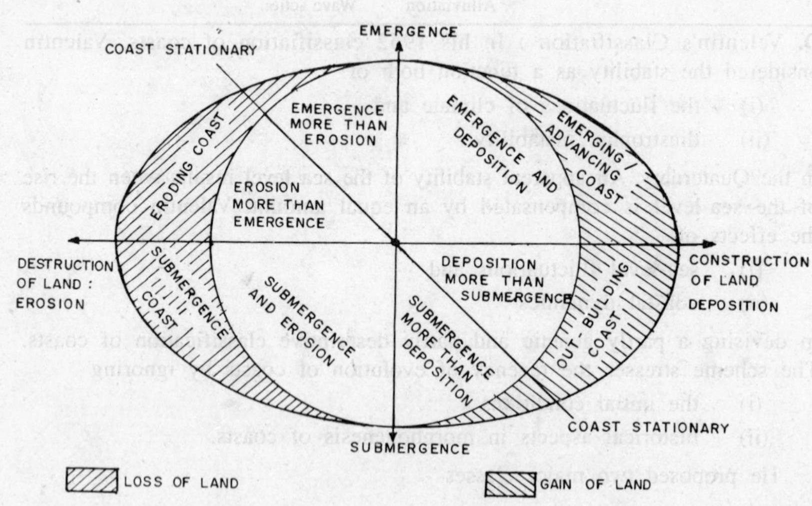

Figure 39 : Valentin's Classification of Coasts

Coasts can be classified into two types that (a) retreated landward due to submergence or erosion of land, and (b) have advanced either by emergence or progradation.

B. *Advancing Coasts*

a) Coasts of emergent submarine floor

b) Coasts of progradation

(i) Organic progradation

Coral coasts (*e.g.* Lachchadive coasts)

Mangrove coasts (*e.g.* Ganga, Godavari and Krishana deltas)

(ii) - Inorganic progradation

- coasts of marine deposition
- coasts of fluvial deposition
- volcanic coasts.

This classification is of great applied significance. It

i) takes note of the gain or loss of area of countries,

ii) draws attention to the problems of (a) protection or (b) reclamation in coastal areas,

iii) also focusses attention of the offshore topography which matters so much in coastal and international navigation,

iv) may be related to geopolitical problems particularly territorial waters,

v) may be of use in economic utilization of offshore submarine wealth.

Such classification suffers from over-simplicity. Increasing investigation of coastal processes and features have emphasized their complexity. A fully genetic classification should therefore include consideration of wider variety of factors; such as

a. *Geological factors* : These would include

(i) the nature of the rock type forming cliffs and shore platforms, and

(ii) the nature of the beach materials.

They would also include

(iii) the form of coastal relief, which is related closely to the immediate past history of the area in terms of

- tectonic movements
- isostatic movements
- subaerial processes acting on the coastal zone and
- changes in sea level.

b. *Marine processes* : The dominant marine processes are those associated with

(i) wave action leading to erosion

(ii) transport and

(iii) deposition

in addition

(iv) tidal environments and

(v) a range of coastal currents, also have effect.

c. *Climatic conditions* vary throughout the world, giving rise to

(i) different processes acting on coasts above the tidal levels and

(ii) different types of sediment input.

They are also more widely important, since they influence a number of other variables : storms zones are

(iii) the main zones of wave generation, past climatic changes have resulted in

(iv) changing sea-levels and

(v) affected in different parts of the world in different ways (*e.g.* glaciation of humid temperate areas);

and they also effect

(vi) types of organism living in the coastal zone.

d. *Biological factors* are important especially in the evolution of smaller scale features.

e. *Scale* : A classification must be related to features at a particular level of size, since genetic features change in emphasis. Thus an adequate classification of coasts related to origin should have a series of levels in terms of scale.

For a classification on the world scale it would seem that a climatic basis could be most satisfactory (Table IX. 41) but within these divisions there could be subdivisions in descending order -

(i) on global tectonic features

(ii) tidal environments

(iii) protection and transport conditions and

(iv) rock types

E. Assuming the surface of the earth is a series of lithospheric plates on which the continents and ocean basins are of only minor significance, Inman and Nordstrom (1971) recognized only four categories of coasts *viz.* those

TABLE IX. 41. : THE MAJOR WORLD COASTAL TYPES

(after Davis 1972)

World climatic zone	Processes and forms
1. Low latitude coasts	a) Reduced wave energy, but with consistent directions of wave approach. Few rocky coasts and weak cliff forms, horizontal shore platform development.
	b) Constructional forms dominant : Supply of sediment abundant-but rarely contains pebbles, massive-barrier-type forms with lithification of beaches. Coastal dunes seldom form.
	c) Biological constructional forms common : coral and algal reefs.
	d) Major subdivisions between coasts where it is hot and wet all year-where muds and mangrove are common and those where there is a major dry season-where there is sandy sediment and local salt flat development.
2. Middle Latitude coasts	a) High wave energy in the zone of frontal storms. Hard rock coasts are common with strong cliff development and sloping shore platforms. Beaches are less common than in land barrier development weak, but pebbles are common in such deposits and sand dunes are formed at the back of sandy beaches.
	b) Biological constructional effects are insignificant.
	c) Major subdivision between coasts in areas which were glaciated during the Quaternary, and those which were not. This affects sediments and sea level change.
3. High latitude coasts	a) Shores are frozen for a part of the year, and wave energy levels are low. Cliff forms are weak influenced by mass movement.
	b) Constructional forms are common, including barriers and include many pebbles. There are few coastal dunes.
	c) Biological constructional effects are insignificant.
	d) Major subdivision coasts where the annual freeze up is short-so that marine processes have some effect on the coast and those where it last almost the whole year. In the latter case there are few marine effects to be seen and ice action becomes dominant.

1. at the edge of active diverging plates (could be block faulted and rift coasts, such as Red sea coasts of Africa and Arabia)
2. at zones of plate covergence (include arc-systems such as Indonesia, Japan and mountain coasts, such as the west side of South America and north side of Mediterranean sea)

3. along major transform faults (such as southern California, Venezuela and north side of New Guinea)

4. on continents embedded within large plates, far from any zones of convergence, spreading or translatory plate motions typified by the Atlantic Ocean coasts, Indian Ocean coasts of India, Africa and Australia.

Coasts of mobile regions have been subject to

(i) Quaternary earth movements and

(ii) eustatic changes.

However, the tectonic components of coastal classification include such issues as

a) the direction and

b) rate of past and present vertical movement

c) the volcanic activity at the coast or offshore, and

d) the trend of deformational axes relative to the coast

F. According to the model now, lithosphere is spreading from ocean ridges, which is being consumed in the building of mountain and ocean trenches, providing a mechanism for the lateral movement of rigid plates. It is of fundamental significance to the evolution of coasts.

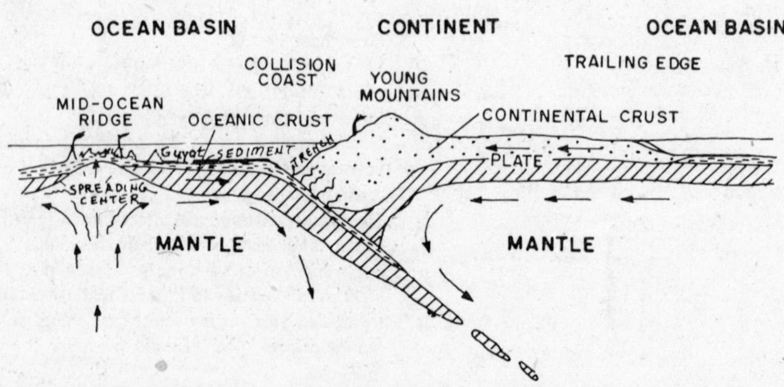

Figure 40 : Inman and Nordstrom's Classification of Coasts : Plate Tectonics, Foramation of Collision and Training edge Coasts

Coasts are classified into three size scales as first, second and third order landforms.

(i) The development of first order landforms features under the influence of plate movements such as coasts, continental shelf and water above the continental shelf of dimension 100 km or more in length. First order coasts may be classified broadly as
- collision coasts
- trailing edge coasts, and
- marginal sea coast.

(ii) Second order features are upto 100 meters long 10 meters wide and 1 meter high above the ocean floor developed on modification of first order coastal features and include features of glacial erosion, deltas major rivers, coastal sand dunes, barrier beaches and barrier lagoons.

(iii) The third order features are function of wave activity, and refer to processes and deposits and landforms in the nearshore environment such as beach face and longshore bars.

F. *Climato- Genetic Classification* by Davies (1980). The scheme takes into consideration the relationship between

i) wave energy and
ii) the morphology of coasts

and identifies the significance of

(i) historical inheritance and
(ii) geological aspects in the evolution of coasts.

These are categorised as

low latitude coasts	{ at 40^0 north and south
mid latitude coasts	{ separating line
high latitude coasts	the general separating limit is bounded by general shore of ice.

Low-latitude coasts : These coasts are characteristic of

(i) low wave energy in swell and trade-wind environment, and
(ii) coastal sediment is abundantly available in tropics,
(iii) few rocky coasts exist because of weak wave energy, cliff and wave.

cut platform development are inconspicuous. Some of these coasts are built by coral and algae. Variations in the coastal morphology are related to rainfall studded with

a) mangroves regime in areas of hot-wet tropical climate while
b) in the hot-dry environment of the tropics with abundant coastal sand dunes

b. Mid-latitude coasts : In contrast they are typically developed in a high energy wave environment. Owing to powerful

i) wave activity and

ii) marine abrasion

cliffs and associated forms are impressively developed in contrast to less conspicuous beaches. Rocky coasts are extensively developed. Imprints of the Pleistocene glacial epoch are well displayed in some sectors.

c. High latitude coasts : Situated as they are in a low-wave energy environment, the high altitude coasts remain frozen for a part of the year. Cliffs are developed greatly, influenced by large scale mass-wasting and beaches dominantly bear shingles and pebbles.

SEA LEVEL FLUCTUATIONS IN INDIA

The Bay of Bengal in the east and the Arabian Sea in the west of the Peninsular India came into existence sometime during the late Cretaceous or early-Tertiary period as prolongations of the original ocean after foundering and breaking up of Gondwanaland, whence the configuration of the Indian coast assumed almost the present form, although the shape underwent modification during the late Pleistocene and early Holocene fluctuations. Plenty of evidences are on record to demonstrate that marine transgressions have taken place, some even in historical times. The existence of submerged forest at depths of 6 to 12 meters below the present sea level on the eastern side of the island of Mumbai is the most well-known examples of subsidence or transgression in recent times. Marine archeologists have recently discovered the ancient city of Dwarka lying under shallow water of the coast of Saurashtra. Examples of eustatic changes of sea level are well documented by the shores of Kathiawar and the Gulf of Kutch, where the raised coral reefs and oyster beds extend in land upto the high level, indicating the elevation of land or fall in sea level. Likewise littoral concrete of pure reef or calcareous beds of well cemented shells and corals are also recorded at many places on the west coast, especially in the city of Mumbai. Similar raised oyster beds near Calcutta too are reported. There are such examples from other parts fo the Indian coast also.

The Rann of Kutch presents an example of repeated sea level falls giving rise to the now extensive marshland. Striking proofs of the eustatic change evidences are provided by coastal dunes along the northern edge islands in the Gulf with

(i) typical sea cliffs,

(ii) sea causes within the plains, and

(iii) raised beaches of littoral concrete on the islands fringes.

An overall fall of 26 meters is estimated in the sea level. The Luni River used to drain into the Arabian sea through the Korie estuary crossing the entire length of the Great Runn, before it because dry and the Western Banas flowed over the little Runn to discharge into the Gulf of Kutch.

The two 'Swatches of No Ground' on the continental shelf fronting the Ganga and Sindhu deltas are thought by some to be the submerged portion of mighty rivers. The shape and size of some islands in front of the Ganga delta changes with the sea level fluctuations and instances of new islands emerging and old ones disappearing in course of time are known. The emergence of New Moore island in the Ganga delta emerged only during the last few years, is an illustrative example.

The Andaman and Nicobar Islands also afford evidence of eustatic rise and fall of the sea level. Raised coral beaches, marine conglomerates and sea caves in uplifted platforms much above the high water mark provide proof of this.

Yet another evidence of submergence is the existence of drowned valleys of several coastal rivers flowing transverse to the coasts.

Some islands are believed to be mesas and buttes on coastal plains were isolated to form islands off the western shore along the Arabian Sea because of submergence of the low-lying areas around them.

The land bridge and shallow platform of Palk strait connecting the southern tip of India with Sri Lanka further shows the fall in sea level around the Peninsula.

CLASSIFICATION OF INDIAN COASTS

Thus far, the Indian coasts have not been explored enough because of which the information data bank of their descriptive and genetic aspects is rather poor. Ahmad (1972) proposed a descriptive classification of the Indian coasts, of seven types with their morphological attributes.

1. *Anadaman and Nicobar type :*
   (i) The coast is rugged
   (ii) moderately indented in the northern group of island, and
   (iii) extensively built of coral

The islands probably represents promontories of the Arakan-Yoma system having escaped Pleistocene submergence.

2. *Deltaic coasts type* : These comprise the major delta of the Ganga, Mahanadi, Godavari, Krishna and Cauvery rivers. In general the Indian coasts are prograding but for that of the Ganga which has remained static or prograded over the last 200 years is debatable. However, the

morphology of the deltaic coasts is subject to modification by strong monsoonic currents and cyclonic storms.

3. *Depositional Plain type* : The eastward sloping coastal plain from the Bay of Bengal to Cape Comorin, except the beaches affected by deltaic deposition, is a product largely of fluvial deposition.

4. *Malabar type :*
   (i) with high relief and irregular terrain,
   (ii) it is a coast of submergence and
   (iii) has undergone recession in the recent times.

5. *Maharashtra type* :
   (i) The coast is submergent with an irregular outline having many steep rocky prominences and embayments.
   (ii) it is marked by
      - dissected terrains
      - terrestrial and marine erosion with drowned valleys, rias and cliffs, spits and barriers.

   The submergence has been caused by faulting.

6. *Gujarat type :*
   (i) The coasts with shallow tectonic depressions
   (ii) marked by gully erosion fronted by estuarine shore mud flats and marshes suggests drowning,
   (iii) besides wave and tidal activity, in protected waters.

7. *Kathiawar coast type :*
   (i) dominated by basalts and its faulted nature
   (ii) the coast is compound one and represents features of both, emergence suggested by barriers and coastal dunes and, submergence indicated by cliffs, estuaries and inlets

CHAPTER X

LAKES

Lakes are inland bodies of standing-water-filling depressions having no direct connection with the sea. They occupy 1.8 percent of the earth's surface. Their different aspects-formation of lakes, physico-chemical conditions and the organisms inhibiting them constitute the subject matter of studies in *limnology* (from the Greek *limne,* lake marsh) - a special branch of hydrology. Conditions necessary for their formation are :

a) basin-like depressions, and

b) sufficient water at least partly to fill them.

The latter *i.e.,* the water again may be either

i) fresh or

ii) salt water

Almost all freshwater lakes always have outlets. The size of the lakes vary from the smallest pond upto (i) Lake Superior, world's largest freshwater lake, occupying nearly 7,75,000 sq km of area, and (ii) Caspian "Sea" the largest saline lake with an area of 436,400 sq km. The great lakes are situated above sea level but few like "Salton Sea" in California and "Dead Sea" of Palestine are below sea level at- 78 meters and 400 meters respectively. Lakes are known to vary in depth from a few centimeter to a maximum of 1870 meters in Lake Baikal, Siberia. Lake Titicaca located in Peru and Bolivia is usually considered to be the highest lake in the world, situated at an altitude of 4000 meters, except for some lakes in the Himalaya which are certainly at still higher altitude such as Devtal lake in Garhwal Himalaya, Mansarovar and Gaurikund *etc.* which are relatively smaller. The Chad lake in Sudan is only about 2 to 6 meters deep but no other lake in the world is marked by such great seasonal variation in area. In dry season it is only 25,000 sq km but swells to 12,50,000 sq km in area in the rainy reason.

Lakes occur in different parts of the world, and several of them are direct results of glaciation. This seems to be the reason for more lakes located at higher latitudes and altitudes. In humid regions, the lakes usually have sufficient water to cause them to overflow the lowest parts of their basin rims. As such, these lakes are freshwater. In arid regions lakes usually do not have outlets, because of both

i) the scanty volume of water and

ii) the high rate of evaporation.

Many depressions in dry regions contain no water at all, while others called *playas,* hold water only temporarily after rains. Arid region lakes with no outlets, almost invariably contain salt (or alkaline) water.

Looking into the time framework, but for some possibly the largest and deepest ones, the lakes, are of shorter duration in relation to the history of the earth. This is so as by nature, they are just temporary obstructions to drainage into which river borne sediments and in some favourable places peat beds and bog iron ores are deposited. Thus they are (i) soon destroyed by processes, such as sedimentation, crustal deformation *etc.* and (ii) are sites for future iron ore deposits (*c.f.* Lake Superior iron ore deposits) and for coal seams (*c.f.* those of the Gondwana coalfields)

Some general statistics of notable lakes of the world have been tabulated :

TABLE X. 42 : GENERAL STATISTICS OF NOTABLE LAKES OF THE WORLD
(source ?)

Lake Name	Area, thousand *in square kilometers*	Mean absolute elevation above sea-level, *in meters*	Greatest depth, *in meters*
Caspian (Sea)	395.0	- 28.0	980
Aral	65.5	53	69
Baikal	30.5	455	1,741
Geneva(Switzerland)	0.6	376	309
Koko Nor (Asia)	4.2	3,205	38
Victoria (Africa)	69.4	1,134	80
Tanganiyika(Africa)	32.9	773	1,435
Superior (N. America)	82.4	183	393
Michigan(N. America)	58.0	177	281
Erie (N. America)	25.7	174	64
Ontario (N. America)	19.5	75	237
Chad	25,000	280	273
Huron	55,800	193	232
Nyasa	35,500	500	732
Great Bear	28,000	130	90
Titicaca	8,150	4165	232
Dead (sea)	900	-428	432
Crater	62	2058	665

TABLE X 43 : NOTABLE LAKES OF CENTRAL ASIA, TIBET AND HIMALAYA
(source ?)

Region	Name	Area *in sq km*	Altitued above sea level *in meter*	Maximum length *in km*
Tianshan	Issikkul	5,000	1,766	184.0
	Chadirkul	200	3,732	22.4
Tarim Basin	Lob Nor (north)	375	865	104.0
	Lob Nor (south)	550	865	41.6
Pamir Region	Greak Karakul	350	4,476	22.4
	Sirikul	75	4,465	22.4
Tibetan	Tso-moriri	115	5,000	25.6
	Pangong	300	4,666	64.0
Kashmir	Wular	110	1,729	17.6
	Dal	20	1,733	6.4
Kumaon	Nainital	a group of small lakes	2,133	
Nepal	Palgu Tso	100	5,000	16.0
Himalaya	Tso Motretung	100	4,666	24.0
Assam Himalaya	Tso Tign	125	5,166	22.4
Tibet	Montealm	750	5,404	76.8
	Mansarovar	330	4,966	25.6
	Kokonor	40,900	3,566	107.2

Economic Utility of lakes :

The lakes are in the service of mankind and for his needs in several ways :

a) *Scenic beauty* - Most lakes in Kashmir, Switzerland and N. America are well known for their scenic charming beauty to the visitor as well as for food, sport and means of communication on which much trade has been built up

b) *Climatic aspect* - They are important modifiers of climate by way of (i) increasing humidity and (ii) decreasing seasonal range of temperature rendering the winters less severe and summers less warm

c) *Water supply* - They act as reservoirs to surplus waters in stream particularly in flood times and thus restrain flood occurrence(s). Such water holdings promise water supply to urban population for domestic purposes

d) *Sediment trapping* - They arrest flood waters sediment brought into the basin by trapping and allow them to settle down

e) *Future sites of economic interests* - They hold promise or prospect of locating bog iron, coal and salts in their basin.

CLASSIFICATION OF LAKE BASINS

Lakes can be classified in variety of ways. The following classification is simple and comprehensive enough to include practically the various possible types met in nature and account for the development of lake depressions.

I. *Basins formed by crustal movements*

a) Rift valley or fault block - These are great linear depressions formed by down dropping of local parts of the earth's crust

b) Broad downwarp of river basin - Gentle subsidence of a segment of a river system results in flooding the area and development of a lake

c) Sinking of valley alluvial fill due to minor faulting (earthquake activity).

d) Upwarp of an isolating arm of the sea. Long inland extensions of the sea get isolated by gradual rise of the intervening area.

II. *Basins formed by streams*

a) Meander cut off , ox-bow type - Basins may be formed by meandering rivers. They are called ox-bow, horse-shoe or cut off lakes such as those of Kashmir valley *e.g.* Anchar, Manasbal *etc.*

b) Valley border lakes located between valley walls and natural levees *i.e.,* ridges formed by flood-time deposits in flood plains

c) Stream dammed - A river may be blocked by a landslide across its valley. The Bundelkhand lakes as also Gohana and Birahi Tal lakes in Garhwal are examples. The Gohana Lake was formed by a landslide across a tributary to the Ganga in 1893.

d) Stream quarried - These lakes occur at the foot of the hill as a result of the impact erosion of the waterfall on a river course. River erosion may also form small lakes in pot-holes developed by whirling action of stones in river eddies. Obviously such lakes are surfaced after the water-falls or rivers have died away.

III. *Basins of Marine origin*

a) Lagoons developed by barrier beaches - In coastal regions seawaves, sometime build bars or spits (structures produced due to arrangement of sediments parallel to the sea- shore is called bars or spits) across the coast or the mouth of a river. These structures detach the back water from the main water body and thus convert the locked water bodies into small lagoons which are small lakes. They are formed in coastal regions. The kayals of Kerala, the Pulicat lake of Tamil Nadu, the Chilka lake in Orissa are examples of this type.

The Caspian Sea, world's largest saline lake, was once a part of the sea and then cut off.

b) Irregularities in the newly uplifted seafloor.

IV. *Dissolution Basins*

Basins are sometimes formed in sinks, surface depressions, chemically etched by the underground solution on rocks like limestone and rock salt. The outlets of the basins may be clogged with impermeable rocks and turn these sinks into lakes as observed on the top of the Salt Range, Pakistan and some of the Kumaon lakes.

V. *Basins of Glacial Origin*

Majority of the lakes are of glacial origin, depressions formed by -

a. Scooping - Lake basins are formed by glacial erosion, as Pir Panjal lakes are.

b. Glacial debris valleys blocking - Sometimes morainic waters laid down across river courses are piled up obstructing the flow to form lakes. Some of the Kumaon lakes are of this nature.

c. Irregular deposition of glacial debris. These may be due to enclosure of waters within moraines.

VI. *Basins formed by Wind*

a. Shallow depressions isolated by dunes - Three or four salt lakes of Rajasthan (other than Puskar, Debari *etc*), typified by the Sambhar lake, represent depressions in an arid country. When full during the monsoon the Sambhar lake, a little over 1 meter in depth, spreads areally over 225 sq km The lakes receive inland drainage and their outlet have been obliterated or blocked by advancing desert sands. Their salinity is attributed to (i) excessive evaporation as also to (ii) the dissolution of particles of salt and saline substances by scanty rainwaters brought alongwith blown wind into Rajasthan from the Rann of Cutch by the monsoon winds. These salts are washed into depressions and get concentrated. Holland and Christie estimated about 13,000 tons of salt are borne like this annually. So collected salt and saline concentrate is evaporated during the summer and the saline efflorescene in the bed is collected and purified. It is observed that the mud of the lake bed is also highly saline upto considerable depth. Other causes may be the former connection with Arabian sea, marine transgression, chemical solution and precipitation in these lake basins. Several small lakes occurring in depressions amidst sand hills of Sind and Gujarat referred to as *dhands* (these are alkaline and saline lakes), contain different sodium salts in their waters.

b. Basins produced by wind deflation - Basins are also produced by wind deflation, *i.e.* blowing away of fine particles particularly from the surface in desert regions and this produces basin. They occasionally receive water and are fed only after heavy showers of rains.

VII. Basins of Volcanic Origin

a. Valleys dammed by lava flows - At times lava flows may block a river when oriented across the valley.

b. Lakes occupying extinct craters or calderas - The Lonar lake is a deep crater-like depression 100 meters deep, 1.6 km in diameter in the Deccan Traps of the Buldana district, Maharashtra. It contains at the bottom a shallow lake of saline water. In the hot weather, the water in the lake practically disappears and yields encrustations of Na_2CO_3 with some NaC1. A subsidence of a tract along circular fractures due to emptying of lava and gases from below at a later Deccan Trap activity is also postulated as an alternative explanation for depression- genesis.

THE ORIGIN OF LAKE WATER

Another, equally important aspect is the origin of lake water. Besides, some grouped as relict type of lakes which owe their water to their earlier connection with World Ocean and later subsequently were out off, others are atmospheric precipitation fill up in relief depressions. Such a closed sea-lake is the Caspian which was earlier connected by a strait with the sea of Azov through a valley. Hydrologically (supply and discharge) the lakes fall into the following two principal groups :

a. Drainless lakes fed by river or atmospheric precipitation which have no drainage and they lose water only through evaporation (examples-Caspian, Aral, Issy-kul *etc.*)

b. Running water lakes amongst which are recognized those -

1. fed by rivers, which loose water bothways by

(i) evaporation, and

(ii) also discharging the water into other basins or rivers by drainage; surface or underground, and

2. with intermittent drainage which takes place during high water regimes.

GEOLOGICAL FUNCTIONS OF LAKES

Some of the most important geological functions of lakes are the following :

a) Lakes especially with no outlets act as perfect settling basins at the floor of which the sediments transported by streams are deposited. Likewise even lakes with outlets, in as much as they too provide the dropouts to the settling filter as reflected by the remarkable clear water on the emergence from Lake Erie of the mighty Niagara River, which drains the four upper Great Lakes, indicating that practically all the sediments brought into the four upper lakes are left back as fallouts. As such outlet streams tend to be relatively much freer with grinding tools and are therefore less effective as erosion agents. Large amount of material accumulate at lake bottoms, of course, much less than that on the ocean floor and the process continues until the basin is wholly filled with the brought in sediments when the lake ceases to exist.

b. Lake basins are storage reservoirs for surface drainage, which regulate the stream flow. Its role is well perceived when it tends to flow out of them, furnishing a check to prevent or dilute, otherwise, destructive floods and thus, reduce the rate of erosion too. Streams are most effective in flood times with additional load and voluminous water.

c. Lakes tend to be an effective environment agent too, in as much as larger lakes or numerous smaller ones together moderate the regional temperature as summer coolers enhancing precipitation possibilities and winters milder.

d. Lakes support abundant, both in terms of amount and varieties of plants and shell-bearing animals. Their shells, skeletons or partly decayed remnants (i) accumulate in the lake basins, thus helping to fill them. Such material later (ii) found in fossil form in old lake deposits, often yield valuable clues which help to discern the geological informatics about it. The residuum of the biogenic materials is one of the principal substances accumulating on the lake bottom periodically-may be annually or seasonally. Such bottom include :

i) bodies of floating organism,

ii) plant and animal remains of the littoral zone,

iii) wind-blown materials both organic and inorganic,

iv) silt, clay or coarser sediment introduced by streams, glaciers, from slope wash or eroded form shore, and

v) marl (largely $CaCO_3$) produced by plants and animals, and mixed with clays, and tend to modify the lake basin geometry and induce the shallowness of water, *i.e.* reduction in water depth.

e. Lake margins are eroded and the area enlarged considerably by wave cutting and the depth reduced consequent upon the spread over of the eroded material on the basin floor.

f. Lakes with no outlets in arid regions on evaporation, precipitate various salts held in solution which may be economical and suitably extracted.

INDIAN LAKES

A. Lakes of Peninsular India

Lakes of this part of the country may be grouped as below :

a) Coastal lakes - These lakes are developed by enclosing lagoonal sea water with bars or spits built by sea waves close to the coasts. Examples of such instances are :

(i) Lake Chilka - in the Ganjam district of Orissa

(ii) Pulicat (or Pullicat) Lake - 4.5 to 16 km (wide) x 2 meters (deep) in Tamil Nadu.

(iii) Kayals - in Kerala State and along Malabar coast.

(iv) Nal - formerly it was along the sea, and lies between Gujarat and Kathiawar. Its area is 90 sq km and the water contains salts in solution.

b) Lower lake of Buldana district, Maharashtra 1.6 km (diameter) x 100 meters (depth) occupies an explosion crater. It is dry during the summer.

c) Lakes of Rajasthan - These are the products of inland drainage basins in arid region. Their basins are being silted up, water levels lowering down and because of constant evaporation they are becoming rich in saline matter. The Sambhar Salt Lake is the largest of such lakes. It lies between Jodhpur and Jaipur cities and is spread over about 225 sq km, at an altitude of 65 meters above sea level with maximum length of 30 km. Pachpedra, Didwana, Lunkraransar are other salt-water lakes of Rajasthan

d) Dhands - these are alkaline or saline lakes of Sind (Pakistan) and

Western Rajasthan containing carbonates, chloride and sulphate of sodium. They are believed to be of aeolian origin.

B. Lakes of Extra-Peninsular India

a. Lakes of Kashmir - Noted among the lakes of Kashmir are Pangkong, Tsomoriri, Wular and Dal.

   i) Wular and Dal are fresh-water lakes and are considered to be hollows in alluvium of the Jhelum river.

   ii) The Pongkong lake (in Ladakh, 65 km long, 3 to 7 km broad situated at an altitude of 4600 meters) and the Tsomoriri (in Rupsu, 24 km long, 3 to 8 km broad situated at an altitude of 5,000 meter.) owe their origin to the damming up of the main valley by the sediments brought by tributary rivers. Smaller lakes in Kashmir and neighbouring areas are called *torns.*

b. Lakes of Kumaon - Noted among these are Nainital, Bhim Tal and Khewan Tal. They are variously believed to be due to (i) blocking of streams (ii) crustal movements, and (iii) bed rock solution.

C. Lakes of Tibet (beyond the boundaries of India)

a. The largest lake in this region is the Kokonor with an area of 4,000 sq km

b. other noted lakes are

   Mansarovar (500 sq km in area) - fresh-water lake

   Rakchas Tal (350 sq km in area) - fresh-water lake

   and

   Gunchu Tso (24 to 48 km in length) - saline lake.

Tibetan lakes have no outlets and receive inland drainage. Their water is becoming more and more saline every year on being loaded with additional stream discharge. They are also shrinking in area as a result of increasing desiccation due to the obstruction of the monsoon wind by the Himalaya and consequent absence of rainfall. Borax is, however, an important constituent of the salts of these Tibetan lakes.

The origin of these lakes may be due to piling of sediments across river valleys either by (i) tributary rivers or (ii) by glaciers. Instances of

(i) glacially scooped bed rock occupied

(ii) upliftment of streams by crustal disturbance resulting from upheaval of Himalaya,

occupying lakes are also forthcoming from the region.

LIFE HISTORY OF LAKES

All lakes are generally youthful with the exception of fluvial lakes that occur on flood-plains, deltas and valley flats. They are also indicative of a youthful topography. The great majority of all lakes on the earth today, however, are of glacial origin formed during Pleistocene or Recent time, and much time has not yet elapsed to fill their basins up with sediments brought or vegetation growing therein or for streams to drain them. Lakes of arid regions may be localized in topographic environment of any age, yet youthful types are more common.

Processes Modifying Lake Basins : The form of lake basin depends upon the forces that operate at the time of its formation. Whatever the original outline may be, it does not last long for gradational process(es) are constantly at work, modifying the lake shores, filling the basin with sediments or drainage. Sooner or later most lakes are destroyed.

Even the largest and deepest existing lakes would ultimately disappear with advancing time which may even be rather longish depending on (i) their own characteristics, and (ii) extraneous attending conditions.

A. *Lowering of Outlets* : Outflowing streams from lakes tend to deepen their outlets, thus draining lake basins whose bottoms are above the level of the stream-cut valley.

B. *Filling of Basins :* Lake basins are filled in various ways and consequent thereto water level therein rises leading to greater outflow which aids in deepening the outlet. Thus filling is complementary to lowering the outlet in draining the lake.

   a. *Filling by Sediment* - Streams sediments collected through its transportational history is deposited in the lake basin. The sedimentation amount varies greatly depending upon the river and terrain characteristics through streams flow. The lake filling this way is well demonstrated by the present day deltas that are being built by many (not all, of course) streams. Slopewash from the lake borders further add to the sediment supply in many lakes.

   b. *Filling by Wave Action* - Waves cut the shores of large lakes and most small lakes much of the time. Thus loosened debris is washed into the basin, gradually shallowing it.

   c. *Wind Filling* - In arid regions large amounts of dust and sand are blown into lakes by the wind. Certain lakes, both ancient and modern, contain volacanic dust, in some cases in incredibly great quantities, thereof blown into the basins.

d. *Filling by Vegetation* - Vegetation accumulates on the shallow borders of many lakes which gradually grows inwards into peat bogs, marshes and muskeys particularly in glaciated terrain. In many places the filling has been complete and new land formed. Also some plants, especially algae, help to precipitate calcium carbonate from solution.

e. *Filling by Animals* - Numerous lime - secreting shell animals live in lake waters which accumulate on the lake floor on their death helping to fill the basin.

f. *Filling by Landslides, Avalanches etc.* - Much materials-heaps of boulders, split off by frost action from adjacent cliffs are brought in the lakes in mountains and plateaus at the foot of steep cliffs by landslides, mudflows, avalanches *etc.*

C. *Obliteration of Basin by Glaciers* : Advancing glaciers are capable to remove morainal dams and destroy lake basins. This evidently has happened in several cases in the past and is of common occurrence in regions of modern glaciers.

D. *Warping of Basins* : The unmistakable evidences of earth movements has been recorded along shorelines and terraces of many ancient lakes. The entire lake may disappear by sufficient upwarping.

E. *Destruction by Evaporation* : This is yet another very important process in arid regions where lake destruction is due to excessive evaporation.

LAKE EROSION AND DEPOSITION

SEA AND LAKE EROSION AND DEPOSITION COMPARED : There is strong commonality between lake erosion and marine erosion but for the effects produced, because of the variance in the erosive power of the two geomorphic agencies. Lake erosion is obviously less conspicuous in relation to the mighty erosion by the sea because of gigantic volume of available water. In larger lakes, however, the resulting erosional features, such as wave-cut terraces, cliffs, caves, coves, stacks and arches, draw a good parallel with similar features produced by sea erosion.

Like-wise, lake deposition too, bears, a strong semblance with similar shallow marine deposits such as beaches, barriers, bars, deltas and wave-built terraces. However, practically there is no equivalence of deep-sea deposits to lake deposits except some accumulation of shells of certain tiny plants (diatoms). The lake floor is entirely composed of terrestrial sediments brought by the streams. Delta growth in lakes is particularly strong because of

(i) absence of very appreciable tides, and

(ii) very powerful wave and undertow action in the usual lakes (except in some very large lakes)

Strong similarity in the cycles of shoreline developments between lakes and the sea is also remarkable.

A characteristic feature of some fresh lakes is the deposition of light gray mixture of carbonate of lime (often in the form of shells) and clay, termed as *marl*. It often occurs in 1.5 to 7 meters thick beds. Yet another characteristic of several lakes and swamps in humid-climate regions is the inward growth and subsequent decay of vegetal matter. These materials are often further decomposed to form rather extensive so-called *peat* beds.

LACUSTRINE DEPOSITS

Lake deposits commonly referred to as lacustrine deposits include deposits which may be grouped according to their nature into three types as again in the case of marine sediments :

a) Mechanical formed (i) chiefly by abundant detritus transported in by streams, and also (ii) those obtained as a result of lake abrasion;

b) Organogenic, formed as a result of accumulation of various organisms on lake floors and

c) Chemical, derived on precipitation of various dissolved salts from lake waters.

Mechanical deposition is most characteristic of most freshwater lakes and when required concentration is attained, chemical precipitation assumes its significance

A. *Mechanical Deposits*

Characteristically coarser sediments like gravels, pebbles *etc.* are laid down along the lake periphery and finer sediments such as fine sandstones and shales grade down towards inner central part. Talchir deposits of Lower Gondwana age is typical example of this peculiarity. The gradation of coarse and fine sediments each other dovetail and overlap because of the inflow of coarse debris, that envelopes the fine sediments below strong storms, or by flooded streams especially in the close proximity of the border. There is a distance limit to which gravel and sand are carried in relatively standing water body. Thus in larger deep lakes, much of the greater part of the bottom is covered with the finest clay and marl. The lake currents form accumulation of the coarse substances and develop shoals, spits, and the like.

B. *Organogenic Deposits*

These are not as important in large lakes, as they are in small ones. Water plants grow inwards from shores and bottom, and thus the lake shrinks and chokes. The restricted environment converts it into a bog in which peat or partially decomposed carbonized matters are accumulated. *Diatoms,* microplants live in the sea and in brackish water also. They secrete beautiful tests or shells of silica, multiply fast and spread over lake bottoms as relatively accessible deposits of a fine white powder, variously called *Tripoli* or *polishing powder*, but are made of the shells of pond snails and other freshwater molluscs of considerable thickness and areal extent. These are called *shell marl*, when loose and incoherent and the lower part of it is without any organic structure being disintegrated by the water. Shell marls often occur under peat bogs and suggest that the bog was originally a lake, which subsequently shrunk and ultimately got filled up by vegetation growth. Bones of dead animals, too, are often met in the marl. These marls on cementation become a firm rock to form a freshwater limestone.

C. *Chemical Deposits*

a) In fresh water lakes - Chemical deposits (various salts) are seldom of importance in fresh lakes. Except some calcium carbonate precipitated and more abundant iron ore limonite ($2Fe_2O_3.3H_2O$) of chemical origin in small freshwater lakes, chemical deposits in larger freshwater lakes are rather uncommon. The iron is brought dissolved in water as ferrous carbonate ($FeCO_3$) containing CO_2 (*c.f.* $CaCO_3$). The $FeCO_3$ begins to precipitate out. But there is a limit to the solution soon on attaining its concentration or saturation in locked lake water, which is subsequently oxidized and precipitated in the form of Fe_2O_3 and $2Fe_2O_3.3H_2O$ (limonite) partly by

   (i) the action of bacteria, called *iron bacteria*, and

   (ii) its contact with air.

   However, at the lake bottom, due to limitation of oxygen availability, iron is precipitated as $FeCO_3$ itself. The extent of bacterial aid in this process is yet hard to estimate.

   Such valuable iron ores are called Bog Iron Ores deposited in lakes.

b) In Salt and Saline Lakes - A saline lake is rather a more comprehensive term, which includes all salts in solution, such as common salt, caustic soda, borax, potash, alum *etc*., in contrast to salt lakes in which common salt (NaC1) constitutes the principal ingradient of the dissolved materials. Saline lakes of all kinds are confined to arid regions where -

   (i) rainfall is little, generally below 50 cm annually and

   (ii) evaporation high.

Such lakes may be formed in either of the two ways :

i) separation of salt water bodies from the sea, or

ii) the long continued concentration of water in basins with no outlets in regions where surplus water is removed by evaporation.

The size of the lake is a function of

- the balance between influx and evaporation, and
- the rise and fall of lake corresponding to high and low water in the inflowing tributaries level and strong evaporation.

Chemical precipitates of lakes, probably the most interesting, characteristic, are important of all lake deposits. Amongst the most abundant of many minerals in the water of salt lake are common salt (halite), sulphate of lime (gypsum), sulphate of soda (glauber salt), sulphate of magnesia (epsom salt), chloride of magnesia and carbonate of lime, and their concentration in the solution goes on increasing with the increase of evaporàtion that removes the surplus water, shrinks the lakes, and with continued evaporation to dryness, all minerals in solution are deposited. Their nature and relative abundance largely depend upon the nature of parent rocks at the periphery of the lake basins and the terrain from which the inflowing materials are brought into the lakes by streams, but for those found in some relic lakes.

The salt deposition is an extremely complex process, and its complexity multiplies with the increase of number of ingradients present in the solution. In general, however, precipitation follows the simple rule of inverse solubility, that is, the least soluble salts are earlier to precipitate, others follow the que and ultimately the most soluble salt takes its turn as the last. No chemical reaction of any significance seems to take place in these lakes as practically all the salts, are the same as brought into the lake, is in very dilute form by springs and streams and laid down on evaporation of saturated solutions. A mixture of chemical and mechanical deposits clearly indicates the slow precipitation rate, but with rapid precipitation thick and almost pure salt masses settle down in their inverse solubility order of the concentration processes. With the entry of flood discharges during a rainy season or a series of rainy seasons in lake waters slow down precipitation rate from the solution, and instead, land-derived sediments accumulate on the lake floor, perhaps, faster. Changes in temperature also affect their solubility. It is for these reasons alternating layers of chemical and mechanical nature are often met in old lake deposits. Thus lacustrine deposits are quite sensitive to the prevailing environmental conditions. Calcium carbonate ($CaCO_3$) and ferric oxide (Fe_2O_3) are the earliest and in moderately saline lakes, this is about the limit of precipitation of dissolved materials.

The next substance to be precipitated on further concentration is gypsum or calcium sulphate ($CaSO_4 . 2H_2O$) if the water is cold which is not readily soluble, yet much more so than calcium carbonate. Anhydrite ($CaSO_4$) is formed from hot water, and the two minerals may occur in succession. Once calcium sulphate is separated out in any form as above, further concentration of salt solution is necessary for subsequent precipitation. After that point has been reached the process continues with the progress of concentration. At an advanced stage the salt is mingled with magnesium sulphate ($MgSO_4$) - should that be present in water.

The "soluble chlorides" of potassium, magnesium and calcium (KCl, $MgCl_2$, $CaCl_2$), readily dissolve in the water they take from the air, in an ordinary room, are always present in the brine; their solution forms the 'mother-liquor' in salt-making which are drained off the salt crystals, and washed, to liberate them of all traces of these bitter compounds. They are deposited only when the water is evaporated to dryness. They form various compound salts with one another and with the sodium chloride, the latter practically concludes the precipitation process from the salt solution. Because of their potassium content, these compound salts are of great economic significance.

Besides as above, there are several smaller lakes which hold sodium carbonate (Na_2CO_3) in solution and in some cases the solution is so concentrated as to cause precipitation. Borax or sodium tetraborate ($Na_2B_4Oa_7$. $1OH_2$ O) is contained in some other lake waters which is deposited in very dry climates.

Gypsum and rock salt are the most abundant and common of chemical deposits made of salt lakes and they frequently occur in association of each other. Concentration may not have proceeded far enough to precipitate common salt. Gypsum alone would be deposited only when vast bodies of rock salt may be preserved, lying below the level of circulating waters as indicated by brine springs suggestive of some salt bodies which are in the process of removal by solution action.

A salt-body is a definite proof of aridity of climate at the time and place of deposition. It can be very safely and confidently asserted that the salt has been precipitated from solution in water through evaporation in arid regions.

BOG LAKES

General Characteristics

Some lakes or ponds show special characteristics as below:

a. small in dimension, generally only a few hundred meters in diameter, however, in literature even some upto 3 km too, are reported

b. characterized by semifloating of such vegetal mass of which a mat extends inward towards open water from the shoreward peripheral margins, to develop a false bottom,

c. characterized by being largely peat, and

d. extremely low salinity waters which range from alkaline to rather strong acidic.

Such lakes are termed "*bog*" and obviously it is a loose conotation with several possibilities and best developed in regions of abundant precipitation in freezing temperature practically round the year. These are fresh water lakes of glacial regions whose topography remained uneffected by stream action. These special types of lakes are called '*bog*', '*moor*' or '*muskeg*'.

The False Bottom

Of all the features characterizing the bog, the false bottom is the most typical. It spreads over from the present shore into the open water where it ends with a zone floating hydrophytes or water lillies. Towards the land margin its boundary is really not sharp. However, the association of peat soil upon which grows a thick mass of shrubs such as coniferous trees *etc.*, help in demarcating its limit. Locally, tamarack and black spruce and some pines grow upon this vegetation mat which during spring months is thoroughly saturated with water but in the late summer is dry and hard. Lakeward there is an increase in (i) saturation and (ii) a general spongy quality to the saturated mat mass below which is a layer of water which extends downward to the true bottom of the lake may be meters below. The observer is also aware of clearly marked zonation of vegetation from shoreward. The extension of the surface is discernible inwards upto which the mat growth is observed. A study of this plant distribution has revealed the lakeward progressive extension of the false bottom until the entire true lake bottom is totally covered leaving no evidence and no indication of the underlying water. In time, likewise, additional mat of false bottom go on developing on anchoring of further plant growth, which increase the thickness of the mat until the false bottom approaches the actual bottom, over which ultimately remains a thick mass of water-soaked peat suggestive of the lake existence once upon a time.

Biologic conditions

Specific biologic characteristics that help to distinguish the bog from most other lakes are :

(i) abundant micro-organisms and snails inhabiting the outer limits of the vegetation and poorly developed benthonic fauna.

(ii) well-developed peripheral flora along the lake shore producing the vegetal mat

(iii) plant debris decomposition to form peat.

Water

In "bog" the water, of strong tea-like colour, underlies beneath the floating mat of vegetation. The water is clear and transparent with variable acidity and the colour is because of the vegetal stain imparted to it during peat-forming process. Streams draining these lakes are typically dark waters which is often said to be due to high iron content, incorrect of course. These waters have been found to analyse rather in mineral matter but, for in some instances calcium carbonate, where the finally comminuted carbonate is present in the surrounding glacial sediments.

Temperature stratification is observed in the deeper and larger bog lakes. But in shallow bodies very marked temperature variations are recorded beneath the false bottom both during a single season and for succeeding years. The cause underlying the temperature variations is not yet reasoned out. However, the possibility of periodic ejections of water from the extensive vegetation mat may not be totally insensitive to temperature variations.

SWAMPS AND PEAT

Marshy grounds saturated with water are called *swamps* of which the Sundarbans, in Ganga delta in West Bengal, is a notable example. The swamps abound four regions -

(i) floodplains
(ii) deltaic regions
(iii) recently glaciated areas
(iv) coastal tracts.

They assume significance because they are the sites of peat formation. Wax, pitch, creolin, phenols *etc.*, are some of the products obtained on peat processing at special chemical plants. In addition, such areas also supply at times *diatomaceous earth* from the shells of minute plants called *diatoms*. This diatomacous earth is used in the manufacture of dynamite and as an abrasive. Swamps are very fertile, particularly when dried

Based on

(i) the supply system,
(ii) vegetation and
(iii) topography

swamps are classifiable broadly into three types

a. *Lowland swamps* : These swamps develop in
   (i) depressions and
   (ii) their surface is flat or concave

Such swamps receive waters besides

(i) from meteoric waters, also from
(ii) surface runoff (rivers) and
(iii) subsurface (underground)

and because of

(i) low relief runoff,

(ii) interaction with atmospheric, hydropheric and lithopheric constitutes

is free, which enriches it with mineral-nutrient-elements they derive through course of precipitation and flow, a favourable condition which stimulates the autotrophic vegetation (like birch, alder or even green mosses) to develop. These vegetation on burial give rise to peats generally of high ash content but low in calorific value.

b. *Upland moors* : Generally they occur at higher altitude at watersheds and thus surface they make is convex whereto the meteoric precipitate, generally poor in mineral salts, is the main source of water supply. Instead, autotrophic oligotrophic vegetation is developed under such conditions. Oligotrophic vegetation is not very much demanding on mineral matter. In course of time, on peatification, the product developed is of high calorific value with low ash content.

c. *Transition moors* : These swamps support mesotrophic vegetation which is relatively less demanding than autotrophic vegetation but more demanding than upland moors, *i.e.* its requirement for mineral-nutritive-elements is small for its growth and accordingly the peat derived from them, too, is of intermediate type.

Peat is dark brown, spongy, fibrous and matted mass resembling tobacco when least altered but a fine black, granular mud when most changed. It is sometimes used as fuel.

Peat, composed visibly only, partially changed plant materials and contains about 2 % of N_2, which it often contains, may be converted to ammonia. A moderate pressure peat is sufficient to transform into coal-like substance. In time-framework it takes about 8 million years to convert peat into bituminous coal. Peat gives the following composition at 100^0 C :

C = 56 - 66%

H_2 = 5 - 9%

O_2 = 18 - 33% ; N_2 = 2% ; Ash (inorganic salt) = 1 - 6%

Formation of Peat

Peat is generally formed under

i) humid and temperate climatic conditions, and

ii) in stagnant and shallow water bodies, such as lakes and lagoons.

In such environment with oxygen, the plant debris cannot be changed to higher ranks of coal. Vegetal materials of various kind from mosses (peat bogs) to trees (forest peat) fluorish in marshy conditions. These materials go on accumulating and in time, because of bacterial attack, they decay in as much as the basic cellulose substance of the plant tissues, disappears soon on account of the presence of sun's rays and oxygen yielding CO_2 and water

$$C_6H_{10}O_5 + 6O_2 = 6CO_2 + 5H_2O.$$

Thus some of the

i) hydrogen is removed as water (H_2O), as also

ii) carbon-dioxide (CO_2), and

iii) some of both the contents as marsh gas (CH_4) (methane)

thus $2\ C_6H_{10}O_5 = C_8H_{10}O_5 + 2CO_3 + 2CH_4 + H_2O$.

The resultant decayed product is enriched in carbon on expulsion of hydrogen from the original plant debris. Humic and ulmic acids produced on plant decay and water containing oxygen, later destroy the bacteria considerably reducing the rate of decay of plant materials. Thus slowing down of plant debris decomposition take place and peat, a black to brown carbonaceous material is formed on partial decay of vegetable matter. The peat formation is abundant in the temperate zone generally between 35-60^0 N latitudes where the temperature is usually between 5 to 15^0 C.

Depending on

i) heat

ii) nature of accumulation basins and

iii) plant materials, due to increased burial

carbon and hydrogen separately combine with oxygen (hydrogen combines more readily than carbon). This produces gradual enrichment as the process continues.

Occurrence of Peat in India

Peat-like substance formation is currently an ongoing process in the layers of

a) the Sunderban area of South Bengal in the Ganga delta on decomposition of mostly the *sundri* trees and

b) the Nilgiri hills in South India. In some depressions bryophyta form a few peat bogs.

In Kashmir, the alluvium of Jhelum and few localities in Nichama and Shatiganga, aquatic vegetation, grasses and sedges have entered into its composition. Peat is being formed in some valleys of Nepal too.

The peat in Ganga delta is low level or lowland peat because of its formation close to the sea level. Those at Nilgiri hills, are referred to as high level or upland peat due to the locale of its formation at an altitude of 2,000 meters. These swamps are fed mainly from atmospheric precipitations, usually low in mineral salts. Oligotrophic vegetation are of high calorific value and low in ash content. Intermixed with sand and loam, peat also occurs 6 to 10 meters below the city of Calcutta. These are characterized by autotrophic vegetation rich in mineral nutrient fed by ground or river waters in addition to atmospheric precipitation and are therefore generally high in ash content and low in calorific value.

CHAPTER XI

GROUND WATERS AND KARSTS

Karst, a Germanized form of an ancient Slovanian term, is a "terrain with distinctive characters of relief and drainage primarily from a higher degree of rock solubility in natural waters than is found elsewhere" (Jennings 1971). It is rather a comprehensive term for solution topography region and is coined after the narrow limestone plateau strip along Adriatic coast in Karst district, Yugoslavia and adjoining portions of Italy borders. Within its ambit are included

i) the landforms

ii) operating processes and

iii) guiding structural controls.

More often *karstification* (a process by which karst topography is developed), is related to subterranean drainage, but indeed the latter is the effect and *not* the cause of solutional activities on tne carbonate rocks - particularly, the limestones and dolomites. But for some isolated instances, some of the best developed karstic landforms are met in humid regions. Factors that contribute the development of solution features include

a) humid to subhumid climate providing adequate rainwater for relatively rapid rate of dissolution,

b) carbonate rocks (*e.g.* limestones) exposed on the surface, ready for dissolution,

c) numerous joints and thin bedding planes facilitating water passage, and

d) major streams entrenchment permitting ground water descent through soluble rocks to emerge into streams.

As such, not all limestone show karst development. In dense limestone for example, jointing and bedding play more significantly than high porosity. Karst features poorly develop in areas that are underlain by porous chalk.

Limestones and dolomites constitute between 5 to 15 percent weight of total sedimentary bulk to which additional 5% evaporites may also be added. They together constitute a set of lithology for karst development.

HYDROLOGICAL CYCLE

Oceans, a gigantic reservoir, is the ultimate

i) storehouse on one hand, and

ii) source of origin of the various types

of water met with, on the other hand. Besides, in liquid state water is locked in solid state in ice and snow, and in vapour form in gaseous state. Water evaporates from the surface of the ocean leaving all salts in the reservoir itself. The evaporated products, the vapours, more through the atmosphere following the pressure gradient, until appropriate conditions are met, when they (the vapours) condense to from droplets which may directly fall on the sea or may reach the sea through its journey on the land, wherefrom evaporation may draw them out again. This entire process of evaporation, condensation and precipitation is referred to as *hydrologic cycle*. This cycle combines in itself the processes of -

a. evaporation including evapotranspiration,

b) condensation,

c) precipitation,

d) surface runoff,

e) infiltration, and

a. *Evaporation* -

i) *In evaporation* the water molecules, from the surface or even when associated with moisture in soils, are driven out through the energy available from solar radiation into the atmosphere.

ii) Plants, too, contribute water molecules into the atmosphere through the process of *evapotranspiration,* and

iii) Under specialized conditions some vapours may as well be obtained directly from solid state - snow and ice, without the intermediate liquid state referred to as *sublimation.*

b. *Condensation* - The mass of air with less than 100 percent relative humidity when cools down without loss of moisture, around dew point temperature, becomes saturated. At this position, condensation is likely to be initiated at a surface or on a nucleus on which it can settle. The morning dew or frost is the result of such condensation on plants, grass or other surfaces. Rain or ice requires 0.1 to 10 μm ranged nuclei. Particles serving as nuclei include clay minerals, salts and combination products.

c. *Precipitation* - Moving mass of air may be saturated on changes in air pressure and temperature for which necessary conditions are :

i) humid air must be cooled down to the dew point temperature,

ii) condensation or freezing of air nuclei must be present,

iii) droplets must coalesce to form raindrop, and

iv) the size of the raindrops must be sufficient to leave the clouds without being totally evaporated until reaching the ground surface.

Air cools while ascending higher altitudes where atmosphere is less dense, and as consequence thereof adiabatic expansion, in response to the lower pressure takes place minimizing heat exchange between air molecules, lowering the air temperature following the thermodynamic laws.

d. *Surface runoff* - The precipitated water on reaching the ground

i) moves downslope, as river or stream or streamlet,

ii) at a rate faster than percolation,

if iii) the lithology is dense or less porous, and

iv) the gradient is more.

e. *Infiltration* - On a

i) less dense or fractured/ jointed lithology, and

ii) more or less horizontal surface

the infiltration is likely to be more and the water moves down to form underground or subsurface water.

The surface topography determines the contact time between the precipitated water and the ground. The contact time is less on steeper slopes and therefore, the time available for percolation underground is limited and as such, the surface runoff is more effective. On flattish ground, the contact time for water to find its passage downward is more, and therefore, the infiltration is more.

f. *Discharge into sea*- The subsurface water too, has sea as its ultimate destination and therefore, it is also referred to as *delayed runoff*, as against surface run off which travel to the sea/ocean as *immediate runoff*.

Thus the entire runoff immediate or delayed reach back to the ocean from where the hydrologic cycle was initiated.

KARST HYDROGEOLOGY

Water occurring beneath the earth's surface is called "*ground water*" or "*underground water*" or "*subsurface/subterranean water*" in contrast to "*meteoric water*" obtained directly from the atmospheric precipitation. This is the most important source of underground water. Snow or lakes in some regions also add to it. Sometimes a small proportion of *primary or juvenile water* may be incorporated, often in the form of steam, into the groundwater below, from hydrothermal solutions and steam liberated from magmatic activity or volcanic eruptions particularly, in regions of contemporary or recent volcanic activity of young mountains. Sedimentation, too, may cause some fresh (river or lake) or saline (lake or sea) water trapping between the sediment pore-spaces alongwith which, they may be buried. Such water is termed *fossil* or *connate water*. The sea water in the formation of underground waters may precipitate-

(i) as entrapped on sea floor between the sediments, *i.e. syngenetic* (*syn* means together and *genesis* stands for origin, creation or birth)

(ii) by infiltration into earlier formed sediment, *i.e. epigenetic* (derived from the Greek word *epi* meaning at, near or with)

The connate water may be squeezed out on further burial and/or sediment compaction, though in some cases thin monomolecular film of water may remain enveloping rock particles/sediments. The latter is referred to as *hygroscopic water*. A temperature of about 105^0 - 110^0 C tend to liberate such waters.

Three main factors determine the distribution of rain precipitation water

A. *Climate* - Aridity induces evaporation, but humidity reduces it and induces surface run off and ground water percolation. Through very fine openings water tends to rise upwards more effectively in arid conditions where because of limited percolation the water table is rather deep and surface runoff usually meagre, but for the period(s) of heavy rainfall when surface runoff is for a short span of time, may be visibly effective.

B. *Topography* - High relief terrain(s) is (are) conducive for rapid runoff, whereas in flattish region the surface runoff is slow which may facilitate, under favourable lithological conditions, domination of downward water percolation.

C. *Lithology* - It determines to a large extent the relative proportion of water distribution in an area. The following rock characters influence most in this respect:

a) Porosity - Groundwaters beneath the earth's surface is accommodated into pore spaces between the sediments or they occupy open rock spaces such as fractures, joints *etc*. The amount of intergranular or open space in rocks is defined by porosity which is numerically expressed as percentage of open or void space in a given volume of rock material *i.e.*

$$\text{Porosity} = \frac{\text{volume of pore space}}{\text{total volume}} \times 100$$

total volume = total volume of rock aggregate + volume of pore space

Imagine a rock sample in which the ratio of volume of solid matter to available open space is 6:4. Then its porosity is

$$\frac{4}{6+4} \times 100 = 40\%$$

Generally, hard and compact rocks like those of igneous and metamorphic origin, often have only very little open space in them to accumulate water. In terrains of such rocks the surface runoff is greater. Sedimentary rocks, as also joints and fissures in any rock provide open space to house in water percolation, reducing immediate run off. In sedimentary rocks too, porosity may be

(i) original, as developed during sediment deposition or

may (ii) secondary, modified due to

authigenic growth or cementation (reducing the original porosity) or

decementation, dissolution (increasing the original porosity), *etc.*

Pores or pore spaces may or may not be interconnected and therefore rocks of equal porosity may not have equal permeability (it is the faculty of allowing a passage of fluid flow through rocks). Thus, *effective porosity* is different and always less than the total porosity. Porosity depends upon several factors, among which are -

(i) size, shape and arrangement of the component grains in the given material-uniform and well sorted sediments have higher porosity in contrast to irregularly arranged homogenous grains and

(ii) size, shape and distribution of fractures and fissures in compact and consolidated rocks.

Direct link between porosity and depth is though not yet established, in general the porosity is found to be decreasing at depth attributable to

a. decrease in the interstitial size on increased pressure due to rock column above and /or

b. increased particle cementation.

Typical porosity for some common rock types are -

Rock type	*Porosity(in %)*
Granite	1.5
Shale, Slate	4.0
Limestones	5.05 to 10.0
Sandstones	10.0 to 15.0
Sand and Gravels	25.0 to 30.0
Soil and Loam	40.0 to 60.0

(c) *Permeability* - Rocks allowing the passage or water or solutions are said to be *permeable* in nature and those which do not permit the flow are called *impermeable* rocks. The surface run off would be more when the rock is impermeable, whereas percolation would be more in permeable rocks.

By definition, permeability of a rock is its ability to transmit water under pressure which in turn depends upon

(i) the presence of minute openings or interstices in the soil or rock

as also (ii) the size, shape and arrangement of such openings.

The water transmission capacity of a formation is computed either by its

coefficient of permeability or by

coefficient of transmissibility

Meinzer defined the *coefficient of permeability,* K as "*rate of flow of water in gallons per day through a cross-sectional area of 1 square foot under a unit hydraulic gradient at a temperature of 60⁰ Fahr*".

The *co-efficient of transmissibility,* T is introduced and defined by Theiss as "*the rate of flow of water in gallons per day through a vertical strip of aquifer 1 foot wide and extending the full saturated height under a hydraulic gradient of 100 percent at a temperature of 60⁰ Fahr*". This expresses aquifer ability as a whole to transmit water.

If m is the average aquifer thickness, the relationship between the coefficient of transmissibility and permeability is expressed by the equation

$$T = Km$$

Darcy's Law, expressed as

$$V = KI$$

where, V= velocity in feet per day

I = hydraulic gradient in feet per foot length

and K= coefficient of permeability measured in gallons per day per square foot

is generally used to explain the flow of water through filter sands, and when, as generally is, total volume of flow is needed, the above expression is modified as

$$Q = TIW$$

where, T = coefficient of transmissibility and

W = width of the flow cross-section in square feet

Permeability of materials depends upon

(i) size and shape of the constituent grains, uniformly coarse-grained gravel is highly permeable, but in heterogenous materials composed of sediments of varying sizes, the permeability is much less, as smaller grains tend to act as fillers of materials

(ii) grain sorting, when high, the permeability increases, whereas the permeability is less in poorly sorted materials

(iii) shape and pattern of interstices, especially the fracture continuity in compact rocks.

Interestingly, porosity is not always a contributing factor in permeability estimation, for instance, clays are highly porous, generally 50% and above, but their permeability is rather very low. It is due to very fine grained particles constituting the clays, which because of their mutual pull develop strong ties towards each other that do not permit water passage through it.

In addition, two other factors also determine water flow in rocks *viz* -

D. *Rocks structure* - If the rocks are inclined, the percolating water tends to follow the most porous layers and can reach greater depth, whereas in horizontally disposed layers, the water moving downwards must cross less porous beds, if any, through its journey.

E. *Vegetation* - It exercises strong influence over the surface runoff. Forests produce cooler atmosphere and thus attract greater rainfall. Also vegetation retains considerable moisture. Under the circumstances, the surface run off is greater in vegetation covered terrains; downward percolation is less and practically no evaporation.

Thus, on precipitation the water is distributed into three ways :

Runoff	Direct flow down the surface slopes	Immediate Runoff or Streamflow
P E R C O L A T I O N	1. Superficial flow through the soil and suboil to streams	
	2. Downward filtration into bed rocks to replensih ground water and maintain its circulation. A part of this water again emerges on the surface through springs, artesian, tube or ordinary wells. A considerable part of the underground water reaches the sea through underground circulation	Delayed Runoff or Ground water
	3.Absorption by soil and vegetation subsequently	Evaporation

DIRECT RUN OFF

When all the pore or open spaces are filled with water, the containing rocks become saturated and no further volume of water can be stored in them. The upper surface below which all available space is occupied with water, is termed as *water table* or groundwater level or *phreatic* (Greek word *phrear,* a well) and the section below, *zone of saturation*. The saturation depth is dependent upon

(i) rock permeability,

(ii) amount of water percolation, and

(iii) the surface topography.

Often the water table is not a level but on the contrary, it is generally a rolling surface. In

(i) humid regions, and

(ii) where the rocks are homogenous

it is subdued replica of the landscape, relatively arching up beneath hills and intersecting the landform at the surface of river, lake or spring. In dried terrain water table is highest below the streams emerging from well-watered hills by stream percolation themselves. However, water table is more or less flat in a flattish plain and follows the the surface topography in general. Three zones of state of water may be identified below the surface:

(a) the upper non-saturation zone of aeration where air, instead of water, occupies much of the void space, which are dry

(b) below is the zone of saturation in which all the open spaces are filled completely with water. This is limited to the depth below which porosity ceases to be present for practical reasons and thus is not far down below the surface. It is rather rarely 700 to 1,000 meters below the ground surface, though in exceptional circumstances instances of ground water occurrences at a depth of 2,000 meters below the surface are also on record. Zone of saturation provides dependable water supply system and needs to be handled carefully for prolonged supply life.

(c) in between the two is the zone of intermittent saturation, in which pores are filled during the seasonal rains. A capillary zone is identified at the upper contact of the water table within this intermediate zone, in which some water rises through a fine conduit into the zone of aeration in response to atmospheric temperature and pressure.

Water above the water table, but below the earth's surface, that is, in the zone of aeration, where all the pores are not filled with water, but may be occupied by air, is termed *vadose water*. A condition may develop, above the main water table, in the zone of aeration, where an impervious bed may hold water saturating the bed, which may show a promise of local water supply temporarily. This is *perched water table* in the zone of aeration. This zone, however, is never completely water filled. It provides passage to the water to the lower zone of saturation and its thickness varies from a few centimeters to as much as 300 meters or even more, depending upon

(i) geology,

(ii) hydrology, and

(iii) the topography

of the terrain. In this zone, water is generally under no hydrostatic pressure except temporarily. Two separate belts are identified in this zone

(i) *soil water belt* from which water is lost by

evaporation and

transpiration by plants, and

(ii) *capillary fringe* which overlies the zone of saturation

Influent streams, that flow above the water table, contribute water to the subsurface waters. In contrast, *effluent streams* receive/ supplies from the groundwaters. The term "runoff" includes

(i) the rainwater received in an area on precipitation, collected and flow into the subsurface stream, as also

(ii) the total discharge of any region, flowing into the stream(s). This suggests possible runoff even from rainless or poor rain area

FLOW IN RELATION TO GROUND WATER CONTOUR

With the help of water level measurements of wells or boreholes, water table configuration is ascertained and depicted in water table contour maps showing water positioning at the time of measurements, as the level fluctuates with

(i) precipitation, and

(ii) withdrawl.

It is important to recognize the difference between

(a) water table levels indicating the upper surface (s) of the zone saturated by free water, and

(b) pressure levels, indicative of the level rise limits of confined water in the ground under head.

These maps depict infiltration or recharge areas. Water table level may be

(i) below the ground surface, whence the surface water penetrates down below till the water reaches the water table (a condition of *influent* seepage) or

(ii) may intersect the ground surface, as it may be in valleys, feeding or providing them water (condition of *effluent* seepage) or

(iii) coinciding with the ground surface, a condition which permanently moists the soils and induces swamp development.

Swamps, lakes and streams, may however, be at times perched above the water table, are influent when they loose water to the ground.

Water table (contour) maps also illustrate the direction and gradient of groundwater movement. The latter varies inversely with the ground permeability through which it flows. The water table gradient or slope is less in more permeable lithology. Water table slopes vary according to the percolation velocity in a lithology of given permeability.

WELLS

Structures punctured by way of drilling, boring or dugging into the ground to encounter water bearing strata of reasonable water yielding capacity are referred to as *wells* - which may be shallow or deep. Amongst more common types of wells are -

(i) *Gravity well* - In such types of wells the surface of water in the well is essentially at the atmospheric pressure. They are vertical or nearly vertical penetration in the saturated subsurface.

(ii) *Pressure* or *Artesian* wells - These are also of almost vertical penetration in water - holding aquifer, but at a much higher hydrostatic pressure due to which the water rises to release the pressure.

(iii) *Galleries* - In contrast to either of the above, these are horizontal tunnels/openings through water bearing aquifer.

Wells could also be grouped on the basis of the depth of water bearing formations-

(i) *Shallow wells* - These are 2.5 to 5.00 meters in diameter, constructed to encounter the top-most water bearing strata.

(ii) *Deep wells* - In contrast, these 2.5 to 50 cm dia, wells are designed to penetrate much deeper through the top-most water bearing strata to second or even lower such beds.

Artesian Wells - Struck in French province of Artois in 12th century, the prefix artesian connotes wells in which water under pressure in the aquifer of permeable beds, is in confined state, overlain and underlain by impervious layers. The conditions for artesian flow include

(i) presence of an aquifer or permeable bed,

(ii) with relatively impermeable basal or top covering beds within which the water bearing aquifer is confined,

(iii) sufficient precipitation in the catchment area or intake for adequate water supply to the aquifer,

(iv) converging dip of aquifer bed, at an angle to impart hydraulic gradient,

(v) lack of any opening/outlet to prevent escape of accumulated water.

A sandstone bed supported over shale bed below and capped above also by shale in structurally a basin like form, would be a good situation to develop artesian pressure. When the precipitation takes place in the catchment area, water enters into sandstone pores from the outcrop receiving rain water, and moves down to a lower position until all the pores there become fully saturated. Thereafter, any further addition of water from the outcrop generates hydrostatic pressure, but being capped above and supported below by impervious layers, water cannot rush out by itself, and thus it is under artesian pressure/condition. Now, if a bore-hole is punctured at or close to such pressure locale, water would rush up through the hole. The height attained by the rising water column above the water level is termed the "*hade*". This rising water would automatically flow over the surface, if the catchment above is the punctured well site, but if it is below, water would ascend upto a height below the surface and would need to be lifted up further by pumping for surface reach.

In many deserts, oasis results due to artesian conditions.

When water is drawn from an ordinary gravity well, the surface of water-table around is lowered and it assumes temporarily the form of an inverted cone, called the *cone of depression* or *exhaustion.* The steepness of its sides depends on the aquifer permeability-flatter in highly pervious or 'open' materials like gravel, and steeper in less pervious kinds of rock where a cone is developed, the pressure gradient so developed induces the necessary flow to level up the depression. Rapid withdrawl and greater amount thereof widens the extent of cone of depression. When a hole has been drilled and a cone of depression, established on well pumping after a rest period, the gradient of the water table wellwards is steepened and water has to be lifted at this *pumping level.* On cessation of pumping, water will adjust itself to *rest level* and if left undisturbed, to the original level of water table. As such the wells may be pumped alternatively, lest the overpumping may consequently lead to gradual extending of the cone of depression, flattening its sides and falling water table level ending up with the drying of the well. Within limits imposed by aquifer thickness, the supply may generally increase by well deepening. Water may be obtained from several permeable beds alternating with impermeable beds successively by boring penetration. In such conditions, a complete cone of depression is not developed for each bed.

The distance from the centre of the cone of depression, where the water table level on water withdrawl is lowered maximum, to the place of least effect in water table, is measurable as *radius of influence* and the area encompassed as *area of influence* and varies with the rock characteristic into which the well is dug being larger in coarse grained, loosely packed and permeable materials and small in dense, less permeable finer materials.

Radius of influence in some sediments is

	meters
Silty	30 - 100
Fine to medium sand	100 - 200
Coarse sand and fine gravel	200 - 400
Coarse gravel	400 - 800

Fresh and Salt Water Relations (Ghyben- Hertzberg Balance) in Coastal area :

A particular case of cone of depression in coastal regions and in certain islands assume special significance, where the strata generally composed of sand, loam, corals and limestones through which rain waters infillter and get stored as fresh groundwater. From the sea, the salt water also seeps through the substratum, but being heavy, causes the freshwater float over it. However, because of only small density difference, the freshwater extending down to rather surprising depths below, balances its column against salt water column as icebergs float over sea water, forming parabolic cones at the boundary of which diffusion between the two waters is negligibly minimal. Usually the freshwater level occurs at an altitude above the sea-level. It is noted that for each foot (0.33 meters) of fresh water that extends above the sea level, it continues 40'(12.12 meters) down below the sea level, from which it follows that 41 ft.(12.42 meters) of fresh water column is balanced against 40 ft. (12.12 meters) of sea-water column. This state of hydrostatic equilibrium between the fresh and salt waters is termed as *Ghyben-Hertzberg balance.*

A relationship of elevation (t) of freshwater column (H) above the sea-water column (h) of specific gravity 'g' may be derived to express as -

$$H = h+t = hg$$

$$t = hg - h$$

$$= h\ (g - 1)$$

$$\text{or} \quad h = t/(g - 1)$$

This determinable and definite relationship between fresh and salt waters may be utilized in salt-water contamination or intrusion. It indicates existence of 40 ft.(12.12 meters) of freshwater for each foot (0.33 meters) between the watertable and seawater. A salt water intrusion steeper cone is developed when freshwater is drawn out in coastal areas. The latter cone is rather shallower. Accordingly a lowering of water table by 1 foot (0.33 meters) involves about 40 ft (12.12 meters) rise of salt water.

Ground Water Recharge

Indiscriminate use of water for drinking, agricultural and/ or industrial purposes leads to its depletion unless the take out is evened by the intake. This balancing is referred to as *ground water recharge*. A formation that yields water, can take it up as well. Generally the introduction of water is faster than the lateral percolation as a result a *water table mount* or reverse cone of depression is developed on injection of water into the ground, if the water table is deep enough below the ground surface. Water in the ground may be introduced by

(a) leading water over permeable surface or water spreading (natural) and

(b) introducing water into "inverted" or recharged wells (artificial)

Natural recharge water spreading may be effected by

(i) ditches that convey water from a surface reservoir or stream to furrows ground plots (as is done in irrigation)

(ii) the "basin" method in which harrows are surrounded by retaining dykes and a shallow sheet of water is maintained over a dyke area (surface basin) or

(iii) feeding water down trenches, shafts or wells, and operation that requires a preliminary desilting of water.

However, if

(i) the amount of recharge exceeds the reservoir capacity,

or (ii) the transmissibility is inadequate,

or (iii) the reservoir is already filled,

the phenomenon of rejected recharge comes into being and the formation cannot accept any further volume of water.

SPRINGS

An aquifer

(i) is a layer of permeable strata

(ii) through which water readily transmits

such as sands, pebbles or cobbles (unconsolidated), sandstone, conglomerate, *etc.* (consolidated). Five broad principal groups of aquifers are identified -

1. Loose coarse grained sand and gravel surface deposits constitute *one group* of *aquifers* and are primarily porous strata in alluvial deposits and glacial drifts.

   (a) Alluvial aquifer consists of silt, sand and gravel composed of mountain outwash of

(i) considerable thickness with

(ii) great amounts of water in storage.

Valleys may be underlain by hundreds of meters of alluvial or possess alluvial deposits as alluvial fans and cones

(b) Glacial drifts include some outstanding pervious members as

(i) glacial outwash, often most of the terminal moraine close to the glacial front or in channels of old melt water

(ii) sand and gravel in eskers and kames

Under certain circumstances

(c) some lakes, and

(d) wind deposits like dunes or coarser (*i.e.* sandy variety of) loess may also constitute aquifer.

2. Consolidated sedimentary rocks, like sandstones and limestones constitute *second aquifer group.* Sandstones provide important artesian conditions. Limestones are good water reservoirs, when

   (i) they are included between impervious rocks

   (ii) solution channels, formed by the dissolution of calcite, the primary constituent of limestone, extend from higher to lower level and get clogged at some point, and

   (iii) the formations are young and thus porous.

While clays and shales are no good as aquifers, limestone-stored water may find its outlet through springs.

3. Fissured or disintegrated *intrusive* igneous rocks *such as granite or schistosity planes of metamorphic rock* form *yet another group of aquifers*
4. Weathered or partly weathered zone over the unweathered parent rock mass too, hosts ground water. It is the *fourth group of aquifers*
5. Occasionally faults may as well store considerable quantities of water.

A layer which is neither porous nor permeable is termed as *aquifuge*, but if it can hold water (*i.e.* if it is porous) but does not readily transmit, it is called *aquiclude*. Clays, shales and mudstones can hold water into the fine pores they possess but do not transmit to furnish an appreciable amount of water. Interbedded permeable and impermeable sequence provides conducive situation for development of secondary water table or perched water table wherein aquiclude inhibits downward water percolation. When an aquifer is intersected by erosion jointing or faulting, or igneous mass blocks the water passage, water tends to ooze out on the surface to release the pressure. Such natural openings through which water flows over the surface is called the *spring*. A host of geological conditions lead to the intersection of ground with water table -

(a) an incline pervious bed underlain by impervious bed, when eroded, is a condition suitable for spring development.

(b) water table generally follows surface topography in a hilly region, which is raised on precipitation, such risen water table may be cut by the hill side(s), causing a spring.

(c) an inclined pervious bed, when faulted, develops artesian condition at the contact, and if the intake of the pervious layer is at a level higher than the surface trace of outcrop, spring is developed

(d) likewise, a discordant igneous intrusion or any other barrier may act like a wall to prevent down water flow in an inclined flow and develop artesian pressure at its contact with an aquifer where water gets stored and subsequently forced up on the surface that form spring.

Four major tracts along which most of the springs in the Indian subcontinent are located -

(i) Bihar belt comprising a series parallel to the coalfield boundaries as in Rajghir area, Munger district,

(ii) West coast belt - including those in Ratnagiri, Thana and Colaba areas in Maharashtra and in Surat area in Gujarat,

(iii) Sind-Baluchistan belt, and

(iv) the Himalayan belt

besides, two minor belts are

(v) in the Mahanadi valley of Orissa, and

(vi) Chittagong district, Bangladesh.

In addition, springs are recorded from Birbhum and Darjeeling districts of West Bengal, from Sikkim, as also from parts of South India.

MINERAL SPRINGS

Springs charged with dissolved mineral salts, in their hot or cold waters are referred to as *mineral springs.* Dissolved salts may be of great variety. The juvenile waters with calcium carbonate in solution, deposit thick accumulations of white lime carbonate in solution, forming *calcarous tufa* or *calc sinter* when loose or *travertine* when crystalline. Mineral waters may have sulphuretted hydrogen, CO_2 and other gases, alongwith Na_2,K-, and Mg - compounds. Limestone regions yield hard water in wells and springs and soft water in igneous terrains. Often mineral water find medicinal applications such as remedy for gout, in intestinal disorder, *etc.*

Silica rich solutions when emitted on the surface form *siliceous sinter* or *geyserite.*

HOT (THERMAL) SPRINGS

Waters issued from hot springs are at temperature appreciably higher than the atmospheric temperatures. They generally range from 36° C to 100° C and waters therein often issue froth on the surface. Many causes are attributed to account for heat of these springs, such as

(i) due to volcanic activity in the region

(ii) sometimes waters descending down below, on contact, with magma receive heat and subsequently issued on the surface as thermal springs

(iii) chemical action, radioactivity, sliding of strata against one another also provide heat to the issuing spring water.

GEYSER

Geysers are special type of hot springs which issue gases and hot vapours at intervals. Water columns in them may rise from a few to hundred meters and both

(i) the intervals between two consecutive eruptions of water, and

(ii) the duration of eruption

may be variable, but for the particular geyser, both the variables are remarkably regular. In view of faithful water emission at average intervals of 65 minutes, for several years, the most illustrious example of geyser in the Yellowtone Park-group of geyser is named Old Faithful geyser which recently is noted to have become a bit irregular. Variations in

(i) heat,

(ii) water supply, and

(iii) nature of the feeding pipe

provide the characteristic of an individual geyser measured in terms of

(i) the interval between consecutive eruptions,

(ii) the duration of eruptions, and

(iii) the height of water column rise, *etc.*

Deposits from geyser water formed at its vent are called *geyserite* which in some cases may even be mere encrustations.

A narrow zig-zag pipe explains the periodicity in water emission in geysers, as such pipe would be composed of some straight arms joined with each other at some angle to form a zig-zag outlay. Water reaches almost at boiling point at higher temperature than 100°C being at great pressure at depth. Steam so produced is convected upward in an arm which in course of time, effectively develops pressure at its union with the other arm, more or less maintaining the uniform temperature in each arm intercepted with pressure buildings at bends. The pressure is relieved at the surface with the ejection of water and until take off pressure is again developed, the ejection is in waiting for the next issuance and thus causing periodicity. Geysers commonly occur in areas of hydrothermal activity.

GROUND WATER PROVINCES OF INDIA

Regions characterized by a general similarity in the mode of groundwater occurrences, underlain by even rocks of diverse type, are termed *groundwater provinces*, which may or may not be contiguous. Subtle differences in their hydrologic and hydraulic properties are often ignored in classifying these provinces. Tayler (1959) grouped various stratigraphic units of India into the following eight groundwater provinces :

A. Pre-Cambrian Crystalline province	
1. Pre Cambrian to Archaen	Delhi system Dharwar-Aravali system
B. Pre-Cambrian	
2. Sedimentary Province	Vindhyan and Kurnool systems, and equivalent Raipur and Indravati series Cuddapah system, Kaladgi, and Bijawar series
C. Upper Carboniferous to Early Cre-taceous	
3. Gondwana Sedimentary Province	Gondwana system
4. Deccan Trap Province (Upper Cretaceous to lower Eocene)	Deccan Trap; basalt flow sand inter-trappean beds
5. Cenozoic Sedimentary Province	Palanaand Nagaur sandstone, in Rajm-uhnudry and Athgarh sandstone
D. Eocene to Pliocene	
6. Cenozoic Fault basins (Pleistocene to Recent)	Narmada and Tapti river alluvium (gravel, sand, silt and caly)
7. Ganga- Brahmputra Alluvium Province	All in the foredeep between the Himalaya and Peniusula (boulder, gravel to sand, silt and clay)
8. Himalaya highland Province	varied rock types

The chemical quality of these provinces is influenced by

(i) lithological units
(ii) topography
(iii) climate, and
(iv) human activity

The entire Himalayan belt, the Kashmir lacustrine sediments, springs, Doon gravels, Tarai springs, Kathmandu lacustrine aquifer, the NW-NE-East sandy-alluvial, tract of the Sindhu (Indus), Ganga and Brahmputra are common areas of great fresh ground-water reservoir.

Over 13 crores cubic feet of water is estimated to be available at depths within 300 meters in the Sindhu-Ganga alluvial region and in the coastal sedimentary strips of rocks with practically (not necessarily) interconnected aquifers in confined to semiconfined state suitable for agricultural and municipal purposes. The formations in Rajasthan- Bikaner, Jaisalmer and Jodhpur areas are important. The yield varies from 45- 150 m^3/hour with TDS 1,000-5,000 ppm, tastes saline to brackish in Bikaner and < 2000 ppm in Jodhpur terrain. In Gujarat, Kutch and Kathiawar sandstones constitute good aquifers occurring at variable depth of 60-300 meters yielding 10-120 m^3/ hour having TDS ranging from 1,000-5,000 ppm.

In Madhya Pradesh and Central India, Bundelkhand granites, Baghelkhand limestones, Malwa Plateau Basalts, Narmada Valley and Satpura range constitute the basinal units carrying good amount of groundwaters. Water occurs in joints and openings of hard rocks, whereas the carbonate rocks of the Vindhyan, Cuddapah and Bijawar regions, themselves as also structures therein, are good passageways for groundwaters. These are actually karst topographic areas.

Gujarat, Neyveli lignite mines in South Arcot in Tamil Nadu, Pondicherry and East and West Godavari districts of Andhra Paradesh offer artesian conditions. Chances of striking similar artesian conditions in the rocks at moderate depth in Coimbatore and Central districts of Tamil Nadu are not much. Thermal and mineral springs are noted in many parts of India-Maharashtra, Panjab, Bihar, Assam and in the foothills of the Himalaya and Kashmir.

Good aquifers are located in the vesicular basalts and in the weathered zone of the Deccan Trap covered regions of Maharashtra and Madhya Pradesh, the deltaic regions of Tanjore. Tiruchirapalli, Madurai and Tinnvelly districts of Tamil Nadu yield fair quantity of good water in the clayey or soft rocky subsoil. Good yield of groundwater is expected from the lateritic subsratum in the coastal tract of Kerala and Karnataka.

HYDRO-GEOCHEMICAL PROVINCES OF INDIA

Groundwater similarities in chemical composition lead to identification of hydrogeochemical provinces. These characteristics are influenced by

(i) physiographic,

(ii) orographic,

(iii) geologic, and

(iv) climatic,

parameters of the terrain. Hydrochemistry is helpful in evaluating and assessing quality rating of the ground waters for various uses.

Three broad hydrogeochemical provinces (Handa 1964) of India have been recognized :

a. **Bicarbonate province** - It includes primary and softened primary water characterized by

(i) more than 90% of the total ions as that of bicarbonates,

(ii) chloride and sulphate < 0.5 mg/litres,

(iii) wide variation in Na, above 3,

(iv) Ca: Mg ratio >2, and

(v) EC 750 micro mhos/cm.

Narmada valley and hilly terrain in western and central India, as also Tarai-Bhabhar belt and contiguous zone stretching along the foothills of the Himalaya and adjoining alluvial tracts are examples of this ctegory.

a. High rainfall (100 to 400 cms),

b. permeable strata, and

c. steep slopes

favours flushing out salts from the soil profile in these regions.

b. **Bio-carbonate-chloride province** - In this province

(i) HCO_3 is still dominant,

(ii) chloride and Na ions assume more significance,

(iii) alkali earth/Na < 1,

(iv) EC increases to 1,500 micro-mhos/cm, and

(v) sulphate: chloride ratio falls as low as 0.1

Its examples are the southern part of the Sindhu-Ganga alluvium and coastal Cenozoic groundwater province.

a. Moderate rainfall (100-200cm), and

b. low terrain slope

facilitates salt to stay to accumulate in groundwater.

c. **Chloride province** - The province shows

(i) > 50% of total anions,

(ii) Na > alkaline earth ions, and

(iii) EC > 1500 micro-mhos/cm.

This province is further sub-classified on the basis of sulphate content in the ground water :

(a) south and south-west Panjab, Haryana, Western Rajasthan, the alluvial tract of Gujarat and some parts of Coramandel and Kerala coasts, characterized by

(i) low rainfall,

(ii) high evapotranspiration, and

(iii) poor drainage

which are responsible for salt accretion in ground water. Some of the well known saline lakes like Didwana, Pachpedra and Kuchaman lie in this subprovince.

d. **Mixed type (HCO_3-Cl) province** - It is of uncommon type characterized by depthwise variation in the chemical composition of water in aquifer, particularly in bicarbonate province. However, in chloride province groundwater occurs essentially in brackish to saline condition and better quality of water, if at all present, is met only in a thin layer over saline water, causing pronounced salinity stratification. Salinity distribution in such water column may increase/ decrease with depth or may be altogether irregular from which freshwater pockets can be delineated with the aid of electrical logging.

The Central Ground Water board has brought a publication of a hydrogeological map of 1: 50,00,000 scale showing different ways of salinity distribution in vertical profile.

(i) Fresh water overlain by saline water : sedimentaries between the rivers Cauvery and Vellar along the coast between Calcutta to Bhubeneswar and eastern half of the Cambay basin

(ii) Freshwater underlain by saline water : coastal tract south of Bhubaneswar, inter-stream areas between the rivers Krishna-Godavari; and central Cambay basin

(iii) Local freshwater pockets in different saline ground water levels; coastal areas

(i) basins of Krishna and Pennar rivers

(ii) Saurashtra and Kutch;

Agra-Delhi and Gurgaon areas, most of the part of Western Rajasthan excluding some platform areas with freshwater.

KARST GEOMORPHOLOGY

As other geological agents do the subsurface/underground waters too, perform three types of work, *viz.* erosion, transportation and deposition. Mechanical erosion in this case is subordinate, as chemical erosion predominates.

EROSION

A. **Mechanical Erosion** - Feeble velocity of underground water movement, makes it rather a weak agent of mechanical erosion. However, it saturates the loosened soils and debris and lubricates the surface, which facilitates the overlying rock masses or soils to slide down under the action of gravity by overcoming the force of frictional resistance. This is particularly notable when the surface on which the sliding takes place is highly inclined. The rocks slided down rest on the foothill to form a conical feature, with wide base and tapering apex, called *talus*. Thus it aides in producing *landslides*. Besides subsurface streams flowing in underground channels may also perform some mechanical work.

B. Chemical Erosion - Solutional activities of the groundwaters are of great significance, as it provides a variety of landforms-minor and major particularly, on carbonate rocks. It is most effective when the pores are opened and the rocks are jointed or fissured, by way of enlarging them. When the underground water is charged with CO_2, it acts as a good solvent of carbonates particularly, limestones, through the process of formation of bicarbonate of calcium 32 times more soluble than the carbonate itself-

$$CaCO + H_2O + CO_2 = Ca(HCO_3)_2$$

Variety of topographic features are developed this way.

A. Minor Karst Landforms

a. on Exposed Surface

Solution pits, facets, flutes and runnels are some features included under this head. *Solution pits* are developed on a horizontally exposed surface (rainpits), whereas larger features are termed *solution pans*. Solution facets are produced by sheet underground water runoff on a gently inclined planar surface, but when the surface is steeply inclined *solution flute (rillenkarren, lapies)* comprising of straight, parallel-sided grooves separated by sharp crests are formed. Sometimes gently inclined bare surface display networks of branching channels, *solution runnels* (*rinuenkarren*), which probably is an establishment of imprints of runnels initially formed under soil cover. Extreme surface pitting may end up in producing sponge-like framework of rocks. Surface pits etching equidimensionally beneath an algal film, termed *phytokarst*, contrasts well with other solution forms which generally elongate downslope or even vertically.

On part dissolution of horizontal or gently sloping surface rocks by rain water and subsequent downward percolation, red clayey soil particles called *terra rossa* are met, both on the surface, as also along channel ways in joints and other openings. Several meters thick, it resembles lateritic soil and is conspicuously absent on steeply inclined surfaces.

b. on Covered Surfaces

Effectiveness of solution activities multiplies manifold when limestone surface has a cover of soil and organic mat over it. These softer materials support rounded and smooth pits as against the sharp knife-edged divides between the intersecting pits on solid rocks.

A *solution notch* is developed over the soil underneath the solid rock cover, attesting the karst process effectiveness below ground level. Likewise *solution slot* or *grike* (*kluftkarren*) intersect as vertical, circular pipes under soil cover as widened joint openings by solution action. At the foot of the limestone cliffs in humid tropics, acid swamp water erodes a meter or more narrow horizontal slit, termed *swamp slot,* over residual clay or organic detritals, which checks further down cutting, especially when

(i) the swamp surface and water table are at the same level,

and (ii) the percolation is rather slow.

Swamp slot constitutes a basic component of *towerkarst (turnkarst mogote, pepino hill*) characterized by corrosion round the basal swampy plains, along the margin of steep-sided karst hill (Figure 41). In *kegelkarst (cone-karst)* the landscape is almost equally shared by star-shaped or valley like conical dolines among conical residual hills which are generally hemispheric rather than conical. Towerkarst and kegelkarst are uncommon in temperate climate. It is developed in hot and wet regions in tropics in Java, Jamaica and South China *etc.*

B. Major Karst Landforms

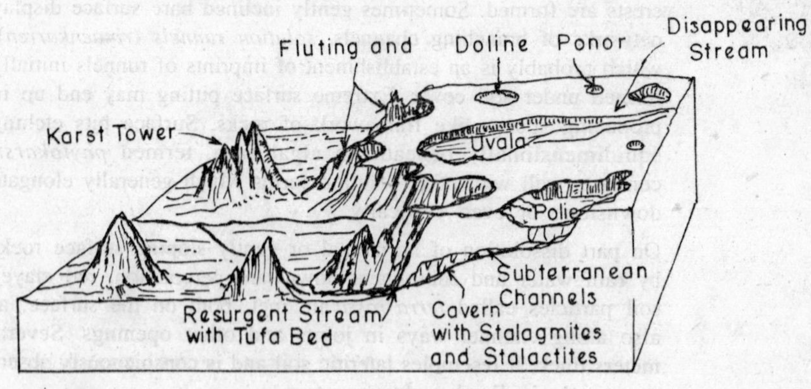

Figure 41 : Some Karst Region Features

Continuum of landforms from the above minor ones to large scale forms, like *dolines etc.*, that can be mapped on topographic sheets and identified on aerial photographs, are observed. These include sinks, sinkholes, solution valleys *etc*. However, caves and caverns cause practical problem in placing on aerial photographs.

Caves and *Caverns* - In vadoze zone, small caverns are formed by solution action along joints and fissures which are enlarged and spread laterally until in contact with insoluable rocks like shale or clay below,desists further corrosion. The roofs of these caves or caverns (if of small size) do not collapse. Their lengths are variable and may extend upto even 50 kms as in the case of Mammoth caves in Kantucky, USA. In Sahastradhara (Dehradun, Uttar Pradesh), the caves generally do not extend beyond a couple to 4/5 meters.

Dolines (*Sinkholes, Sinks, Swallow Holes, Cock Pits, Blue Holes, Cenotes*) - Dolines or limestone sinks are localized basin-shaped, hollows or depressions of different sizes (a few meters in diameter to a meter deep to several kilometers in diameter to hundreds of meters deep) that constitute fundamental geomorphic component of karst landscape, formed (Figure 42) due to

(i) joint control most common ones,

(ii) differential solubility, or

(iii) random vents,

resulting on the collapse of roofs of honey combed and cavernous/porous limestone. Solvent action of surface water can also but rarely form such features. Generally the inward basin slope is 20-30^0, but in collapse sinks, they may be vertical or subvertical. The dissolving water traverses underground.

Based on their geometry four major classes of dolines are recognized

(i) funnel-shaped solution doline

(ii) steep or cliffed collapse doline

(iii) subsidence doline

(iv) subjacent karst collapse doline

The first and second types are not so easily discernible in the field, but for the process they result from. Subsidence and subjacent karst collapse types are characteristic forms in nonsoluble rocks and caused by solution of subsurface buried karst. In stream beds, frequently small dolines called *swallow holes*, *ponors* are formed. These are completely concealed below the alluvium. Intersection of adjacent dolines in tropical region that forms their rapid growth and enlargement, or fusion of smaller dolines to form compound, honeycombed, reticulate or star-shaped, give rise to *cock-pits*.

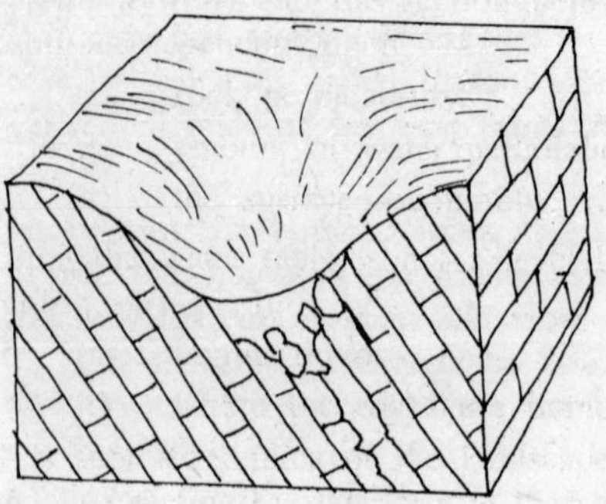

A. Solution Doline

B. Collapse Doline

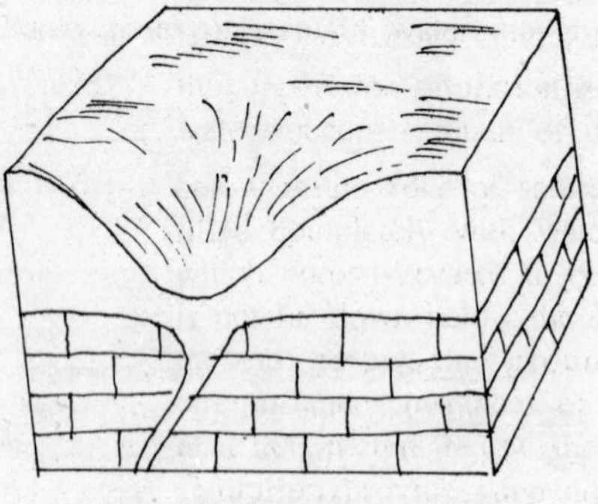

C. Subsidence Doline

D. Subjacent Karst Collapse Doline

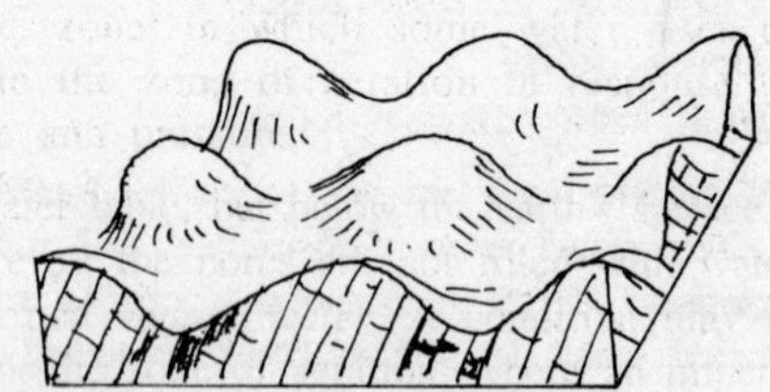

E. Cokpits/Intersecting Star Shaped Doline

Figure 42: Major Doline Classes

The floors frequently swarm at the water table.

Uvalas (*Compound Sinkholes, Karst Windows, Blind Valleys*) - A compound or chain of intersecting dolincs, *uvala* (a Serbo-Croatian term) refers to relict surface forms of network of dry valleys wherein surface runoff is progressively abstracted into groundwater circulation. Similarly a *karst window* is developed when an unroofed subsurface stream emerges as surface valley in a window and thereafter traverses back into subsurface situation.

The term *blind valley* is often used when a normal valley disappears - downstream into a swallow hole. In a flat interbedded limestone, shale and sandstone country, *karst valley*, a special form of blind valley is common, where a surface drainage initially developed on nonsoluble rocks, later erodes into soluble rock. This is possible only when the subsurface drainage overpowers the surface drainage, not so common. Mammoth Cave area, Keutucky, USA illustrates a good such example. Large blind or karst valleys may be in folded or faulted rock usually elongated along tectonic axes, abuting against steep enclosing soluble or insoluble rocks are called *poljes*.

Solution valleys - Such features as sinks, swallow holes, caves, caverns and underground channels may on dissolution result in the formation of solution valley distinguishable from a river cut valley by the following facts:

	Solution valley	*River valley*
Transverse section	Steep-side U- shaped	V-shaped
Width	variable	less variable
Water	no water above the water table	contains water
Bed	may have water when bed strewn due to materials of heterogenous solubility	generally the bed is smooth and devoid of heterogenous materials
Tributaries	when present, tributary valley above are hanging over the main valley	tributary streams meet the main valley at the same level
*Natural bridges**	some rocks remanents in the form of natural bridge over the valley may be seen	no natural bridge

*a natural bridge is (a) left over segment of unroofed cave or (b) formed by progressive underground abstraction of surface flow, *either*

(i) across the rock of an entrenched stream meander, or

(ii) from a high-level stream into a lower adjacent stream

Stylolites - These are products of solution action of underground water along the planes of separation of adjacent carbonate beds, one over another in which less soluble vertical columns project into each other forming an irregular zig-zag junctions resembling "suture joints".

TRANSPORTATION

Unlike by other geological agents, the only mode of transportation by the underground water is to carry the dissolved substances in solution by percolating waters. They may

(i) reach to the seas or lakes by underground percolation,

(ii) add to the stıeam water to be carried on to the seas or lakes, consequently increasing the water salinity therein,

and (iii) be deposited, at least partly, at times into the interstices of sediments to act as a binding cementation works.

DEPOSITION

The following factors control the deposition of dissolved material from the subsurface underground water:

(a) *Loss of CO_2 and other dissolved gases* - The dissolved gases are lost when subjected to

- heating,
- evaporation or
- reaction

on decrease of dissolution power. Like waters charged with carbon - di- oxide are active as an acid underground but on reaching the surface the gas escapes rapidly and carbonates released out from the solution.

(b) *Fall of pressure* - But for few exceptions, the solubility of substances in general lowers with decrease in pressure. As such when hot solutions from below ascends upwards towards the surface, their solubility diminishes and the solute is thrown out of the solution.

(c) *Decrease in temperature* - With the exceptions of a few a hot solvent ordinarily holds in it more dissolved substances than its cooler equivalent. Thus the solubility decreases with falling temperature and when the hot subsurface saturated solutions (but for carbonates), loose temperature on reaching the surface as a result of heat loss, deposition takes place.

(d) *Water evporation* - On evaporation of water at or near the surface, the dissolved solute tends to concentrate and saturates the solution, until the solution is over saturated when the mineral compounds are left out as deposits

(e) *Chemical reaction* - As a consequence of reaction between

(i) mixing of two percolating solutions,

(ii) gases and solutions, and

(iii) solid particles/bodies and solutions deposition may be caused.

Depositional Products and features -

A. *Cave deposits* - These deposits are formed only in the vadoze zone, above the water table, where air too, is filled in the voids. The vadoze water charged with CO_2 percolates down the surface with dissolved $CaCO_3$ in it and emerges at the cave roof in limestone region forming droplets clinging on the ceiling. Some water and CO_2 being gas escapes from the water on

(i) pressure release,

(ii) temperature loss, and may be

(iii) little evaporation (?) too.

Thus a deposit is formed, over which another depositional covering succeeds similarly with the addition of more water, forming a long "icicles" of calcite growing from the roof downwards which hangs as pendant-like masses. Such deposits are called *stalactites,* which when cut transverse, show their successive growth shells.

Part of the solution falls down under the action of gravity alone on the cave floor, where accumulations grow as above upwards from the floor, also showing shelly nature when crosscut. These deposits are termed *stalagmites.*

Continuance of growth of stalactites from the roof downwards and stalagmites from the floor upwards may end up with pair coalescing turning into development of columns called *dripstones* which chamber the caves. Such cave-ornamentation of fantastic shape is curious to the visitor.

B. *Geode* - Completely or partially filled cavities by deposition from underground solution are known as *geodes.* Generally silica, but rarely calcite form such filling substances. Often amorphous silica is so deposited layer by layer producing characteristic "liesegang banding" typical in agates. This silica is generally deposited from colloidal solution. Crystals when developed grow centreward of the cavity resembling the teeth of a comb. The structure so developed is called *comb structure.*

C. *Replacement* - Underground water carrying some dissolved substance replaces host substance volume by volume and even molecule for molecule without disturbing the pre-existing fabric but effecting change in chemical composition. In doing so, the host substance is removed by the solution and the substance brought in by it is deposited, perfectly maintaining the morphology. Such forms are *psuedomorphs* (psuedo=false; morph=face or from) for instance, $CaCO_3$ be dissolved and silica deposited in its place. In this way sometimes the woody tissues of tree- trunks are replaced by equal amount of deposition material, and the cellulose is transformed by silica into a hard rock like material, but still preserving the woody structure. This process is called *petrtification* and the transformed material as *petrified* wood. Fossils shells may also be replaced likewise and the replacing substance may be silica, calcium carbonate, iron pyrites, oxide of iron, calcium or barium sulphates, *etc.*

D. *Concretions* - Odd-shaped nodules occurring in sedimentary rocks adequately demonstrate the capacity of water to dissolve substances and redeposit by their precipitation around some foreign organic or inorganic materials. They are called *concretions* and the Indian "kankar" is a typical concretionary deposit of calcium carbonate over a silica/ siliceous nuclei.

KARST (GEOMORPHIC) CYCE OR KARST CYCLE OF EROSION

There is probably no type of landscape to which the concept of cyclic evolution can be applied with less complication than a karst landscape (Thornbury 1966). Here

(a) the nature or type of rock involved is one, *viz.* limestone
(b) generally the structure is simple, but for faulting in Yogoslavian region and folding in the Great valley of Appalachian terrain which introduce some minor departure from simplicity, and
(c) dominance of only one geomorphic process, *i.e.* solution.

Karst geomorphic cycle of landscape begins and ends up with surface drainage and it is for this reason Davis contended himself by considering it as a special and transitory phase marking the mature stage of normal fluvial cycle over lithology favourable for (Figure 43) solution action. Widely accepted and recognized karst cycles includes four stages of evolutionary development.

(a) *Youth stage* - In this stage surface drainage is inaugurated on
  (i) original bare limestone uplifted above the base level, or
  (ii) one that has been entrenched over limestone uplifted above the base level after removal of insoluble coverings of sandstone and/or shale by erosion.

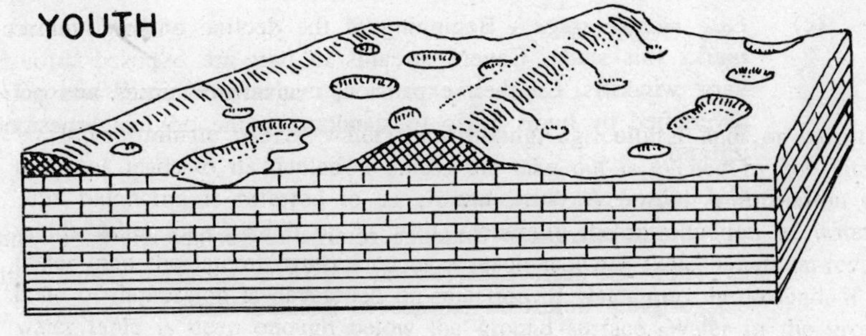

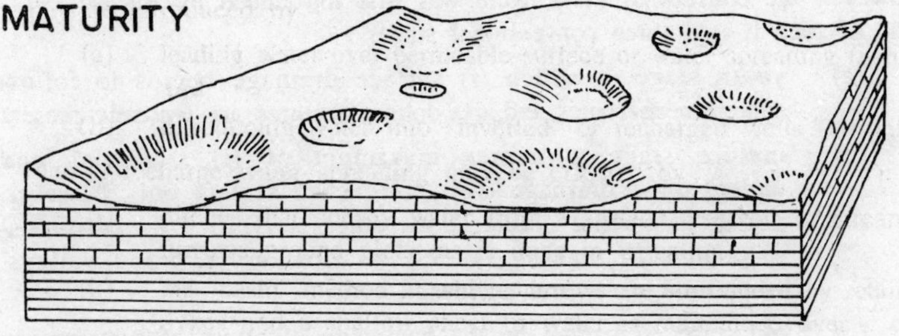

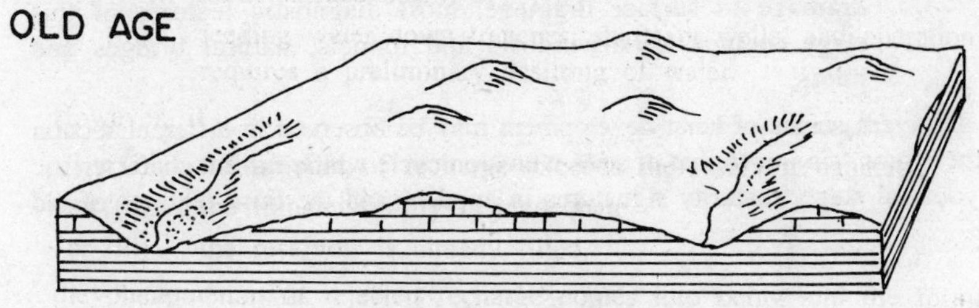

Figure 43 : Karst Erosion Cycle

This surface drainage on progressive expansion sinks down below as underground drainage, though is rather far from complete. In the absence of large caverns, the stage is characterized by the development of sink holes, lapies and stream dolines.

(b) *Mature stage* - This is the stage of maximum karst development, characterized by

(i) maximum subterranean drainage, and

(ii) limited surface drainage only as short sinking cracks which end abruptly in shallow holes or blind valleys,

(iii) extensive development of cavern networks and formation of travertine deposits.

(c) *Late mature stage* - Beginning of the decline of karst features marks this stage. Cavern streams in part are exposed through karst windows. On their expansion, uvalas are formed and polje diversified by hums begin to standout on the original limestone upland

(d) *Old stage* - At this stage

(i) the subterranean drainage returns to surface drainage, and

(ii) now the surface expression is like a peneplain dotted with original limestone surface as isolated hums.

However, the concept of karst cycle was first introduced by Beede (1911) who divided it into three conventional stages -

(a) youth stage in which (i) surface drainage begins to follow subterranean routes and (ii) dolines features are typically present

(b) mature stage witnesses maximum of (i) sinkholes and underground drainage well entrenched stream only however, persist as surface drainage; the stage is typified by numerous-thousands of sinking creeks, swallow holes, blind valleys, compound sinks and dolines.

(c) old stage is marked by the beginning of return of subterranean drainage to surface drainage; most diagnostic features of this stage include karst-windows and tunnels, natural bridges and hums.

Different stages of karst development may be observed in different section of the region, similar to that of stream erosion cycle which exhibit characteristic of youthful stage, maturity signatures in another, and the third features of old age.

CHAPTER XII

DESERTS

The factors and thus the processes operative in humid or arid climates are often the same and yet the topographic forms in arid rigions are not identical because of their different

(i) intensities, and

(ii) conditions.

The deserts are typically noted for high temperatures. The most diagnostic features include

(a) paucity of precipitation, and

(b) its sporadic distribution.

Daily (i) high temperatures and (ii) low precipitation cause the rate of evaporation rather high to the extent that the amount of water lost by evaporation is appreciably greater than the amount of precipitation. Evaporation, being in excess over precipitation, leads to the growth of only sparse vegetation in arid regions. As such weathering, particularly of chemical nature and consequently soil development is rather limited. Largely streams are intermittent and they are usually active with the precipitation only during rainy periods when the channel valley is filled partly or wholly with water.

ORIGIN OF DESERTS

World's deserts are distributed

(i) in mid-latitude regions,

(ii) in the lee of high mountain ranges, and

(iii) along cold-water coasts

Accordingly, deserts may be genetically grouped as -

(a) *Polar deserts :* These are limited to areas blanketted with great ice-sheets of Antartica and Greenland where moisture available is insufficient in quantity, but trapped as frozen, and thus not available for plant growth, simulating desertic conditions. Such droughts in polar regions are termed *"physiographic drought"*

(b) *Coastal deserts* : Coastal regions may or may not be affected by ocean currents-cold or warm. Desert conditions often prevail in regions affected by cold currents, *e.g.* the west coast of South America washed by the northward Humboldt's Current through which cold water moves equatorwards. Moisture-laden wind blowing over the cold current is cooled causing condensation. Subsequently, the wind, now drier, moves landwards, relatively becoming warmer being away from the cold currents, whence the ability to hold water, whatever little still remains unprecipitated, increases.

(c) *Middle-latitude deserts* : Also referred to as "*topographic deserts*" because

(i) they are situated far in the interiors of large terrestrial masses that constitute the continents, or

(ii) of the positioning of high mountainous regions normal to the paths followed by the prevailing winds.

Of the above two, the latter, (ii), seems to be more common cause of such desert development. In such a situation, the warm moisture - laden wind is deflected upward in its attempt to pass the high mountain ranges and in doing so, it gets an opportunity to expand and rarify, which it does, thus cooling adiabatically which causes the condensation of the air-moisture as rain or snow. Moving further, the leeside of the mountain range, however, the relatively moisture free descending air is heated again adiabatically, whereby the moisture laden capacity of the air increases further. However, after condensation the cooling rate lowers simulating conditions conducive for desert development. For each 1,000 feet (303 meters) rise, air cools by 5.5^0 C until condensation occurs down to 3.2^0C/1,000 feet (303 meters). The temperature rises 5.5^0 C for each 1,000 feet (303 meters) air descend. Accordingly, the descending air from an elevation of 10,000 feet (3030 meters) down to 5,000 feet (1515 meters) in the lee of the mountain range crest would warm upto 5x5.5^0 C or 27.5^0 C enhancing its moisture holding capacity manifold.

(d) *Low - latitude or Tropical deserts* : The air circulation in the world is a function of

(i) solar radiation and

(ii) earth's rotation.

Air temperature rises due to solar radiation near the equator and expands, becomes lighter as it rises up at higher altitude over relatively denser air, and moves poleward. Through its journey, it descends down to the surface from high altitudes near the "horse" latitudes (about 30 latitudes) of the subtropical higher pressure belts. During its downward descend the air is compressed leading to increase in molecular collisions which causes the air to be heated adiabatically. As the moisture holding capacity of the air bears linear relationship, increase in temperature enables the air to hold moisture with it, warm descending hot air leaves a dry effect. Consequently, the area beneath receives rather a low rainfall and the rate of evaporation becomes higher.

THAR DESERT

Almost of the same age or slightly younger than the Sindhu(Indus)-Ganga alluvium (of Pleistocene to Recent age) are the wind - borne wide expanse of aeolian sandy accumulation lying in the west of the country is the great Thar desert tract of India extending between 31^0 N, and 21^0 N, and between 69^0 E, and 76^{0}E stretching from the west of Aravallis to the basin of Indus, and from the southern confines of the Haryana and Panjab plains, the basin of Sutlej to down south latitude of 25^0 N occupying an area of 650 km long x 250-300 km wide i.e. 295,000 sq km enveloping below much of the terrain's solid geology. Of this, about two-thirds lie in Rajasthan. A small triangular patch between the Jhelum, the Sindhu (Indus) and the Salt Range, referred to as *Thal* desert, occurs in the Pakistan Panjab. It is believed to be a recent desert essentially of Holocene origin, developed due to

(i) post glacial climatic desiccation, as also

(ii) presumably by unmindful desertification activities induced by man. The latter could be of significance only for a part of the vast expanse of the Great Indian Desert. Historical documents indicate that at the time of invasion by Alexander, the Great (4th century B. C.), the region north of Kutch and south of Haryana and Panjab was fertile and forested and supported cities and township of sizeable population. As of now, evidences are forthcoming to support that desertification of the Thar is possibly referrable to late Pleistocene, before which it was under the sea right from the Permo-Carboniferous until its rise sometimes in Upper Tertiary. Small microlithic tools in good quantity scattered over some dunes, indicate their stability presumably since the mesolithic times when the early man dwelt over the terrain.

The northern part of the vast Thar Desert is classified as sand desert of subtropical zone and the southern part lies in the tropical zone. This enormous Thar tract comprises

(i) the proper desert or *Marusthali* with 200-250 meters high sand dunes above the sea level. Not a single river flows through this desolate land. It is fringed by

(ii) the semi desert known as *Bagar* which includes some sand free patches, may be alluvial fans to the west of Aravalli with steppe vegetation. It indeed is a semi-arid plain with flat surface enclosing in some parts salt-soaked *playa* depressions, locally referred to as *sar*

The sandy monopoly of the desert is diversified by

(a) low elevation hilly projections, as also

(b) sand dunes and hills

The deserts generally are elevated upto about 325 meters, reducing to about 150 meters towards India's border with Pakistan. Great antiquity is evidenced through the desert. In some cases, the magnitude of denudation, dissection shown by deep ravine, developed by the water action through dune attests the prolong operation for long time-span during which such distinctive features could produce even when the rainfall is rather scanty. Features like

(i) calcification,

(ii) deep gullying, and

(iii) marked weathering horizons,

bear testimony to the desert stability for long periods of time, as in Gujarat. The older rocky prominences above the sands are curiously worn down to give rounded appearance. They present sand-blasted topography striking desert-erosion illustrations. The commonest type of the dune is U-shaped parabolic with a rake-like form which occurs transverse to the dominant early south-west monsoon winds. The ridges are 10-15 meter high 150-200 meters wide and 5-10 km. long. These dunes strike

(a) parallel to the wind direction in the southern part where the force of the wind is great, and

(b) transverse to the prevailing winds elsewhere in more distant parts of the deserts where that force has abated.

The windward slope is long, gentle and undulatory, while the opposite slope is more abrupt and steep. In the southern part these ridges, locally called *bluts* are of larger size 50 to 40 meters high. The dunes by and large are slowly progressing. The crescentic type (*Barchan*) of dunes (200 meters long and upto 20 meters high occurs in groups) are migratory in nature and more common in the interior.

Within *seif* chains are large number of depressions, salt lakes or hollows three meters deep, elliptical in shape and aligned parallel to the ridges of the longitudinal sands, locally called *dands* or *dhands*. Those dhands without outlets become huge evaporating pans, which after floods increase in salinity.

Composition of the desert sand - The well sorted and uniformly rounded by wind attrition during their drift, quartz is the commonest mineral constituent of these sands in which feldspar and hornblende also occur with a fair proportion of calcareous substances as clasts of marine foraminiferal shells which offer a clue to the site of origin of the sands with which they are intimately mixed. But for their high degree of rounding, these sands are indistinguishable from the sea shore sand. These desert sediments include two populations

(i) a saltation population, and

(ii) a suspension population.

Weathered sand or reddened/brownish sand contrasts with fresh sand in its increased content of silt and clay-18% in Rajasthan and 33% in Gujarat. In addition, widespread kankarized sand is found in inland parts of the Thar. The sand is held together in it by calcium carbonate nodules. This sand is lighter in colour and its $CaCo_3$ content may reach as high as 50-60 % in comparison to 27% for most sand dunes in inland situation. In the eastern and north-eastern parts, in the semi-arid region, the desert is covered by shrubs and stunted xerophytic vegetation. The tract with thin sand covered, supported some cultivation on the availability of water which indeed is a problem in this region. Some, now abandoned villages and towns are manifestations of such activities in the near past.

The subsoil water quality varies from locality to locality. In some more eastern parts at times the water quality is tolerable for

(i) human consumption, and

(ii) agricultural use,

but in some other parts the well waters are distinctly brackish to saline over which, of course, freshwater may afloat during the monsoon on percolating down from the surface temporarily because of inland drainage in the area although a view, of course, untenable now, attributes it to transporation of salt from the Gulf of Cambay and Rann of Kutch by the southwest monsoon. Recent investigations support, if at all, the wind carries very little saline matter from the sea at least in the more interior parts of Rajasthan.

Seemingly, the aridity in the region dates back hardly to last 3000-4000 years for Sind and Baluchistan as also the neighbouring parts of Rajasthan were forested and, therefore, experienced at least relatively humid climate in the pre-and early-historic times, a fact well borne out by the use of burnt bricks for buildings. Well designed drainage systems in the settlements of Mohanja-daro times further attest the prevalence of more humid climate. History of the region is also on record to indicate that wild elephants lived in Rajasthan deserts which have been more moist than those obtaining at present to sustain such population. However, the situation was much more different during the 4th century when Alexander the Great, invaded Panjab and Baluchistan. They had already transformed into arid regions. Rivers of that time, while traversing through Rajasthan, fought rather a loosing battle against the desert inroads, reminiscences of which are now traceable only through dry, forsaken channels with rather a flat-floored, wide (to carry massive bedloads) and steep-walled (because of the strong vertical or columnar jointing in dry soils valley geometry). Their longitudinal profiles are steep. When dry, which most of the time they are, floored by alluvium. In Spanish speaking regions, these distinctive channels are called *arroyos, (wadi* in Arabic or *onadi* in French).

Origin of the Rajasthan Desert : The Indian desert owes its origin

(a) primarily to a long-continued and extreme degree of desiccation of the region

(b) in addition to the sand transportation by the southwest monsoon winds that prevail over the region most of the year though subordinately to

(c) its location in the neighbourhood arid land areas of Arabia and African horn, as also

(d) relative narrowness of the sea and thus cannot be the source of moisture- laden winds (on one hand, and on the other hand, the Bay of Bengal branch of the southwest monsoon winds virtually dies up at final leg of its journey through wide coverage across the country)

(e) prevalence of upper air westerly currents in the region that is strong enough to cut any ascending air, whatever little they may be

(f) mass of warmer air originating in the drier northwest region occupies and establishes stability in the upper atmosphere of the region that prevents any ascent of the air and condensation, and

(g) to the non - precipitation of the moisture these winds carry because of the alignment of the Aravalli mountains to the wind direction and thus lack of any impediment in the way to force the moisture to condense down as rain.

As such, in the absence of any moisture, the dynamic wind is capable of blowing over tons of dust and sand-particles, like dark clouds, received in huge quantities from the sea-coast from the Rann of Kutch as also from Lower Sindhu basin. Such sandy accumulation is an ongoing process in the present situation because of virtually no rains being much above 12.5 cm. and therefore, the thick pile of sand remains where deposited, for they are not in a position to be transported by water to the sea. In addition the rocky prominences of the tract, on weathering under the influence of great diurnal and thermal variations, which may be of the order of 40^{0}C in some parts, further contribute to the depositing sand. The seasonal variation is greater. Such thermal weathering leading to a mechanical disintegration and *desquamation* of rocks produces huge quantity of loose debris that remains as such in the absence of chemical or organic action to transform it into a soil, particularly in the absence of vegetal matter.

Thus, "the Thar offers instructive illustrations of the action of aeolian agencies" (Wadia 1984). There is strikingly sudden contrast of the topography on passing over from Gujarat or even Madhya Bharat to the country west and south of the Aravalli punctuated with bare and bold hill, masses and the peculiar sand-blasted treeless landscapes for several kilometer around under a cloudless transparent sky. The terrain is blanketed with thick pile of mechanical debris produced by powerful insulation, the disintegration of the bare rock-surface by desquamation, the saline and alkaline efflorescences of many parts, and the general absence of soil and humus. The growing salinity of its lake basins (by wind-borne salt dust from the sea-coasts) in the Rajasthan desert is more subtle and less understood phenomenon so far.

The climatic analysis supplements the following physiographic observations -

(i) the rivers are misfit and soon die out due to alluvial suffocation with no permanent flowing river draining through but for the Luni, the origin of which is located in high rainfall terrain that enables it to carve its own valley,

(ii) lack of integrated drainage network system and absence of controlling level to which the landscape is lowered (the rivers rather lay their load and sink down into ground alluvial surface)

(iii) gullying through, and dissection of mountain yield debris which are accumulated in the intermont depression, and

(iv) lateral confluence of alluvial sands and formation of sloping *bajadas* along mountain fronts are characteristics of an arid climate. Annual rainfall is less than 500mm (500 mm isohyet is the scientific basis for cartographic expression to delimit the desert)

It is observed that Pleistocene climatic oscillations caused the development of middle and low latitude true deserts. Subsidence of the Ganga delta and subsequent subrecent rise of Aravalli-Delhi axis led to the gradual Eocene seafloor rise of the Thar, as a consequence of which, the Yamuna, hitherto a tributary of the Saraswati (the two together constituted the Ghagghar) of the Sindhu drainage system, shifted to join the Ganga drainage network. The increased aridity is attributed to the

(i) rise of the Himalaya, and

(ii) lowering of the Aravalli

may be responsible for the shifting and desiccation of rivers like Saraswati, and gradually making the desert unhabitable.

The Planning commission of India reported (1952) "The great Indian desert of Rajasthan" is spreading northwards to non-desertic areas of Haryana, Delhi and Uttar Pradesh. Desertification around the periphery of Thar is still rather speculative. Computation of rate of about 0.8 km/year of spreading of the desert and "is approaching approximalely 125 sq km of fertile land" is unacceptable to Roy and Pande (1970) who opined that "the desert is neither expanding nor contracting" but has been occupying essentially at the same position". Later Manu *et al* (1974) expressed "at least tentatively hard to support the popular myth of the deserts spread".

Recent geomorphic studies brought to light that the present scenario is the outcome of the prevailing wind regime superposed over the fossil longitudinal dunes and sand piles and developed to its present existence in about 10,000 years B. P. , through more than one cycle of Quaternary climatic changes and associated desertification and de-desertification (Table XII.44). Integrated prior drainage network in the central Luni basin, in the desert plains in Pali, Jodhpur and Nagaur districts are well documented, which deposited thick layer of alluvial materials over much of today's desert within the stratigraphic column which indeed acts as sweet water reservoir, is indicative of past wet phase in the region. Thick caliche deposits at varying depth with the alluvial layout are also recorded.

Physiographically, the Aravalli constitute the eastern limit of the Thar and stands majestically high to prevent penetration of wind blown sands eastwards, but for several breaks, like

(i) the Kantali-Sota Sabe break (which does not reflect any penetration across it)

(ii) the Sambhar gap (an ineffective one)

(iii) the Luni gap, through which wind blown sands of Alwar and Jaipur districts could have been deposited, and

(iv) the Jawai gap or the Sirohi saddle through which wind blown sands penetrated along the Banas river

that could allow wind blown sand accumulation eastwards. However, the penetration through these gaps meets the following doubts:

(a) dynamically, the wind can transport sands of less than 0.01mm in diameter. The annual rainfall near Jawai gap is 630 mm increasing eastwards to more than 700 mm This high precipitation is expected to moisten the windward sand for deposition much before the Jawai gap or Sirohi saddle crossing

TABLE XII 44 : GEOMORPHIC EVIDENCE OF DESERTIFICATION/DE DESERTIFICATION DURING THE LATE QUATERNARY PERIOD IN WESTERN RAJASTHAN

(after Bimal Ghose, Surendra Singh and Amal Kar 1977)

Period in years B.P.*	Major Geomorphic Events	Geomorphic Processes	Climatic Phases	Desertification/ de-desertification
Early centuries A. D.	Increasing fluvial activities but increasing biotic stress on the land has resulted in salinity hazard along prior drainage channels and more deflation and deposition of sediments in the sandy plains and duny tracts	Biotic activities	Slight amelioration of climate	Smal lscale amelioration of climate but biotic desertification is present
Early centuries A.D-3,800	Formation of low active dunes like barchans, shrub, coppice *etc*., reactivation of the flanks and crests of the stabilized dunes. Inland basins turned into saline depressions	Aeolian	Arid	Climatic desertification
3,800-10,000	Major dunes stabilized drainage channels revived, sandy inland basins filled with fresh water and fluvial sediments	Fluvial	Humid	Amelioration of climate to de-desertification
10,000 -?	Formation of major dunes disintegration of drainage system by aeolian sand; creation of sandy inland basins	Aeolian	Arid	Climatic desertification
?	Well interated drainage system; vast alluvial plaines	Fluvial	Humid	No evidence of desertification

*B. P. means Before present: Alluvial material (dated by C^{14} method).

(b) total dearth of loosened sand particles near the vicinity of the gap along the prevailing wind

(c) on the contrary, the deposits are away from these gaps and the velocity of the prevailing wind is questionable in as much as to spread sands over such a long distance

(d) sand characteristics in the Banas river basin in the east do not match well with aeolian origin.

These together with studies of Survey of India sheets do not support the penetration of wind blown sands eastwards. On the other hand, studies indicate an increase of desert sands within the arid regions and reveal

(a) the ongoing fresh sand deposition,

(b) the increasing trend towards N. W. in Panjab and Haryana, as also in Jodhpur district (Rajasthan),

(c) strong desert creep westwards towards the desert itself rather than in the semi-arid neighbourhood

(d) decrease of desert sand in an area is closely associated with an area of increased sand nearly resulting in fertile land deterioration, and

(e) inaccessibility of the presence of Aravalli gaps as sole cause of wind blown sand penetration eastwards, particularly in view of the ground truth.

The deterioration of the desert may possibly be summarised as below :

(i) The uplift of the Himalaya and lowering of the Aravalli led to shifting and desiccation of the Saraswati river, which became fatal to increase aridity. The Pleistocene rise of Potwar plateau probably adversely effected Himalaya's drainage system, causing shifting (the river Saraswati) and capturing (Yamuna by the Ganga river system) of some streams which affected the conditions within the Indian desert;

(ii) inevitable lowering of the water table as a consequence due to increased aridity,

(iii) increase in the groundwater salinity,

(iv) physiographic changes in recent times,

(v) practice of cutting down trees and woodlands, uncontrolled grazing and uneconomic exploitation of natural resources,

(vi) cultivation of dunes, responsible for sand shifting leading to deterioration of adjoining fertile lands.

Extension of desert-like conditions at the outskirts of a desert boundary or their increase or intensification within the peripheral limits of a desert, are triggered by a set of processes that involve mutual interactions of

(i) climatic,

(ii) edaphic, and

(iii) biotic

factors.

Desertification may be defined as the spread of desert-like conditions, outside a desert boundary or accentuation of them within a desert, triggered by a package of processes that involve an interplay of

(i) climatic,

(ii) edaphic, and

(iii) biotic

factors, which may be

(a) operating with the start of a climatic accident until resulting finally in the desert formation,

(b) leading to an accentuation of the xeric conditions within an already existing desert regime, or

(c) causing desert sprawling from within a desert tract outside towards its neighbouring areas.

Accordingly, the term "*climatic desertification*" is coined for the desertification process induced by the climate, in contrast "*biotic desertification*" basically accentuated due to biotic interaction. Aridity takes its roots

(a) soon on climatic accident,

(b) which

  (i) weakens precipitation, or

  (ii) the moisture-laden wind

by large scale downward settling of airstreams. Under such conditions evaporation potential gradually built upto offsets the annual precipitation leading to

(i) irregular or erratic groundwater supply,

(ii) lowering of the water tables, and

(iii) disintegration and desiccation of the drainage systems

which contribute to the initiation and furtherance of an "*arid cycle of erosion*" causing

(a) misfit streams that short-lived and die out due to excessive desert sands or alluvial suffocation,

(b) scouring on the hills through wind-driven sand causing caves, cavities and fluted channels along the hills-Zugen, mushroom form and Yardang formation,

(c) formation of depression or "playas" which are converted into salt lake,

(d) formation and movement of sand dunes,

(e) pediplains and inselberg topography due to slope retreat of hills, plateau and intermont basins developed on prolonged wind erosion,

(f) typical vegetation with xerophytic tendency and aridosols of deserts.

Human activities including wrongful and excessive use of natural resource of a desert set inequilibrium in the desert eco-system and upset balance between

(i) rainfall, and
(ii) landuse

and thus accentuate the desertic conditions leading to "biotic desertification" by way of

(a) over and uneconomic exploitation of resources,
(b) overgrazing by domestic animals and wild ungulate ,
(c) deforestation and indiscriminate chopping of trees, and
(d) uneconomic landuse practices - cultivation of sand dunes and marginal lands causing sands to shift and deposit on the adjoining fertile lands.

These human interferences cause

(i) soil erosion,
(ii) creep, and
(iii) expansion of desert

DESERT TYPES

On the basis of

(i) surface form, and
(ii) composition

deserts may be described as

Rocky Deserts - In this type, a group of small hills alternate with rough crested mountains enclosing valleys with steep sloping fissured walls composed of hard crystalline rocks that look like pitted, wooden logs on account of their comby and cellular structure. Because of the desert varnishing at times, talus debris and outcrops appear like pig iron casts. The pedestal here are produced by the transport of the torrential rain descending on bare slopes where its lead (proluvium) is deposited. They are less towering in humid regions. Valleys are covered with rock waste. Vegetation grows on some of them. Some have springs which disappear soon on rock waste.

Stony Deserts- Comprising *either* rock waste (hammada) or pebbles (serir) over the gently undulating "varnished" surface of which lies a sandy-clayey soil, these are residual products of deflation, found extensively in Sahara.

Sandy Deserts - Include loose sand areas, barchans and dunes, generaly poor in vegetation, but when present they are called steppe type. Bed rock exposures with shrub growth are met in the intervening short valleys.

Clayey Deserts - They cover limited areas. The flat shores in inland basins and in some large lakes are clayey deserts (Mediterranean, Aral,

Caspian). Their surfaces formed by the deposition of fine silt by floods, are even and cracked (polygonal sections) in which vegetation, if present grows. The soil may sometimes contains salt. They are called *takyrs* in Central Asia.

GEOLOGICAL WORK OF WIND

Considering generally, wind, a draft of air in motion, is of low density and consequently a weak geomorphic agent, but it assumes a great significance in arid and semi-arid environments where its efficiency in modifying the land-surface is remarkable . Its work includes erosion, transportation and deposition of sediments which is a function of its

(i) kinetic energy, and
(ii) the rate of energy utilization per unit area,

EROSION - Three processes

(i) abrasion,
(ii) attrition, and
(iii) deflation

are included in the erosional work done by wind.

Wind Abrasion

Armed with sand-sized grains acquired during its motion, the wind acts as a powerful abrading agent. These sediments act as a tool in shaping the rock bodies. The sculpturing is achieved in several ways. Of these polishing is recognized as a natural effect of abrasion or sandblast action. Being of uniform character homogenous rock surface tends to be smoothened and polished whereas because of unequal responses in inhomogenous heterogenous rocks, softer constituents are more easily cut and their surfaces grooved and fluted. It is for this reason the relatively soft cementing materials are abraded faster than the harder gravels and nodules in conglomerates and are affected rather slowly resulting in their beehives appearance and thus present *honey-combed surfaces,* though differential weathering also has a role to play. Coarser materials like pebbles and gravels are worn and polished by wind abrasion, and show facetting on their one or more surface meeting in sharp edges. Such features are called *ventifacts* or *dreikanter*. For their development a combination of

- strong winds
- abundance of sand, and
- absence of vegetation

is necessary, a condition so uncommon and as such ventifacts or dreikanter do not occur commonly.

As the sand is seldom lifted more than 2 meters above the ground, the wind abrasion is conspicuous within a few meters of the earth's surface. The maximum effect is limited below this upper limit, but on the lower side, it is little above the ground level, as the wind velocity is impeded due to surface friction close to it.

With this limited operational height, little above the ground, but generally below 2 meters undercutting of rocks is a marked feature leading to the development of *pedestal* or *pivotal* rocks carved out on granites of Madan Mahal, Jabalpur or in western Rajasthan near Jodhpur as mushroom shaped blocks.

Wind Attrition

The sand particles in the transporting wind system not only strike the upstanding rock that they meet on their way, but also collide with each other, because of their different size and shape. All the fragments contained therein do not travel with the same velocity and because of their differential motion in the same wind system, they strike one another mutually and breakdown further with size reduction to be ultimately converted into sand and dust particles which cover large parts of the deserts. This process of mutual striking of particulate material with one another is called *attrition*. The particles in this process tend to become rounded too, because of smoothening out of angular corners due to mutual striking impacts developing into *millet seed grains,* so called because of its morphological resemblance to millet seed.

Wind Deflation

Deflation can

a. (i) winnow silt and fine sand
   (ii) reducing the surface level, and
   (iii) concentrating rocks and coarse sand in a protective surface lag
b. scoop blow-outs, the largest of which may be kilometers across upto 100 meters deep,
c. carve out sharp, sinuous ridges (yardangs) and parallel depression in dry silts with a local relief of upto 9 meters, and
d. sand blast rock surfaces, sometimes undercutting rock outcrops within 30 to 60 cm of ground.

Of these, lag surface mantles most stony deserts. Lags of crude, mechanically weathered regolith constitutes the *hammada* or rocky desert, with pavement of rounded gravel and coarse sand produced by deflation of alluvial deposits to form *serir* or pebbly deserts. Blow-outs and yardangs have rather limited occurrences.

Deflation is the process of scooping away the sediments or chemically weathered materials and thus prevents closing up of depressions for filling in with detritus. The material is transported to be heaped up elsewhere. This process may produce great erosional basins in the desert to form *oasis* such as the one of the Libyan Desert.

The wind erosion intensity is the function of

(i) the nature of the terrain - made up of resistant rocks or otherwise protective vegetative cover present or absent,

and, (ii) the wind velocity

FEATURES OF WIND EROSION

Features of wind erosion enormously vary in size from as small as grooves, a few millimetre in length to as large as deflation basins measuring several hundred square kilometer in area as a function of

(a) relief,

(b) relative hardness of rock, and

(c) wind regime

They develop stable streamlined forms primarily due to the fact that the aerodynamic shape offers least resistance to the movement of the wind. Some such landforms are listed below (Table XII 45).

TABLE XII. 45 : LANDFORMS OF WIND EROSION

(source ?)

Land-feature	Size-class	Process
Grooves	small to large	Corrasion
Ventifacts	small to medium	(abrasion)
Yardangs	large	Corrasion and Aerodynamic erosion combined/
Pedestal rocks	medium	Corrasion, weathering and rainwash
Deflation hollows (blowout depressions)	small to large	Deflation
Lag deposits	mediu	

Grooves

Minor depressions, grooves on resistant rock walls or projections related to wind abrasion (corrasion) are attributable to

(i) differential weathering caused by abrasion, and

(ii) rainwash

presumably developed in drier climatic phase of late Pleistocene. Active impressive grooves are noted in association with the present day desert environment.

Ventifacts/Driekanter

Wind erosion induced polished rock fragments, faceted by combination of several flat surfaces meeting at sharp angles in vegetation-free hot deserts, depict pits and striae parallel to the wind direction and their long axis vary a few millimeter to upto 3 meters. The ventifacts are believed to be a product of abrasion by salting grains. The rock fragment or stone on the desert floor is abraded on its windward side to produce a flat surface until it turns or rolls whence the surface now exposed to the wind action is subjected to abrasion producing a new flat surface (Figure 44). The process may go on to produce several such flat surfaces on the rock fragments whose morphological attributes may be explained by wind flow dynamics around the ventifacts. Certain aspects of windflow dynamic observed in relation to the ventifacts geometry and surface features thereon include

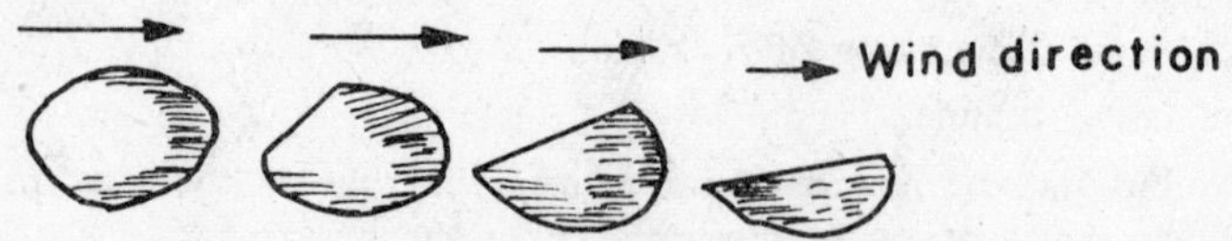

Figure 44: Ventifact Development

(a) negative and positive flow including
   (i) basal burnishing and
   (ii) strongly concave basal surfaces
(b) vorticity and local velocity - a gentle vorticity is set up which erodes a basal concavity destroying most of the basal margins. In due course the contact between basal surface and ground is lost facilitating a turn or roll of the stone, bringing a new impact face against the prevailing wind.

Pedestal Rocks

These curious features with wide rock caps balancing on narrow columns are conspicuous structures produced by wind erosion on upstanding rock masses. Such sculpturing is produced because the blown particles are not carried higher more than 2 meters, only the lower regions of upstanding rock masses are abraded and thus undercutting the foot regions slender without much affecting the rock mass above this zone, resulting in a wide rock cap fan standing on a slender rock column, a shape matching with the pedstal fan and accordingly they are called *pedestal rocks* of which a good illustration is seen in the granites of Madan Mahal near Jabalpur. These features are found in many semiarid regions. Instances of such features occurring over 1-2 meters in producing such features is questionable and as an alternative Bryan (1926) offered a possibility of differential weathering and rainwash while Blackwelder (1931) attributed the development to wind deflation. However, exfoliation on sheltered surface, is now regarded as one of the causes for undercutting.

Deflation Hollows

Strong winds are capable of deflating sand-sized particles particularly, in the absence or near absence of vegetation from the weakly indurated exposed rocks and desert sands producing *blowouts* of varying shapes and sizes upto several square kilometer in area and their depth is a function of

(i) the protective cover of gravel, or

(ii) the subsoil water depth

The moist sand has a tendency to resist further deflation of the basin floor. Sediments produced on mass wasting and escarpment retreat are involved in the deflation processes (Ball, 1933)

Lag Deposits/Desert Pavement

Selective winnowing of sand and silt particles from the desert floor by wind leaves a cover of a fairly smooth surface of gravels and pebbles called *lag deposits* or *desert pavement* limiting the lower vertical depth of erosion by deflation. The residual exhumed surface known as *hammada* (in Sahara), *gibber plain* (in Australia) and *desert armour* (in the United States) become glossy by wind abrasion on these stones and brightly polished with a coating of silica or oxides of iron and manganese giving a yellow-brown to blackish brown colour to the stone. The coating is referred to as *desert varnish,* the mechanism of such coating which involves oxidation of salts in weathered products and in surface efflorescences is still not well under stood.

It is generally believed that the capillary rise from near the local moisture sources brings the mineral matter to the surface (Eagle and Sharp, 1958) in solution from which water evaporates under the influence of high thermal efficacy, leaving the solute to paint the large and small particulate materials. The soft finer and lighter and material is in time deflated out, leaving behind the desert with more durable and coarser material.

Yardangs

Yardangs are usually sets of ridges

(i) asymmetrical in shape-steep windward fronts, and

(ii) long tapering leeward tails,

and aligned parallel to the prevailing wind direction (Figure 45). In monodirectional wind direction they tend to be more symmetrical. They may be a metre to several kilometer long generally maintaining 3:1 length to width ratio and occur as streamlined wind eroded features in extremely hyperarid environment.

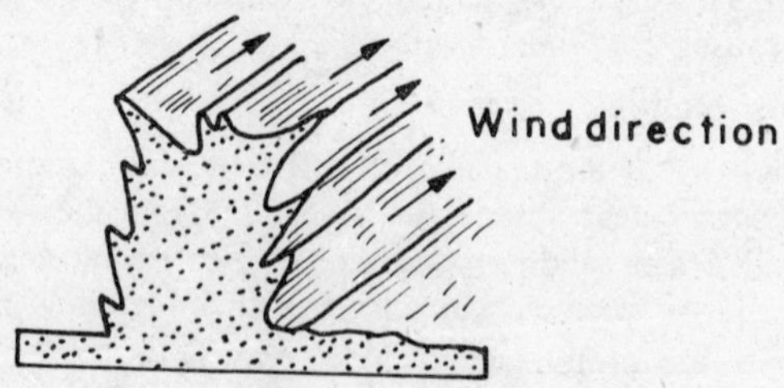

Figure 45 : Yardang and Wind Direction

Whiteny (1983) raised doubts on the efficacy of sandblast action for developing the streamlined form of yardangs. She observed that the wind erosion primarily occurs on the leeward side of yardangs. In monodirectional wind regions the rest of the surface is sheltered by high steep- sided windward face against the abrasion prime by air-borne sand, which necessitates consideration of other wind erosion processes on the leeward face of the yardangs. She demonstrated that the morphology of yardangs is aerodynamically controlled in which the steep windward face has developed due to sandblast by a strong primary windflow, and the long tapering leeward surfaces are because of deflation and abrasion by dust particles held in the interfacial flow in contact with the surface. In competent materials, erosion is, however, mostly dúe to suspended particles moving along beautifully burnished, long flutings on the leeward surface. Convergence of streamlines over the crest line leads to the development of flutes. Yardangs present in incompetent lacustrine clays and lake bed sediments are known as *mudlions.*

TRANSPORTATION -Wind transport is achieved by

a. *Suspension* in which silt-sized particles are removed
   (i) over great distances and
   (ii) to great heights in the atmosphere
b. *Saltation* wherein sand grains hop, and rebound, energizing other sand particles - a motion limited to short distances and heights generally not over 1.5 to 2.5 meters, and
c. *Surface creep* leading forward movement of coarse sand and small pebbles due to the momentum gained from jumping sand particles.

DEPOSITION- Depending upon

(i) the wind velocity,
(ii) sand/sediment morphology (size and slope), and
(iii) the volume of particles in the system,

the wind transports them to varying distances, until

(i) prevented by some obstacle which checks the wind velocity, or

(ii) precipitation takes place

whence the wind is off loaded and deposition takes place, such as sand ripples, sand shadows and drifts, dunes, whalebacks, loess *etc.*

The infrequent rains in the desert regions at times generate peakflows as streamflood or sheetflood and develop features like pediments, bajada (bahada) and playa (Figure 46)

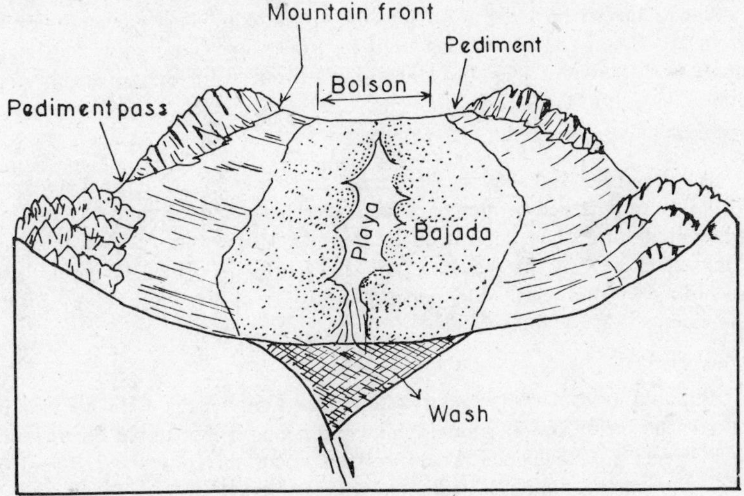

Figure 46 : Mountainous Desert Features

TYPES OF SAND DEPOSITS

The wind deposits by no means are restricted to either desert origin or location. The four non-desert environments in which sand deposits may be fixed are

(i) along shorelines,

(ii) along stream courses in semi-arid regions,

(iii) in areas where loosely cemented sandstones have disintegrated to supply sand, and

(iv) in areas of glacial outwash.

Five chief large scale sand deposit forms identified (Bagnold, 1941) within the desert itself include

- Sand shadow,
- Sand drifts,
- Sand sheets,

- Sand levees or whalebacks, and
- Sand dunes.

Sand Shadow

As the wind velocity is reduced the transported material is dropped behind in the shelter of the obstruction that meets on the way and is accordingly termed *sand shadow*. These obstruction may be boulders, cliff(s)or bushes. Such deposits do not rise more than the obstruction itself, for above the shadow the wind may be still strong. Sand accumulation goes on in the lee of the obstruction until its front siope attains limiting angle of repose and thereafter the sand slides down the front of the advancing deposit along the slip-face.

Sand Drift

A variant sand shadow, called *sand drift* is noted in a gap between two obstructions where the interspace acts like a funnel through which sand is transmitted forward to deposit on the leeside. Usually the sand drift connotes a thin sand deposit for which a term *sand sheet* is coined (Bagnold, 1941). Accordingly, some confusion persists in so far as the terminology usage in this connection is concerned.

Sand Sheets

Bagnold (1941) preferred this term for sandy areas with an extremely flat surface lacing in any topographic expression, except ripple ornamentation. Such sheets are developed where pebbles occur mixed with sand and wind lends the small mounds and is unable to build dunes. These areas are referred to as *sand seas*.

Sand Levees or Whalebacks

These are sand ridges with

(i) flat tops
(ii) aligned in the prevailing wind direction, and
(iii) lack the collapsing fronts found in the seifs.

They appear like the back of whales. They seem to be the residue of dune chains which have moved downwind and the coarser sand has been left behind as a prominent ridge. At times, these features may take any definite form and appear only as undulation. In either case they seem to be areas of fossilized or immobile sand particularly, in contrast to the barchan and seifs where the sand is energized and thus mobile.

SAND DUNES

Ideal outline for dunes is provided by

(a) uniformity of

(i) wind direction and

(ii) sand supply, as also on

(b) absence of rainfall

Their irregular shapes are rather unusual and must owe to some anomalous local conditions. In fact regularity of outline of individual dunes offers clue (s) for interpretation of general prevailing climatic conditions (Table XII. 46)

The principal dune form controls are

(i) wind velocity and

(ii) the nature of the surface over which dune moves, provided

(a) adequate sand supply and

(b) restricted wind direction exist.

TABLE XII. 46 : DOMINANT FACTORS CONTROLLING IDEAL DUNE SHAPES

(source ?)

Dune shape	Wind direction	Wind velocity	Sand supply	Character of floor
Transverse	In *one* dominant direction but range of direction may be high	Upto 40 km per hour. Become irregular (blowouts) at higher velocity.	Large beaches/ Alluvial valleys	Loose sand
Longitudinal	In *one* or two dominant directions but range is less than 30 degrees, generally only a few	Greater than 40 km per hour. Stable strong wind form	Moderate to large	Loose sand or hard floor, often continuous with rocky ridges parallel to wind direction
Barchan	In *one* narrowly restricted direction	24 to 40km per hour	Moderate to low limited	Firm floor, often a rock surface
Irregular	Variable and unsteady	Less than 40 km per hour	Low to high	Not significant

In steady prevailing wind regions three major types of dune forms are recognized

(i) Transverse

(ii) Barchan

(iii) Longitudinal

Modified varieties are

(iv) Blowout shapes

(v) Irregular or complex amoeboid outlines

In regions of erratic directional wind attributes, the dune morphology (Figure 47) too, is modified in various ways. Their forms may be simple and complex, grading into each other. They may be built, destroyed and rebuilt by the operative wind independent of the ground cover. Fossil sand dunes are practically cemented, relatively stable forms and occur in humid parts of Gujarat, Rajasthan and Haryana as also in many other deserts. Dunes display a great variety of forms. Hack (1941) on a semi-emperical basis showed a probable distribution of dune forms (Figure 47)in relation to

(i) sand supply (ii) vegetal cover, and (iii) wind effectiveness

Some major dune types and their characteristic are listed in Table XII. 47

TABEL XII. 47 : SOME MAJOR DUNE TYPE AND THEIR CHARACTERISTICS

(modified after F.J., Pettijohn, *et. al.*, 1972)

Type	Morphology	Relation to Wind
Crescent (trans-verse, barchan, sigmoidal and parabolic)	Heights upto 100 meters	All perpendicular to wind ; barchan has horns downwind, parabolic has horns upwind. Transverse dunes have relatively straight crest
Longitudinal (seif, parallel and whale-back)	As much as 100 meters high, 1 to 2 km wide and 20 to 200 km long. Linear, sinuous, crosswinds increase in height and width. Form spectacular "wind rows" in some continental deserts. May be straight or slightly sinuous Whalebacks are flat topped.	Extend downward roughly parallel to average wind vector
Complex (pyramidal, hooked dune "massifs" heaps, star and reversing)	Many and varied but pyramidal dunes to 200 meters or more	Variable, shifting winds plus merging dune forms
Sheets	Flat to gently undulating. Can cover wide areas	Accretion probably predominates over avalanching

TransverseDunes - These are dunes of common occurrences in deserts and coastal regions. They are usually regularly spaced, asymmetrical elongated ridges, windward sloping generally 1 to 2 degrees, but with steeper distal slope between 16 to 20 degrees, oriented at right angles to the prevailing dominant wind direction. In the Luni basin in Rajasthan they measure 1.5-2.0 km long, 12-15 meters high and about 400 meters wide. It is a feature where loosened unconsolidated sand material is in abundant supply. The feature is intimately associated with *sabakhas* or salt-encrusted depressions, generally near the water table. Being moistened sand, horns are not developed in sabakhas, as developed in drier desert regions. Deflation becomes strong when due to lowering of water table the interdunal sabakhas disappear. Sand assumes a dunal ridge form that migrates across the surface as transverse coastal dunes.

Transverse dunes are unstable forms (Bagnold, 1941). The rate of sand supply is the crucial factor in their stability and with the reduction in supply rate, its transverse nature is transformed into isolated barchans or seifs. However, some (Mckee, 1966 and Glennie, 1970) regard transverse dunes as stable forms because of their commonness in several world deserts.

Barchan Dunes - These are crescent-shaped dunes, asymmetrical in cross-section with a gentle windward convex slope and steep leeward concave slope, with horns or cusps pointing downwind, developed across the prevailing wind as isolated sand bodies or long chain-like series of migrating dunes. They occur in such deserts where strong winds constantly blow in one direction and are characteristically associated with regions where

(i) the sand supply is adequate, and

(ii) the wind is moderately strong

They may be 30 meters in height, and a width measuring 10 to 20 times its height. In the Central Luni basin, Rajasthan barchans are less well common as compared to other types of sand dunes. The dunes may be 3 to 5 meters above the interdunal area with 1-2 degrees windward slope and steeper leeward face around 10-15 degrees.

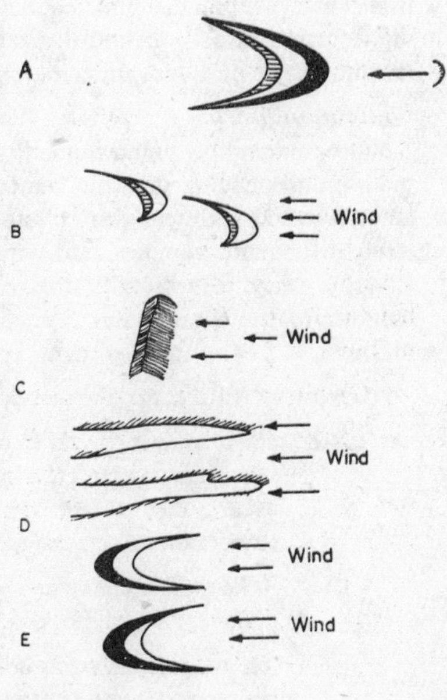

Figure 47 : Dune Types
A. Barchans B. One wing of Barchans not present C. Transverse Dune D. Longitudinal Dune E. Parabolic Dune

Bagnold (1941) considered windflow dynamics over and around a growing sand mound to explain the evolution of barchans. The streamlines pressed closest together about the summit diverge over the margins of the dune where the amount of sand is thin. The drag velocity of the convex windward slope, therefore, tends to maximize over the crest, but decreases rapidly in the flank regions. The flanks lengthen in the down current direction to form horn until the wind drag on sand equals the wind drag on the main sand body. The horns so produced gradually advance into the leeward protected sheltered zone where the wind is stagnant, the drag velocity and thus sand supply rate become lowest arresting further horn growth because of the wind shadow effect. The grains eroded and transported from the upwind slope cascade across the crest into the slip off slope formed in the zone of dead sand and the dunes migrate in the wind direction in this process without loosing its mass. The sand distributed uniformly across a transverse section, is directed outward through the flanks. Therefore, a new dune grows directly behind the slip face. Such a process explains the echelon arrangement of dunes in a barchan field.

Longitudinal Dunes (Seifs) - These are dunes of commonest type, linear features extending uninterruptedly for many kilometer at a stretch, mobile in nature and developed in the heart of the trade wind deserts. They are serrated, sharp-crested features with collapsing fronts. With increasing size linear dunes tend to become complex and compound. Their length, width and in between spacing vary. Empirically the width of seif dunes is about six times the height. In the Thar Desert, the longitudinal sand dunes may measure 30-70 m high, 2.5 kms long and on average spaced 250 m apart.

Opinions differ for the origin of seif dunes.

(i) Seif dunes develop (Bagnold, 1941) from a modification of barchans under the influence of bidirectional wind regime. As such, one of the barchan horns elongates considerably in the direction of occasional crosswinds.

(ii) The seif dunes may normally develop along the vector of two converging wind direction (Mckee, 1966)

(iii) Seif develops from helical wind circulation in a vast level expanse of sand deserts noted for strong winds. Uniformly spaced linear dune ridges, developed on piling up of loose eroded sand produced by the effect of helical wind circulation, perpetuated the helical effect by the growing height of sand ridges in introducing rising convective currents in the wind system.

Parabolic Dunes - A parabolic dune is a U-shaped sand accumulation associated with blowout depressions in which the horns of dunes point upwind. It is partially stabilized. The windwards face gently rises from the blowout, making a steep convex slope to the lee of the downcrest. Its variant include

(i) blowout dunes in arid plains, and

(ii) hair-pin forms near coastal areas.

The parabolic sand dunes are geometrically similar to barchan but for their association with a blowout. Several individual parabolic dunes coalesce to form a continuous chain of dunes traversing several hundred meters. The size of dunes is a function of

(i) the sand supply

(ii) the effectiveness of the prevailing wind, and

(iii) vegetation cover.

Coastal and flood plain areas are perennial source of sand for building parabolic dunes. In the Thar, complex parabolic dunes extend from the Rann of Kutch to Bikaner and Jodhpur. Covering an area of some 1,00,000 sq km the dunes on an average are 2.6km long and 2.4 km wide, suggested that the parabolic dunes of the Thar deserts are fossil forms inherited from a more humid climatic phase of shorter duration some 5,000 years B.P. This postulation is on the assumption that the U-shaped dunes were probably from the material eroded from pre-existing sand shields and obstacle dunes in the relatively low energy wind regime of the past several thousand years.

The morphology and evolution of parabolic sand dunes are probably controlled by the stabilizing effect of vegetation that grows along the margins of the feature where the sand cover is thin. The bare central section of the dune moves more rapidly in the downwind direction, than its flanks. Vegetation may grow on the shifting dunal sand surface but it gets subsequently buried by the collapsing front. In this process, a central blowout develops ahead of the relative dunal segment. If the blow out elongates in the downwind direction at the expanse of the width of flanks, a hairpin dune develops.

Yet another wind-borne deposit, but of finer grade than sand, is loess.

Loess - It is fine grained dust or silt-sized suspended particles that are transported by the wind outside the source area. They settle down on loss of energy. Once deposited the particles considerably resist their further transportation by the wind because of

(i) greater cohesion among the fine-grained particles and

(ii) the protective vegetative cover, the growth rate of which exceeds the rate of sedimentation.

Grey to buff in colour, the loess is homogeneous, not stratified, porous, permeable and unconsolidated. Sometimes, the fine dispersal of abundant limonite imparts yellow to buff colour to its deposit. Depending upon

(i) the nature of the material in the contributing source area, and

(ii) the distance travelled while air-borne,

their grain size distribution too, depicts spatial variance, dominantly between 0.01 and 0.05 mm which approximates upto 50 percent by weight of loess deposits.

The materials constituting loess are derived from several sources :

(i) glacial deposits, in particular, outwash, called "cold loess" generally better sorted, and of a narrow range of grain size distribution, being processed by glacial melt water

(ii) flood plain sediments including colluvium

(iii) weathering of rocks in hot-dry climate, often referred to as "hot"or "desert" loess

(iv) from desert sands

Mineralogically, quartz (40-80 percent), feldspar (10-20 percent), montmorillonite and heavy minerals constitute loess matter. The chemical composition of loess, however, varies in respect of carbonate content of calcium and magnesium from practically over 35 percent, is a function of the location-it is high in dry regions and low in humid terrain.

Loess deposits cover an area of thousands of square kilometer reaching thickness of up to 50 meters or more

Table XII. 48 lists the properties that provide the basis for general recognition of loess.

FLUVIAL FORMS IN DRY LANDS

PEDIMENTS, PEDIPLAINS AND INSELBEGRS

Pediments

There is a sharp break in slope profile between the mountains and the adjoining plains in the arid and semi-arid mountainous terrain, the plains having a gentle slope not exceeding 7^0 in contrast to the neighbouring mountains with steep slope or slopes ranging from 15^0 upwards to practically vertical. The area of gentle slope called *piedment* lying between the mountain and the playa usually comprises of two distinct parts

(i) the *pediments* - the upper section of eroded bed rock surface close to the mountain- these are *plains of degradation* developed at the expense of mountain front retreat,

(ii) the *bahada* or *bajada* - lower section, *plains of aggradation*

The pediment has little or no cover and is parallel to the mountain front. It is flat relatively smooth except for the regional slope and not undulating as *bajada* is. Most pediment surfaces are somewhat concave upward both in longitudinal and transverse profile.

TABLE XII. 48 : LOESS CHARACTERISTICS-GENERAL

Colour		Yellow or yellow buff characteristic, also ash grey, grey buff or brown 'A' horizon often very dark
Texture	a.	Friable
	b.	Dominantly silt but also contains clay and sandsized particles. Ordinarily the fraction, 0.01 to 0.05 mm contains, at least, 50% of the material by weight
	c.	An aggregate of clay particles appears to surround each large silt or sand grain
	d.	Silt and sand particles are angular to subangular
	e.	Shows an open texture with little interlocking of grains under the microscope and many irregular voids
Composition		
	a.	Silt fraction consists of quartz (+ 50%), feldspar, often, as abundant as + 20%, calcite, dolomite, mica less than 10%, locally volcanic ash shards; chert, hornblende, chlorite and pyroxene are minor
	b.	Clay fraction is somewhat variable, but montmorillonite and illite are most abundant
	c.	Chemical composition SiO_2 65 to 71%, Al_2O_3 and Fe_2O_3 3.25 to 4.25%, CaO 3.05% 10.70%, MgO 1.68 to 5.10%, K_2O 1.21 to 2.78%, Na_2O 1.31 to 1.97%, H_2O 0.51 to 2.20% Calcite, an important constituent, although not always present, occurs as silt grains, concretions, tubes, snail shells and cement.
Structures		
	a.	Stands with vertical faces principally in artificial cuts and banks undercut by streams
	b.	Vertical jointing is noted in most exposures
	c.	Tabular structure may reach a maximum of 10 to 11 meters below the surface. They are coatings on rods and often are associated with small rootlet tubes 1 mm or less in diameter
	d.	Incrustations of carbonate appear on fracture faces, around exposed roots, and below overhanging walls, often as stalactite growth
	e.	Concretions, particularly of calcite, are very common and the depth from the surface to their first occurrences is directly proportional to rainfall
Engineering Properties		
	a.	Vertical permeability to water is greater than horizontal
	b.	Dry undisturbed loess will withstand loads which it is incapable of supporting after saturation by rain or canal water.
	c.	Liquid limit varies between 36 to 46%
	d.	Plastic limit varies between 21 to 26%
	e.	Plastic index varies between 9 to 20
Topographic Forms		
	a.	Chimneys or pipes occur along valleys, narrow ravines, and sharp divides
	b.	Blankets the underlying topography
	c.	Intricate drainage pattern especially in the tributary system. Slopes show step-like terraces ("cat-steps") due to minor landsliding or creep.

Thin veneer of alluvium, which typical pediments have, but at their lower margins, advances upvalley mountainward where they merge into *rock fans*. A rock fan is defined as a partial cone or fan-shaped bed rock surface apexing in the mouth of a canyon. The pediments widen mountainward. Once the pediments are formed, it is quite possible that a change in

(i) climate or

(ii) diastrophism

may initiate different stream conditions leading to the development of *fan-topped pediments* on alluvial fans depositing completely over them. On the other hand, fans or bajada-surfaces formed earlier may be altered through the normal processes of pedimentation to true pediments.

Mode of Formation/Origin of Pediments : Pediments are characteristic of both, semi-arid and arid regions. However, their mode of origin is still unsettled for want of adequate answers to several questions -

(i) climatic conditions, under the influence of which they were formed, need to be understood in more details although they are universally believed to be the products of arid-zone erosion.

(ii) the sharp slope breaks between the pediment and the mountain front is to be reasoned out as probably, it indicates a change in the nature of the process, as also

(iii) the processes leading to pediment formation are to be identified. The exact mode of origin of rock-cut plains is thus debatable.

Views expressed by different investigators may be categorized as below:

(a) Resurrected surface - A pediment is considered to be a resurrected surface in this opinion group, which is the normal product of a mountain front recession in an arid region. The pediments "are resurrected surface due to the stripping of the alluvium which once rested upon them" - Lawson

(b) Pediments formed by Sheetfloods - This group of physiographers are of the opinion that rock plains are not formed by lateral erosion but are developed due to

(i) normal desert weathering

(ii) the backwearing of mountain slope, and

(iii) sheetfloods that strip off the debris from the rock floor of the retreating mountain front

(c) Pediments fromed through Lateral Erosion-Johnson visualizes

(i) an inner zone, the *zone of degradation*, close to the mountain highland wherein vertical down cutting of stream reaches its maximum relative importance

(ii) an intermediate zone, the *zone of lateral corrasion*, surrounding the mountain base in which lateral cutting by streams attains its maximum relative importance, and

(iii) an outer zone, *the zone of aggradation*, where-upbuilding by deposition of alluvium has its maximum importance

In this theory pediments are believed to owe their genesis to lateral erosion by short-lived mountain streams of the desert emerging from the mountain front which debouch upon a piedmont slope to form rock fans having their apex towards the mountain front, and as the stream courses shift laterally, the rock fans are extended and result in the formation of pediments.

Various objections have been raised to this theory:

(i) absence of such rock fans in general, in the desert

(ii) lack of mountain front indentation, in contrast actually rather straight mountain fronts are observed

(iii) sheet flood erosion is generally more significant in deserts than stream erosion

(iv) with the cutting back of mountain fronts the erosive power of streams would progressively decline although the pedimentation process seems to continue, and

(v) the presence of inselbergs or residual knobs on the pediment surface is compatible with lateral erosion.

On overview it seems probable that lateral planation by stream is the dominant process in forming the pediments, but sheetwash too, may be an important factor.

INSELBERGS AND PEDIPLAINS

Inselbergs are island-like residual steep sided isolated hills and knobs of extremely hard rock that rise above the neighbouring flat plains. Unlike pediment, this landscape is not related to mountain basins. The actual shape of the inselberg is determined by the nature of rocks. There are extensive plains which possibly represent the last stage of pediment formation and have been called pediplain by L. C. King.

Bajada

A fan shaped smooth depositional slope segment situated between playa and pediment is known as *bajada*. A true bajada is a surface of coalescing alluvial fans. It is concave upward in form with its sloping surface ranging from almost 0^0 in the playa to as much as 10^0 at the pediment (Figures 46 and 48)

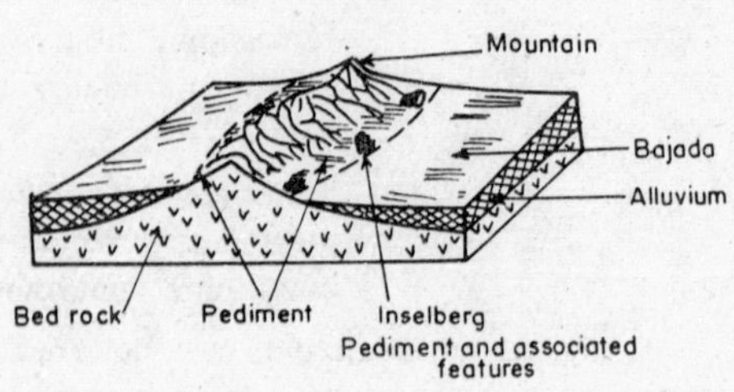

Figure 48: Pediments and associated features

It comprises of a heterogenous assemblage of upto 300 meters thick poorly sorted gravels, colluvium and mudflow deposits formed during or immediately after flash floods. Its dynamics of depositional surface includes

(i) frequent shifting of active channels

(ii) plugging of channels by mudflows, and

(iii) breaching of plugged material of subsequent floods opening new distributary channels.

SALINE LAKES

On the eastern edge of the Thar deserts several saline lakes and flats including

the Sambhar, and

the Didwana

lakes occur. Both the above lakes are commercial source of common and other salts. The Sambhar lake illustrates *bolson* (Figure 47), a term coined for an extensive, flat depression surrounded by hills, characterized by centripetal drainage. The Didwana, Kuchaman, Sargol and Khatu lakes are playas flat-floored, undrained desert basins, smaller in dimension in which water accumulates after rains and thereafter quickly evaporates, leaving behind soluble salt deposits.

A. The Sambhar Salt Lake, about 30 km long x 3-8 km wide encompassing an area of about 225 sq km when full during the rains, trending WNW-ESE, at an altitude of 360 meters in the Aravalli terrain, is located between Jaipur and Jodhpur districts in Rajasthan. During the summer months the area is considerably shrunk. It is an extremely shallow depression in which water is generally less than 50 cm/0.5 meters deep, but for some fetches the depth may be upto 2 meters within the lake periphery. The base of this lake is dated as 10,000 years B.P. The basin is surrounded by aolian sands except in the west where low schists-gneisses hillocks are exposed.

The Kuchaman Salt Lake quite small, covering hardly an area of 12 sq km is situated 10km NW of the Sambhar.

The Didwana Salt Lake only 5.5 km long and 2.5 km wide, fanning over 13.5sq km occurs about 35 km northwest of the Sambhar lake. It is set in desert sands overlying a granitic basement. The playa is rather too shallow and virtually dries up in the summer in contrast to the Sambhar which remains filled at least partly with water at all times.

The Khatu Salt Lake and is situated is rather a very small playa about 2.5 km southwest of the Didwana.

Other small lakes in this region are at Sujangarh and near Degana.

B. Choking or damming of stream channels, active or abandoned, constituted another group of saline basins or flats of which Pachpedra Salt basin stretching over about 100 sq km is an illustration. It is a vast flat north of Luni occupying its old bed in the eastern Barmer district. There are distinct traces of old river bed where Luni takes a sudden bend downstream.

C. Yet another type are those met in Bikaner area where they are referred to as *sar.* There are several depressions surrounded by dunes with saline encrustations. Lunkransar Tal, an elliptical depression of 3.5 km is one such lake situated about 70 km northeast of Bikaner city wherefrom salt is regularly produced. It is completely dry in the summer but after the rains, brine is formed from which salt is extracted after exploitation.

Lake Salinity - The salt genesis in these lakes is not unified and is still debatable. Four possibilities have been proposed in this regard, in which salt

(a) rises from the underlying *in situ* beds or brine spring

(b) is obtained on leaching from the surrounding rocks

(c) is transported by the wind from the Rann of Kutch

(d) is relict of the receding seas

First two factors, in recent drilling and geochemical investigations favour the salinity possibility of in the lakes. Thick concentrations of halite and potash beneath the sand are struck in drilling in the Bikaner area in the widespread gypsum beds. In addition, potash and bi-carbonate, carbonates and sulphates of various elements are found in these lake waters besides the common salt. A mix of gypsum and common salt occurs in the soil around them.

PLAYA

Arid regions are characterized by extremely sparse rainfall. Still water is available to act as an important agent of erosion, transportation and deposition, as also it is probably more effective than even driving winds in desert landscape sculpturing.

Not many streams everflowing through desert regions, find their way to sea, and only a few originate in areas that provide sufficient water to sustain them through their long courses across the desert, if they can withstand agricultural, domestic and industrial uses which effectively drastically affect their flow adversely.

Because of the peculiar climatic and hydrologic conditions, stream beds in most deserts are generally dry for long periods of time. These rivers, often carry water only during floods. Still the flow is short-lived because of

(i) rapid evaporation of water, or

(ii) vanishing into ever absorbing highly permeable rubble and debris of the desert.

At some places, however, broad dry plains slope down from the mountain ranges toward central basins which receive surface runoff from time to time. A central undrained depression, within a bolson or elongaled valley, of varying shapes and sizes, but possessing a flat almost featureless surface is known as *playa*(Figure 47). In tectonically unstable areas, warping and/or faulting exercises, a dominant control in the evolution of the feature (Sharp, 1954).

Some micro-relief is noted on playas which is due to abundance of channels, plunge pools, mineralogical differences in the soil composition and associated features related to inland drainage and self-risen grounds caused by capillary. Water may briefly accumulate on playas, which on evaporation leaves salt encrustation surface called *alkali flats*. Playas having high salt- concentration are known as "*salinas*". These playa lakes usually dry up rather quickly or at best survive as shallow, silty lakes, such as Great Salt Lake.

Floods in the desert are local and for short periods, like rain which produces it, and are unlike floods in humid areas. The runoff moves swiftly and floods rise and fall with great rapidity on the bare ground of the desert particularly, because of the lack of vegetative cover which could offer resistance to the moving water. These *flash-floods* give little warning in which surging wall of debris-laden water may sweep down the stream bed to destroy everything in its path. Sudden flooding of a desert by a cloud burst also moves material to lower elevations, reducing the size of the rock fragments and scouring the bed rock surfaces in the process.

The moisture amount in the deserts does change. Glaciation produce *pluvial periods* (from the L. pluvia, "rain") whence the moisture is arrested in solid ice which certainly is most hospitable in contrast to true arid regions. During any single pluvial period

(i) rainfall is greater,

(ii) temperature is lower,

(iii) rate of evaporation is less, and

(iv) vegetation is more extensive.

It is at such time, large portions of depressions in the region, may be dotted with lakes, called *pluvial lakes.*

The continental glaciers tend to modify the general global atmospheric circulation causing movement of rain-bearing belt and the temperature dropping. In effect, rates of evaporation lowers and that of precipitation increases. With the ice recession, the climate again becomes dry and aridity prevails.

CYCLE OF EROSION IN ARID LANDS

Arid regions are characterized by

(i) poor rainfall and therefore,

(ii) scanty vegetation

The precipitation is in torrents of short durations. Waters from these "cloud bursts" or snow melts cause excessive rainwash, gullying, scouring and erosion on account of their great speeds and large volumes. But for few perennial streams, which are rare such as the Nile, generally the streams are mostly dry with rather wider valleys than the volume of water they handle. Bolsons, an inland basin sometime exist at different heights in desert regions. Subsurface water is low, chemical weathering is little, but mechanical rock wear is most pronounced in desert areas.

Beginning of the Cycle - Imagine consequent stream flowing over a recently elevated tectonically effected by tilting or folding, upland. Some of these would evaporate, others disappear to their base level, the bed rock, or join lakes, but certainly do not flow out. Subsequent streams do not develop because of the poor performance of earlier streams. At the initiation of the arid cycle, river action is more important and in time wind action picks up with the advancement of cycle.

Youth stage - Highland streams cut to erode mountain slopes and transport the debris downward to lay down as valley fills

(a) alluvial fans composed of coarse debris are formed (*bajadas*)

(b) water table is elevated

(c) playa lakes covering valley floors are developed

(d) streams are loosing themselves

(e) seepage waters carry mineral solutions to produce salinas

(f) dunes are developed where sand is deflated from lake borders

Subdued topography with moderate relief is the outcome thus developed on highlands' erosion and raising of bed rocks. In the absence of draining out of water at this stage the wind lowers the average regional altitude. Even in desertic terrains, valleys are degraded on the highland slopes but the basin slope is aggraded by deposition. Although wind may erode the exposed rocks and fan out finer waste mostly in the directon it blows, yet water remains the principal erosional and depositional agent. As such, the wind reduces the average altitude of the area from where no water flows out of the area.

Maturity Stage - This stage is punctuated by

(i) divide destruction

(ii) head erosion, and

(iii) development of drainage channels.

Bajadas are more striking. Flattening and thickenning of inland basins are caused by the merging of stream courses and coalescence of valleys. Conspicuous even faced, broad surface of bare rock, sloping, truncated plains called pediments or rock fans mark the bases of receding mountain. Geologic work done by the wind in low relief area assume relatively greater significance because of lesser erosional stream capacity. In this stage, the basins coalesce and stream patterns integrate. "The change of the total base levels will be fewer and slower and the obliteration of the upland the development of graded piedmont slopes and the aggradation of the chief basins will be more and more extensive" - Davis.

Old Stage - Topographic relief is further reduced due to wearing away of solid rock mass and island-like masses, inselbergs rise boldly through gently sloping pediments that developed against widely separated, now less towering, naked hills, joining broad basin front. They are largely wind-swept and enlarge the areal extent constantly. Their morphology may, of course, be modified in time by some diastrophism.

WIND AND STREAM EROSION - COMPARISON

(a) Careful observations have led to the conclusion that the corrasive action of wind is generally much less effective in sculpturing and cutting away of rocks than formerly surmised. There is, however, no doubt about the local efficacy of wind corrasion when conditions are very favourable.

(b) Stream erosion, even in desert regions, accomplishes much more work than wind erosion because when it rains, it is often in the form of so-called "cloudbursts". In desert areas of high relief, therefore, torrents of water rush down the stream courses carrying heavy loads of rock debris derived from the abundant weathered rock material almost unprotected by vegetation.

(c) Well exposed outcrops in the vast western desert (of USA) show plain evidence of mechanical and chemical weathering, but with little or no evidence of wind erosion.

(d) The greatest geologic work accomplished by the wind seems to be deflation, the picking up and transporting of really large quantities of finer materials which either were laid down through the agency of running water or were produced by weathering.

APPENDIX I

GULF OF KHAMBHAT, GUJARAT, INDIA : AN OVERVIEW OF ITS GEOENVIRONMENT*

Md. B. Islam[1], S.S. Merh[2], L. S. Chamyal[2]

1. **Department of Geology, Rajshahi University, Rajshahi, Bangladesh**
2. **Department of Geology, M. S. University of Baroda, Vadodara. 390 002, India**

ABSTRACT

The Gulf of Khambhat in Gujarat forms an important segment of the West Coast of India and comprises an area marked by a regime of high tide domination, strong currents and low wave energy. The gulf waters are constantly churning up the sediments, rearranging them by tidal currents. The shape of the gulf is structurally controlled; the tectonic framework being the manifestation of the various Cambay Basin boundary faults. Geologically, gulf coast is made up of Post-Mesozoic depositional sequences, marine to fluvio-marine, and mostly consisting of a variety of Quaternary sediments. The coastal landscape reflects the tectonic setting and diversity of coastal marine processes. The coastal deposits comprise tidal mudflats and mud-banks. Beach development is inhibited, and restricted to small coastal stretches only. The mineralogical make up of the mud shows an overall uniformity, and consists of 60 to 70 % silt-size grains of montmorillonite with subordinate illite, chlorite and kaolinite. The beach sands on the Saurashtra side differ on the two sides of Shetrunji river; to its south the beaches contain biogenic carbonate sands (miliolite) while to the north, the grains are mostly quartzose and lithic. The sand grains of the mainland beaches are mostly quartz, ilmenite and magnetite. The gulf offshore is characterised by mud-banks, shoals and underwater ridges and provides a very complex picture of sediment input, transport and deposition, the processes being significantly influenced by the narrow funnel shaped gulf configuration, bottom topography, coastal physiography, tidal prisms, fluvial sediment influx and river water input.

INTRODUCTION

The Gulf of Khambhat which forms an important coastal segment of the West Coast comprises a very unique offshore feature. It forms a funnel shaped entrant of the Arabian Sea sandwiched between the Gujarat Mainland, and the Saurashtra peninsula, opens up into the Arabian Sea rather abruptly, and is located in the widest portion of the West Coast continental shelf. It is 50 km in its widest part along its southern limit and is approximately 135 km long in the north-south directions (Fig.1). The gulf dividing the Mainland Gujarat coast from that of the Saurashtra provides a baffling picture of coastal marine processes. It comprises an area marked by strong and powerful tidal action, and forms a site where the tide-rise is the highest along the West

*Contributed paper

Coast. The high and low tides during different seasons, in combination with the addition of the detritus-laden waters of the rivers from the Mainland side of Gujarat have generated a very complex pattern since the advent of the Quaternary period. The gulf waters are constantly churning up the sediments, rearranging them by tidal currents. Offshore features have originated during the Quaternary times. The various onshore and offshore factors responsible for the diversity comprise (1) tectonic setting (2) eustatic and neotectonic sea level changes, (3) lithology of the geological formations, (4) location, shape and size of the gulf, (5) shallowness of the gulf bottom, (6) trends of the costline segments, (7) rainfall, (8) wind-directions, (9) fluvial regimes of the coastal areas and (10) action of tides and waves.

The various sections of the coastline overlooking the gulf are so diverse in their terrain features that each section presents its own distinctive geological and geomorphic characteristics. Not only the coastline geology and geomorphology is varied, but the geohydrological conditions prevailing in the coastal areas show marked diversity.

The Gulf of Khambhat has received little attention in the past, and only in recent years, some data is available on the geological and geomorphological aspects of the gulf coastline and on the offshore processes. Studies by ONGC (Raju, 1968; Mathur *et al.*, 1968; Chandra and Choudhary, 1969; Eremenko, *et al.*, 1969; Markevich, *et al.*, 1976; Rao and Talukdar, 1980; Biswas, 1982) have shown that the configurations of the Gulf of Khambhat is fault-controlled, and its shape, coastline features, drainage pattern and the Cenozoic depositional history are all related to the numerous regional faults.

Earliest, geological reference to the Saurashtra coast of the gulf is available in the work of Feden (1884). Describing the coastal rocks of southern Saurashtra he included even the Quaternary rocks (except miliolite) into his Gaj Beds of Tertiary age. Subsequently, the officers of the ONGC (Negi, 1955; Babu, 1958; Rao and Jain, 1959 and Datta, 1959) in connection with the oil exploration programme, conducted geological and geophysical studies along the Bhavnagar-Ghogha coast. Shrivastava (1963) furnished a systematic petrographic account of the various rocks of the same area. References to the Saurashtra side of the gulf coast are also found in the works of Mathur *et al* (1968), Raju (1968) and Chandra and Choudhary (1969). The more recent work on the Bhavnagar-Ghogha coast is that of Ganapathi (1981), who has given a precise and integrated stratigraphic picture of the western coast of the gulf.

Whereas the Saurashtra coast of the gulf has been investigated in fair detail, the Mainland coast has not received any attention, perhaps as this part of the coast comprises mostly Quaternary alluvium. The works of Chandra and Choudhary (1969) which described the stratigraphy of the Cambay

Basin is the only source to obtain the geological details of the eastern coastline of the gulf, but their information pertains to the sub-surface only.

Geomorphic aspects have also received very little attention, except for some preliminary studies carried out by Iqbaluddin and Banerjee (1978) on the geomorphic processes operating along the Bhavnagar coast. Recently, workers of the M. S. University of Baroda (Ganapathi, 1981; Ganapathi and Merh, 1982 ; Patel and Merh, 1982 ; and Patel *et al.*, 1985) have studied the coastal geomorphology of the gulf. Ganapathi (1981) has given an elaborate description of the Bhavnagar-Ghogha coast, while Patel and Merh (1982) have described the landforms of the Bhogava -Sabarmati confluence area that marks the northern tip of the gulf. The geomorphic features of the Mainland coastline between the rivers Mahi and Tapi have been described by Patel *et al.* (1985). These authors have attempted to explain the different landforms *vis-a-vis* sea level changes and neotectonism.

Some indirect information on the paleoclimatic changes and eustatic sea-level fluctuations during the Holocene epoch are available in the works of archaeology pertaining to the Harappean port town of Lothal located quite close to the gulf (Rao, 1973, 1979, Pandya, 1985).

Offshore data have appeared in the form of research papers by Rao (1975); Borde *et al.* (1982), Vishwanathan *et al.* (1984), Nayak and Sahai (1985) and Muley *et al.* (1985). Whereas Rao (1975) has described the foraminiferal content of the gulf sediments, Borde *et al.* (1982) have described the composition of the estuarine sediments. Vishwanathan *et al.* (1984) have on the basis of Landsat Imagery studies, given a preliminary account of the sediment plume dispersion and have correlated the phenomena with bathymetry. Nayak and Sahai (1985) also using the Landsat Imagery have provided a good account of the tidal as well as seasonal sea level changes, current patterns and sediment transport, and the relationship of these phenomena with some erosional and depositional coastal geomorphic features. Muley *et al* (1985) have studied the SALYUT-7 space photography and have described the coastal geomorphology as well as the sediment transport in the gulf.

STRUCTURAL SETTING

The tectonic framework of the gulf is, in a way the manifestation of the various boundary faults which have given rise to the oil bearing Tertiary Cambay Basin, and the site of the gulf is marked by a period of subsidence during Pleistocene followed by some uplift during the Holocene. Whereas the Cambay Basin comprises Tertiary deposition, the present day gulf is the Quaternary manifestation of the same tectonic set up. The tectonic framework of the Cambay Basin has been investigated by several workers of the ONGC (Raju, 1968, Mathur and Evans, 1994 ; Mathur *et al.* 1968; Chandra and

Choudhary, 1969; Markevich *et al.*, 1976; Rao and Talukdar, 1980; Biswas 1982). Two N-S major lineaments comprising the east and west basin-bounding faults as well as numerous cross-cutting faults have been found responsible for the existing configuration and geological and geomorphological evolution of the Gulf of Khambhat and the coastal areas facing the gulf. The Western Cambay Basin Bounding Fault (WCBBF) extends along the Bhavnagar coast of the gulf. The extension of the Narmada Fault marks the southern limit of the actual gulf on the Saurashtra side, whereas along the Mainland, the gulf environment extends upto the mouth of the river Tapti. Eastern Cambay Basin Bounding Fault (ECBBF) which lies approximately 50 km inland does have its effect on the coastal configuration. The two basin bounding faults have been shown to continue southward as enechelon step faults. Raju and Srinivasan (1983) have described the Cambay Basin as an intracratonic rift graben between Saurashtra uplift and Aravalli ranges extending in a roughly north-south alignment from North Gujarat through the Gulf of Khambhat to as far south as Ratnagiri offshore.

Tectonically, the Gulf of Khambhat can be divided into three parts. The north-south Saurashtra coast overlooking the gulf marks the site of WCBBF. Ganapathi (1981) has referred to the southern extension of this fault as Ghogha-Sanand fault. The two faults together have considerably influenced the Quaternary history of the Saurashtra side of the gulf area. The northern tip of the gulf which is marked by a narrow east-west muddy coast almost forms the meeting point of Sabarmati and Mahi, and most probably follows a fault line (Mathur *et al.*, 1968). On the basis of LANDSAT-I imagery studies, Powar (1981) has shown a lineament along Mahi- Narmada coast, but existence of such a fault is doubtful. No fault line has been reported by any worker. According to the ONGC workers, the Mainland coast forms a partly submerged graben of Jambusar and Narmada blocks of the Cambay Basin. South of Narmada however, a fault has been shown by Biswas (1982) a little inland and parallel to the Dharwar trend. An interesting tectonic feature of the Gujarat mainland coast is the presence of numerous transverse faults broadly coinciding with the trends of the major rivers like Mahi, Narmada and Tapti. On the basis of Deep Seismic Sounding, Kaila *et. al.*, (1981) have delineated some more basement lineaments which have cut even the Moho boundary (Figure 2). This cross cutting lineaments are reflected in the depositional history of the Cambay Basin during Tertiary and Quaternary Times.

The existing coastal and bathymetric features, especially on the Saurashtra side reveal well marked fault control. In the present study, the authors have postulated a N-S trending fault bordering the eastern side of the Piram Island. It is quite possible that the Ghogha-Sanand Fault which according to Ganapathi (1981) runs in between Ghogha and Piram, in fact, lies to the east of Piram Island. Of course, the possibility of a smaller N-S fault between Ghogha and Piram cannot be ruled out. Considering the various evidences, the schematic section across the Gulf shows the relative positions of the different formations (Figure 3).

STRATIGRAPHIC FRAMEWORK

The Gulf of Khambhat and its environs, comprises mostly Post-Mesozoic depositional sequences, dominantly marine to fluviomarine. Resting unconformably over the basalts of the Deccan Trap, the Tertiary rocks constitute major bulk of the Cambay Basin. Although a large part of the Gulf of Khambhat does not show any significant outcrops of the Tertiary rocks in its coastal areas, their presence, stratigraphy and lithology have been investigated by the ONGC (Table 1; Figure 4)

A large part of the coast is exclusively made up of Quaternary deposits. The various coastal blocks of the Gulf, *viz.* Saurashtra, Bhal and Mainland show a significant geological diversity. Along the Saurashtra coast, within a distance of 15 km inlandward successively Tertiary and Quaternary formations over a trappean basement are met with (Table 2)

In the Bhal area, the raised mudflats which are several meters thick, rest over an older alluvium pointing to the over running of the alluvial area by a Sub-Recent transgressive sea. Though the exact thickness of the underlying alluvium is not known, it may be varying from 100 to 200 meters, gradually merging downward into the marine clays. Geologically, the Bhal presents a good example of Late Quaternary sea level fluctuations; during the last transgression (Flanorian -Early Holocene), the Gulf of Khambhat was connected through the Gulf of Kachchh to the Rann of Kachchh.

The Mainland Gujarat coast is different. Here Recent to Sub-Recent mudflats and sporadic beach sands are seen resting over an older continental (non-marine) segment. Recent studies indicate that these deposits show an age range of Middle Pleistocene to early Holocene. These dominantly fluvial deposits are almost 150 to 200 meters thick north of Narmada; nowhere Tertiary rocks are exposed on the surface. The south of Narmada mudflats and sandy beaches rest over alluvium. Here the alluvial cover has been reported to be relatively thin, of the order 100 meters or less, through which crop out the Tertiaries around Rajpipla, 25 to 30 km away from the coast.

The gulf coastline can be divided into three blocks namely (i) Saurashtra block (ii) Bhal blocks and (iii) Mainland block.

On the bases of geology, physiography and terrain characteristics, the three blocks have been further subdivided into numerous segments as under :

1. Saurashtra block
   (a) Methla - Gopnath (MG) segment.
   (b) Gopnath - Ghogha - Bhavnagar (GGB) segment.
   (c) Bhavnagar -Dholera (BD) segment.
2. Bhal block
   Dholera - Khambhat (DK) segment.
3. Mainland block
   (a) Khambhat - Dahej (KD) segment
   (b) Hansot - Hazira (HH) segment

COASTAL GEOLOGY

The coastal plains of *Saurashtra block* show a well developed Post-Mesozoic sequence comprising Deccan Trap and laterites overlain by Tertiary and Quaternary sediments. The rocks of different stratigraphic ages occur quite close to the shoreline. A NW-SE fault along the Shetrunji river and E-W fault along the Kalubhar river appear to have played a significant role in the Quaternary depositional history and geomorphic characteristics of the various segment (Ganapathi, 1981). Whereas the Shetrunji Fault separates MG segment from the GGB segment the Kalubhar Fault demarcates the GGB segment from the BD segment. These faults as well as the extension of Narmada Fault and Ghogha -Sanand Fault (= WCBBF) have controlled the deposition of the coastal Neogene rocks. Since the advent of Miocene differential uplifts and subsidences along these faults are reflected in the stratigraphic differences of the three segment.

In the M-G segment the Tertiary - Quaternary rocks show a width of about 10 km. The Shetrunji Fault has restricted the deposition of miliolite to its south and on crossing Shetrunji river towards Gopnath a striking change in the lithology and stratigraphy is seen. The trap comes quite close to the shoreline almost 2 to 3 km. only. Laterites and Gaj rocks are exposed along the coast itself; miliolites change over to a non-carbonate facies of Lakhank Formation.

The BD segment is very much different from the earlier two segments. Here, the coast for several kilometers inland consists of unconsolidated Sub-Recent sediments, alluvial as well as tidal marine. Nowhere Tertiary rocks are exposed, and the Recent sediments are seen to abut against the Deccan Trap.

The *Bhal block* comprises a linear NNW-SSW trending depressed zone separating the Gujarat Mainland from the Saurashtra peninsula. Tectonically it forms a part of the western flank of the Cambay Basin lying to the immediate west of the WCBB fault. Surface deposits comprise only the Sub

Recent and Recent sediments, but in the subsurface, Tertiary rocks have been reported by ONGC. Although the main graben to the east of this fault (mostly falling within the Mainland) shows a very pronounced downthrow, the western flank of the basin also points to a distinct subsidence. This phenomenon coupled with late Quaternary sea level changes (Merh 1992) has given rise to the Bhal sedimentary sequence.

Major bulk of the sediments of this block consists of an older alluvial accumulation overlain by tidal mud deposits formed during the Holocene sea level rise. Since then, the sea has regressed to its present level. The late Quaternary exposed stratigraphy could be described as under :

Present day mudflats	-	Late Holocene (Recent)
Raised mudflats	-	Early Holocene (Flandrian)
Fluvial deposits	-	? Late Pleistocene.

As seen today, elevated relicts of older alluvium peep out as bets or island above the saline wasteland.

The coastal *Mainland block* overlooking the Gulf, extending from Khambhat (Cambay) to Hajira (between the Mahi and Tapi rivers) belongs to the Broach and Narmada tectonic blocks of the Cambay Basin structure. From north to south, it shows a very conspicuous tectonic control in the deposition of the Quaternary sedements, as a result of which the exposed geology of it's two segments show well defined differences. Nowhere the Tertiaries are exposed. The Narmada Geofracture marks the tectonic lineament which separates the Mainland coastal block into the two geologically diverse segments *viz.* Cambay - Dahej and Hansot - Hajira. From the point of view of Quaternary - Tertiary stratigraphy also, the segments differ from each other-the northern CD segment consists of a huge pile of about 150 m thickness of Quaternary alluvium resting in the subsurface over Broach and Jambusar Formations. The southern HH segment has a reduced alluvial - thickness, (less than 100 m) and the alluvial sediments are resting directly over Jhagadia Formation (Agrawal, 1984).

The late Quaternary stratigraphy of the Mainland Gujarat reflects the control exercised by the factors of sea-level changes and neotectonism along ENE-WSW fractures. The coastal deposits are all Quaternary and show the following depositional sequence.

Present day mudflats, beaches, Coastal dunes *etc.*	-	Late Holocene (Recent)
Raised mudflats, paleodunes and coastal ridges	-	Early Holocene
Older Alluvium	-	? Mid to Upper Pleistocene.

COASTAL LANDSCAPE

The coastal landscape in the various blocks, being a reflection of the geological diversity and varying responses to the offshore processes during the Quaternary period, differs from segment to segment (Figure 6).

Geomorphically too, the two segments of Saurashtra block are quite different. The MG segment forms a straight N60^0 E trending coast comprising essentially an intertidal rocky platform, with narrow sandy beaches backed by steep miliolite cliffs and dunes and separated by projecting headlands. It indicates an emergent coast with a geomorphic history of fluctuating strandline. Intertidal platform, cliffs and stacks are the erosional features while the depositional landforms belong to three generations, related to three strandline levels (Merh, 1992). The Shetrunji river which marks the limit of this segment consists of an estuarine delta. To the north lies the GGB segment. The coastal morphology towards Ghogha is different and dominated by non-miliolitic deposits and lacking in well defined cliffs. The coastline features comprise erosional landforms like a backshore rocky undulating terrain, and a wave - cut intertidal platform and deposional features like backshore alluvial plains coastal ridges, beaches, river spit-bars and mudflats. The erosional features are on the whole less pronounced. The BD segment marks the merger of the Bhavnagar creek with that of Sabarmati. It is made up of landforms of unconsolidated sediments deposited during the late Quaternary strandlines. The tidal creek of Bhavnagar constitutes the mouth of Kalubhar river and is characterised by raised mudflats, recent mudflats, mangrove swamps, point bars, mud island and offshore mud banks. During high tides the gulf waters enter through the creek and flood vast inland portions. Within the raised mudflats, occur patches of alluvium as 'bets' which support some vegetation. The raised mudflats inlandward merge into the coastal alluvium.

On crossing the Sabarmati river, the coastal segments fall within the Mainland block. The KD segment provides a good example of a drowned alluvial coast, and consists of a thick (exceeding 150 m) accumulation of late Quaternary fluvio-marine to fluvial sediments topped by marine mud. The continental sediments, generally referred to Gujarat Alluvium (Narmada Formation) represent deposits of an ancient super-fluvial system, subsequently dissected by the tectonically controlled new channels of present day Sabarmati, Mahi and Dhadhar (Sridhar *et al*, 1994). The landforms ovserved in this segment are (1) estuarine river mouths (2) foreshore and offshore mudbanks (3) older (raised) mudflats (4) relict alluvial patches within raised mudflats and (5) beaches and sandy ridges. The estuarine river mouths of Mahi and Dhadha are broad, muddy and extend for several kilometers inland. This segment is under the maximum influence of high tidal currents. The Dhadhar river mouth provides a good example of choking of the river due to increasing tidal mud accumulation, so much so that the river meanders, bifurcates and then joins again while flowing within the muddy deposits. Mudbanks of variable dimensions and shapes occur in abundance between the mouths of Mahi and Dhadha river.

The HH segment across the river Narmada is different geomorphologically as well as in terms of coastal marine processes. The southern limit of the segment is flanked by a fault along the course of the Tapi river (Raju 1968, Kaila 1982). Located to the south of the Narmada Geofracture, tectonically this segment has been reported to show much less vertical tectonism during the Quaternary period. The alluvial cover about 100 m at its maximum near the mouth of Narmada river progressively increases seaward, which ultimately goes baneath the ancient and present day mudflats. Landward the thickness decreases and dies out towards east and south. The Narmada and Tapi form two major estuaries. The coast in between is dissected by smaller creeks of river Kim, Tena and Sena Narmada estuary has a well developed river mouth. Aliabet, is an island of about 50sq km area The configuration of this island is seen to be constantly being modified with the passage of time (Nayak and Sahai, 1985), and the present authors have also recorded that during low tide, the Aliabet gets almost connected with the left bank of the river mouth. The Tapi estuarine mouth is comparatively smaller; somewhat 'hook-shaped'; it shows striking heterogeneity of depositional and erosional phenomena. The most significant feature is the mouth bar 'Kadiabet' having an area of 11 sq km; the bet itself is criss-crossed by numerous river channels. The sandy (alluvial) patch between this mouth bar and the adjoining Mindhola creek to the south perhaps forms an ancient mouth bar; on it are now located the port towns of Dumas and Magdalla. Numerous small estuarine creeks are the striking geomorphic features of this segment, the prominent ones being those of Kim, Sena and Tena. The creeks extend for several kilometers inlandward, through which tidal waters are pushed for long distances. The rivers do not carry much sediment load, and the estuarine mud mainly consists of the reworked silts of the littoral and offshore areas, brought by tidal waters. A merging of the tidal deposits of these river mouths and creeks has given rise to broad stretches of 4 to 5 km wide mudflats throughout this segment. Raised mudflats are also present and these become increasingly prominent southward. In comparison to the KD segment, sandy beaches are better developed, which vary in width from 50 to 300 m, the widest beach being at Hajira. The formation of the beach is mainly on account of the destruction of the older dune ridges and onshore accumulation of sand particles by wind-generated waves. An interesting depositional landform of the segment is the presence of older coastal sand ridges, occurring beyond the raised mudflats, or within the mudflats. These sandy ridges belong to several generations (Patel *et. al.* 1985), related to periods of sea stands during rising or falling levels of late Quaternary strand line.

In contrast to the coastline to the north of Narmada, the alluvial plains here are less extensive, much thinner, and form cliffs only in the inner parts of the rivers Narmada, Tapi and Kim. Alluvial islands within the mudflats are common and conspicuous, and easily recognised by the slight elevation and the vegetational cover.

Besides the above described coastal segments, the offshore island of Piram, constitutes an independent geomorphic unit essentially controlled by tectonics. This island shows several landscape features like mudflats, beaches, rocky platform, cliffs and dunes. The mudflats, occur all over and terminate landward against the rocky platform of the Piram Beds. Beaches of coarse to medium sized sands are seen on the southern and eastern fringes of the island. The southern part of the island is marked by cliffs 5 to 8 m high. The central part of the island is capped by 3 to 4 m high stabilised sand dunes.

COASTAL DEPOSITS

Tidal mudflats constitute, by far, the most extensive deposits, and are essentially the products of the strong tidal currents operating with the Gulf. These striking sediments have remained practically uninvestigated. Compared to the tidal mud flats and mud banks, the development of sandy beaches is quite subordinate. On Saurashtra side, beach sands are encountered only to the south of Ghogha and on Piram island. Beach development on the material on the two coastal flanks are quite different.

Samples of tidal mud collected from representative locations were analysed to know their grain size, bulk mineralogy and chemistry and clay mineralogy.

For *grain-size data,* muds from 50 spots were analysed. Following standard sieving procedures, coarser (sand) and finer (silt and clay; <62μ) were further analysed for the estimation of silt and clay percentages with the help of Shimtzu Particle Size Analyzer (Model SA-CP_2).

The major bulk of tidal mud in all segments is made up of silt-sized particles, their proportion varying from 60% while clay content varies from 15 to 30% (Fig. 6). The sediments show a relative richness in clay fraction nearer to the low water line (LWL) as compared to those near the HWL.

Representative mud samples were analysed (XRD) to know the *bulk mineralogy.* Studies were carried out on the Phillips PW-1720 X-Ray generator with Phillips PM 1840 diffrectometer with silicon detector using Ni filtered CuK radiation. The results obtained are documented (Table 3; Figure 7). The clay minerals are montmorillonite (smectite); illite, chlorite and kaolinite and all these clays appear to be land-derived through rivers. The mineralogical make up of the mudflats points to an overall uniformity and this is as expected because the analysed material represents the products of continuous mixing and repeated cycling of gulf sedments and fluvial material by tidal processes. The chemistry of the sediments does not lead to any clear-cut conclusions and indirectly reflects the mineralogy. 30 mud samples from different locations were analysed for major oxides, following the standard procedure adopted by Shapiro and Brannock (1962) and Vogel (1975). The data obtained points to an increase of SiO_2 values from south to the north while Al_2O_3 exhibits a reverse trend. In respect of iron oxides, Fe_2O_3 shows a slight increase from south to north on both the coastal flanks; the FeO values, on the other hand, show an overall decrease on Saurashtra side as compared to the Mainland.The other oxide like CaO_2, MgO, NgO, Na_2O and K_2O do not show any trend.

The provenances and source rocks appear to have been responsible for this variation. Mahi and Sabarmati rivers bring a lot of sediments from the granite and metamorphic areas of North and Central Gujarat, whereas the rivers on Saurashtra side as well as those of South Gujarat drain basaltic terrains. This variation in detritus perhaps explains the behaviour of SiO_2 and Al_2O_3. The increase of Al_2O_3 could also be attributed to the laterite (bauxitic) rocks of Saurashtra and Mainland.

Clay Mineralogy analysis was carried out on 23 representative samples, by using XRD. To determine the abundances of clay minerals, the method of Biscaye (1965) was followed. It was observed that four major clay minerals, *viz.* montmorillonite, illite, chlorite and kaolinite, are universally present; the first one being most predominant (70% to 80%). Nair *et. al.* (1982) have shown the northern part of the shelf area of the Arabian sea (which includes the gulf as well) to comprise a montmorillonite zone. These authors have attributed the major bulk of the clay minerals to the land-derived river detritus. The abundance of montmorillonite is correlatable to the preponderance of the Deccan Trap along the Gujrat coast. The mud samples in the north part near the mouth of river Sabarmati and Mahi show an appreciable increase in illite and marginal increase in kaolinite and chlorite contents.

Compared to the extensive development of tidal mudflats, that of sandy beaches is much subordinate. On Saurashtra side beach sand are encountered south of Ghogha and in the Piram Island. Beach development on the Mainland is restricted to Dahej and to its south. Beaches on the two coastal flanks are quite different and their nature varies from north to south as well.

Along the Saurashtra coast sands appear near Ghogha and continue southward beyond the Shetrunji river. The nature of beach material is different on the two sides of the river mouth. To the north, it consists of coarse to fine sands, mostly quartzose and lithic; onshoreward it is almost made up of rounded to subangular shingles. The minerals other than quartz are chalcedony, agate, ilmentite, beryl, garnet, hornblende, muscovite, gypsum and opaques. Rock fragments are of basalt and Gaj limestone; shells and shell fragments of foraminifera, ostracoda, gastropoda, pelecypoda are present in small proportions (14-22%). To the south of Shetrunji river, the beaches are better dveloped, extend almost continuously from Gopnath to Methla and further northwest. The beach material consists typically of bioclastic grains (mainly foraminiferal and abraded molluscun shell fragments) commonly referred to miliolite sand. The carbonate fraction could be as high 60%, other grains being those of quartz, hornblende, ilmenite *etc*. In Piram Island, beach sands which are restricted to the southern and eastern parts consist of dominantly quartz with some chalcedony, ilmenite, beryl, garnet *etc*.

Sand on Mainland first appear at Dahej and then continue southward showing increasing development. North of Narmada at Dahej and Luhara the beach is seen as an inter-tidal zone made up of very fine to medium grains of quartz,ilmenite, magnetite, beryl, gernet, gibbsite, tourmaline and hornblende; the quartz and ilmenite are the dominant minerals. The heavy mineral concentration is 38 to 45% whereas carbonate content is merely 10 to 15 %. At Hajira beach, the heavy mineral content mostly ilmenite, is as high as 75% to 80 %

The Saurashtra beach has a different mode of origin from that of Mainland. On Saurashtra side, the carbonate sands of Gopnath Methla appear to have been transported by longshore drift generated by wave action and tidal currents. The beach to the north of Shetrunji river reveals the phenomenon of accumulation of coarser land- derived detritus, the finer sediments having been washed away to the offshore parts. On the other hand, the beach deposits of the Mainland point out accumulation of sand in those areas only which are affected by the southwesterly wind. It is also observed that a sizeable proportion of present beach is derived from the erosion and rewoking of the ancient submerged dunal material. In addition, a lot of fluvial sands added by Narmada and Tapi are washed back by waves and transported northward by strong tidal currents parallel to the Mainland coast; southern and eastern parts consist of dominantly quartz with some chalcedony, ilmenite, beryl, garnet *etc*.

Sands on Mainland first appear at Dahej and then continue southward showing increasing development. North of Narmada at Dahej and Luhara the beach is seen as an inter-tidal zone made up of very fine to medium grain of quartz, ilmenite, magnetite, beryl, garnet, gibbsite, tourmaline and hornblende; the quartz and ilmenite are the dominant minerals. The heavy mineral concentration is 38 to 45% whereas carbonate content is merely 10 to 15%. At Hajira beach, the heavy mineral content mostly ilmenite, is as high as 75% to 80%.

The Saurashtra beach has a different mode of origin from that of Mainland. On Saurashtra side, the carbonate sands of Gopnath-Methla appear to have been transported by longshore drift generated by wave action and tidal currents. The beach to the north of Shetrunji river reveals the phenomenon of accumulation of coarser land-derived detritus, the finer sediment having been washed away to the offshore parts. On the other hand, the beach deposits of the Mainland point to accumulation of sand in those areas only which are affected by the southwesterly wind. It is also observed that a sizeable proporation of present beach is derived from the erosion and reworking of the ancient submerged dunal material. In addition, a lot of fluvial sands added by Narmada and Tapi are washed back by waves and transported northward by strong tidal currents parallel to the Mainland coast.

OFFSHORE PROCESSES

High tidal range and low wave energy characterise the Gulf offshore which typically resembles the macrotidal type of Hayes (Figure 8). The offshore environment presents a complex picture of sediment input, transport and deposition, and its mudbanks, shoals and underwater ridges reveal a high-tide domination, strong tidal currents, low wave energy, together with other variables like shape of the gulf, coastal physiography tidal prisms, fluvial sediment influx and river water input. The gulf is an area of highest tides along the West Coast, the tides rising to as much as 12 m (Figure 9). The main causes of very high tides comprise (1) convergence of Saurashtra and Mainland west lines into a funnel shaped gulf (2) augmentation to the total volume of tidal water by the addition of water from the rivers (3) shallowness of the gulf, the maximum depth being 55m only, and (4) action of strong south-westerly monsoonal winds.

The bathymetry of the gulf is quite varied and the bottom is seen to comprise a large number of shoals, underwater ridges and deep channels (Figure 10) and the features are essentially the products of (a) accumulation and distribution of sediment load by tidal currents. The inner part of the gulf is shallow, never exceeding in depth beyond 27m and replete with mudbanks and shoals. The outer gulf is relatively deeper, broader and has varied bottom topography; the floor is made up of undrwater channels and ridges which tend to diverge and open up southward. The deepest part is marked by a median channel 45 to 50m deep, the portion just outside the gulf forms the open flat continental shelf averaging 30 to 35 m deep, dissected by a number of channels intervening sandy ridges, parts of some of these ridges rise above the LWL.

Apart from the phenomenon of the rise and fall of water level, the tides generates very strong tidal currents. The geographic location and gulf configuration, the broad and extensive continental shelf within which the gulf is located and dominant south-westerly direction of the monsoonal winds, all these factors in combination with the irregular floor of the gulf bottom have been resposnible for the complexity of the pattern and behaviour of tides. The tidal current directions as observed on satellite imagery reveal following salient facts :

(i) The current directions during flood and ebb tides have more or less identical paths and follow the bathymetric features.

(ii) The fanning pattern outside the gulf mouth is a reflection of the trends of numerous underwater rhythmic linear ridges which regulates the entry and exit of tidal waters.

(iii) The unevenness of the inner gulf bottom, the numerous mudbanks and shoals, and the obstruction caused by the Piram Island, are also the factors that govern the movement of the tidal waters.

One of the most striking offshore phenomena is the role of the mouths of rivers Sabarmati, Mahi and Narmada. These react differently to the rising and receding tides. During the flood tide, the inflowing river water would experience a resistance thereby slowing down or even reversing the river water flow direction. On the other hand, during the ebb tides, the river water would join the seawater in its outward journey. The tidal currents are thus weak at the river mouths during flood tide, but they are quite strong during the ebb tide.

It has been observed that the tidal water entering the gulf are relatively less loaded with suspended sediments, and the inner part of the gulf becomes heavily charged with suspended sediments during flood tide conditions, mainly due the churning up of the bottom sediments as well as those of the tidal flats, mud banks and shoals. At the peak tide the stagnation of water would cause settling of the sediments, especially sand and silt size particles.

CONCLUDING REMARKS

The present study provides an insight into the geological and geomorphological evolution of the Gulf of Khambhat during Tertiary and Quaternary times and also provides a picture of the rather complex scenario of the offshore happenings in the tide dominated gulf waters. The gulf is obviously a product of the various onshore and offshore processes controlled by the factors of geology, sea level changes, climate and strong tidal currents. Glacio-eustatic sea level changes have also contributed to the evolution of the gulf, which has preserved numerous evidences of late Quaternary sea levels. The submerged milliolite dunes of Methla-Gopnath, offshore extension of Gujarat Alluvium beneath the tidal flats point to the low strandline of the terminal Pleistocene.The major evolution of the Gulf however is related to the Holocene transgression (Flandrian) and its regression to the present level. During the last transgression, which drowned vast areas, the gulf was connected through a shallow bay to the Gulf of Kachchh and joined further to the Little and Great Ranns. The extensive tidal mud accumulation of the inner gulf is thus part and parcel of the vast accumulation of the fine detritus in the sea which over-ran and inundated the then existing alluvial topography. This high sea has now regressed to the present level.

The turbulence of the waters of the Gulf of Khambhat have fascinated and intrigued everyone. Strong tides, sediment laden currents and their interaction with the coastline have been matters of observation not fully understood. This study, for the first time, throws some light on the various geo-environmental factors and processes operating in the gulf during the late Quaternary.

ACKNOWLEDGEMENT

Authors gratefully acknowledge the assistance provided by Shri. H. K. Sing and Shri V. Sridhar in finalising the manuscript of the paper.

REFERENCES

Agarwal, G. C. : Further studies on Tertiary and Quaternary sequences of South Gujarat with reference to their structural style and geomorphic expressions. *Ph. D. thesis, M. S. U. Baroda, (Unpublished)*, 1984.

Babu, S. K. : Report on work done around Ghogha and surrounding area during field season 1957-58. - *ONGC Report* (*Unpublished*), 1958.

Biscaye, P. E. : Minerology and sedimentation of recent deep sea clay in the Atlantic Ocean and adjacent seas. - *Geol. Soc. Bull. Am.* 76 : 803-832, 1965.

Biswas, S. K. Rift basins in western Margin of India and their hydrocarbon prospects with special reference to Kutch basin. - *Bull. Am. Assoc. Petrol. Geol.*, 66, 10, 1497-1513, 1982.

Borde, D. V., Sarin, M. M. and Somayajulu, B. L.K. : Composition of Narmada and Tapi estuarine particles and adjacent Arabian Sea sediments. - *Ind. Jl. Marine Sciences*, 11, 51-62, 1982.

Chandra, P. K. and Choudhary, L. R. : Stratigraphy of the Cambay Basin. - *Bull. ONGC*, 6, 2, 37-50, 1969.

Datta, A. K. : Geological work done in Bhavnagar - Ghogha region, Saurashtra.- *ONGC Report* (*Unpublished*)., 1959.

Eremenko, N. A., Negi, B. S., Kasiarov, M. V., Serigin, A. M., Deshpande, B.G., Sengupta, S. N., Talukdar, S. N., Sastri, V.V., Sokolov, I. P., Povlenko, A. I., Datta, A. K. and Raju A. T. R. : - Tectonic Map of India. - *Bull. ONGC*., 6,1-11, 1969.

Fedden, F. : The Geology of Kathiawar peninsula in Gujarat. - *Mem. Geo. Surv. Ind.*, 21, 2, 78-125, 1884.

Ganapathi, S. : Geological and geomorphological evolution of the coastline between Bhavnagar and Mahuva of southeastern Saurashtra. - *Ph. D. thesis, M. S. U.* Baroda, (*Unpublished*), 1981.

Ganapathi, S. and Merh, S. S. : Geological and Geomorphological evolution of the coastline between Bhavnagar and Mahuva of southern Saurashtra, Gujarat State - In the report on the work based on the UGC sponsored research programme on Quaternary Geology, Dept. of Geology, M. S. U. Baroda 146-168 (*Unpublished*), 1982.

Kaila, K. L. , Krishna, V. G., and Mall, D. M. : Crustal structure along Mehmedabed- Billimora profile in the Cambay basin, India, from deep seismic soundings.- *Tectonophysics*, 76, 99-130, 1981.

Markevich, V. P. : Prognostic assessment of oil and gas prospects on India and main guidelines to future exploration. - *ONGC Report (Unpublished)*, 1976.

Mathur, L. P. and Evans, P : Oil in India. - *XXII International Geol. Cong.* New Delhi, 1964.

Mathur, L. P. , Rao, K. L. N. and Chaube, A. N. : Tectonic framework of Cambay Basin, India. - *Bull. ONGC*., 5, 17-28. 1968.

Merh, S. S. : Quaternary sea level changes along Indian coast. - *Proc. Ind. Nat. Acad*., 58, 5, 461-472, 1992.

Muley, M. V. Shah, V. P. Bokil, K. K. and Manjul, S. S. : Coastal zone studies in the Gulf of Khambhat using SALYUT-7 space photographs. - *Proc. Workshop on preliminary analysis of SALYUT-7 space photographs.* Space Application Centre, Ahmedabad, 127-141, 1985.

Nair, R. R. and Sahai, B. : Coastal geomorphology of the Gulf of Khambhat. - *Proc. Symp. Quat. Episodes*, (*Ed.* Merh, S. S. and Vashi, N. M.). Dept. of Geology. M. S. U. Baroda, 87-96, 1985.

Nayak, S. R. and Sahai, B. : Coastal geomorphology of the Gulf of Khambhat. - *Proc. Symp. Quat. Episodes*, (*Ed.* Merh, S.S. and Vashi, N. M.). Dept. of Geology. M. S. U. Baaroda, 87-96, 1985.

Negi, B. S. : Gravity and Magnetic survey in sub-surface structures in Bhavnagar - Ghogha area. - *ONGC Report (Unpublished)*, 1955.

Pandya, S. H. : Location of Lothal : A multidisciplinary approach. - *Proc. Symp. Quat. Episodes* (*Ed.* Merh, S. S. and Vashi, N. M.) Dept. Geol., M. S. U. Baroda, 259-263, 1985.

Patel, K. F. : Studies on the tidal muds of Saurashtra nd Mainland Gujarat Coast with special reference to their geochemistry. - *Ph. D. Thesis, M. S. U. Baroda (Unpublished)*, 1985.

Patel, M. P. and Merh, S. S. : Landform studies on the Bhogava - Sabarmati confluence area of Gujarat with reference to Holocene sea-level changes. - In the report of work based on the U. G. C. sponsored research programme on Quaternary Geology. Dept. of Geology, M. S. university of Baroda (*Unpublished*), 219-226.

Patel, M. P. Patel, S. G. and Merh, S. S. : Geomorphic evidence of Quaternary sea-level changes in Mahi - Tapi coastal segment of Gujarat. - *Proc. Symp. Quat. Episode* (*Ed.* Merh, S. S. and Vashi, N. M.) Dept. of Geology, M. S. University. Baroda, 49-62, 1985.

Powar, K. B. : Lineament fabric and dyke pattern in the western part of the Deccan volcanic province. Deccan Volcanism, *Geol. Soc. Ind. Mem*., 3, 45-57, 1981.

Raju, A. T. R. : Geological evolution of Assam and Cambay Tertiary Basins of India. - Bull. Am. Assoc. Petrol. Geol., 52, 12, 2422-2437, 1968.

Raju, A. T. R. and Srinivasan, S. : More hydrocarbon from well explored Cambay basin. - *Petroleum Asia Journal*, K. D. M. I. P. E., O. N. G. C., Dehradun, 25-35, 1983.

Rao, P. V. and Jain, S. K. : Geology of Bhavnagar - Ghogha area. - *ONGC Report* (*Unpublished*), 1969.

Rao, R. P. and Talukdar, S. N. : Petroleum geology of Bombay High field, India. - *Glant Oil and Gas fields of the Decade* 1968-1978 (Ed. M. T. Halbouty), 1980.

Rao, S. R.: *Lothal and Indus Civilization*. - New Delhi, 1973.

Rao, S. R. : Lothal. - A Harappan Pont Town (1955-62). *Mem. Archeolog. Surv. India,* 1, 38, 1979.

Shapiro, L. and Brannock, W. W.: Rapid analysis of Silica - Carbonate and Phosphate rocks. - *Bull. U. S. Geol. Surv.*, 1194A, 1962.

Sridhar, V., Chamyal, L. S., Merh, S. S. : - North Gujarat rivers : Remnants fo super fluvial system. - *Jl. Geol. Soc. Ind.*, Vol. 44, 1994.

Vogel, A. I. : *A text book of Quantitative Inorganic Analysis including elementary Instrumental analysis*. - 3 ed., 1975.

TABLE 1: STRATIGRAPHIC SUCCESSION OF SAURASHTRA COASTAL BLOCK

Period	Period/Epoch	North of Shetrunji River: Formation	North of Shetrunji River: Lithology	South of Shetrunji River: Formation	South of Shetrunji River: Lithology
	PERIOD/EPOCH	THE AREA NORTH OF SHETRUNJI RIVER		THE AREA SOUTH OF SHETRUNJI RIVER	
		FORMATION	LITHOLOGY	FORMATION	LITHOLOGY
QUATERNARY	Holocene	Recent Deposits	Alluvium, Beach and Dune	Recent Deposits	Alluvium, Beach and Dune Sands Mudflats and Soils
QUATERNARY	Pleistocene to Early Holocene	Lakhanka Formation	Soft friable ferruginous sandstones and sands with layers rich in agate pebbles and intercalations of gray coloured clays	Miliolite Formation	Biosparites with Intercalations of red coloured clays
			UNCONFORMITY		
TERTIARY	Mid-Pliocene	Piram Beds	Hard and well cemented fossiliferous conglomerates with alteration of sandstones and claystones		
			DISCONFORMITY		
TERTIARY	Lower Miocene	GAJ FORMATION: Bhumbhli conglomerate member	Fossiliferous conglomerates, grits and argillacious sandstones with intercalations of clays		? ? ? ?
			UNCONFORMITY		
TERTIARY	Lower Miocene	GAJ FORMATION: Ratanpur clay member	Gray and yellow coloured clays and marls with gypsum layers at the base, basal conglomerates, and argilla ceous sandstone with cross lamination	Gaj formation	Yellow and gray coloured clays with biosparites.

---	---	---	UNCONFORMITY	---	---
	Paleocene	Lateritic Rocks	Unstratified red, Brown and Yellowish Brown Hard laterites with clay pockets	Lateritc rocks	Red, Brown, Yellow-ish, Brown Hard laterites with clay pockets
	Cretaceous	DECÇAN TRAP	Variety of basalitic lava as well as felsite and rhyolite with dole-rite dykes	Deccan Trap	Veriety of basaltic lava flows, feisite and rhyolite with dolerite dykes (Ganapathi, 1981)

Table 2: GENERALIZED STRATIGRAPHIC SUCCESSION OF CAMBAY BASIN

AREA / AGE	SURFACE: WESTERN MARGIN	SUB SURFACE: THARAD	SUB SURFACE: AHMEDABAD MEHSANA	SUB SURFACE: TARAPUR	SUB SURFACE: BROACH	SUB SURFACE: NARMADA	SURFACE: EASTERN MARGIN
RECENT TO PLEISTOCENE	ALLUVIUM	GUJARAT ALLUVIUM.					ALLUVIUM
	AGATE CONGL.	BUDHANPUR FORMATION	JAMBUSAR FORMATION				
PLIOCENE			BROACH FORMATION				
U. MIOCENE	PIRAM BEDS	ANTROL FORMATION	JAGADIA FORMATION				
M MIOCENE	KUDA / BHUMBALI	DHIMA FORMATION	KAND FORMATION				
L. MIOCENE	RATANPUR	DEODAR FORMATION	BABAGURU FORMATION				
				TARKESHWAR FORMATION			
OLIGOCENE			TARAPUR SHALE		DADHAR FORMATION		
U. EOCENE		WAV FORMATION	TARAPUR SHALE		ANKLESVAR FORMATION	TELWA SH. / ARDOL Mbr. / KANWA SH.	NUMULITICS
M. EOCENE		THARAD FORMATION	KALOL FORMATION	VASO FORMATION	ANKLESVAR FORMATION	HAJAD Mbr. / HAZIRA SHALE	
L. EOCENE			KADI FORMATION	UPPER CAMBAY SHALE		HAZIRA SHALE	
PALEOCENE	LATERITIC ROCKS	BALUTRI FORMATION	LOWER CAMBAY SHALE / VAGAD KHOL FORMATION				
UPPER CRETACEOUS	DECCAN TRAP GROUP						
L CRETACEOUS TO JURASSIC	MESOZOIC SEDIMENTS						
ARCHAEAN	GRANITE						

Table 3 : X-RAY ANALYSES OF MUD SAMPLES

TABLE		SOUTH SAURASHTRA (Methia - Bhavnagar) SAMPLES NOS.												NORTH-SAURASHTRA (Bhavnagar - Vadgam) SAMFLES NOS									NORTH MAINLAND (Khambhat - Dahej) SAMPLES NOS.			
	Minerals	1	2	3	4	5	6	7	8	9	10	11	12	13	14	15	16	17	18	19	20	21	22	23	24	25
CLAY MINERALS	Mo	O	O	O	O	O	O	O	O	O	O	O	O	O	O	O	O	O	O	O	O	O	O	O	O	O
	Ka	O	O	O	O	O	O	O	O	O	–	–	O	O	O	O	O	O	–	O	O	–	O	O	O	O
	Ha	–	O	O	O	O	–	–	O	O	O	O	–	O	–	–	O	–	O	–	O	O	O	O	–	O
	Na	–	–	O	–	O	–	O	O	O	–	–	O	–	O	–	O	O	–	O	O	–	–	–	O	–
	Dl	–	–	–	–	–	–	–	–	O	–	–	O	–	O	–	–	–	–	–	–	–	–	O	–	–
	I	O	O	O	O	O	O	O	O	O	O	O	O	O	O	O	O	O	O	O	O	O	O	O	–	O
	Bi	–	–	–	–	–	–	O	–	O	–	–	–	–	–	O	–	–	–	O	–	–	–	–	–	–
	Mu	O	O	O	O	O	O	O	O	O	O	O	O	O	O	O	O	O	O	–	O	O	O	O	O	O
	Zi	–	–	–	–	–	–	O	–	–	–	–	–	–	–	–	–	–	–	–	–	–	–	–	–	–
	Gla	–	–	–	–	–	–	–	–	O	–	–	–	–	–	–	–	–	–	–	–	–	–	–	–	–
	Ch	O	O	O	O	O	O	O	O	O	O	O	O	O	O	O	O	O	O	O	O	O	O	O	O	O
	Pe	–	–	–	–	–	–	–	–	–	–	–	–	–	–	–	–	–	–	–	–	–	–	O	–	–
	Se	–	O	–	O	–	O	O	O	–	–	O	–	–	O	O	O	O	–	O	O	O	O	O	O	O
NONCLAY MINERALS	Q	O	O	O	O	O	O	O	O	O	O	O	O	O	O	O	O	O	O	O	O	O	O	O	O	O
	Ca	O	O	O	O	O	O	–	–	O	O	O	–	O	O	O	O	O	O	O	O	–	O	O	O	O
	Ar	–	–	–	–	–	O	–	–	–	–	–	–	–	–	O	–	–	O	–	–	–	–	–	–	–
	Ze	–	–	O	–	–	–	O	O	O	O	–	–	–	–	–	O	O	–	O	–	–	–	O	–	–
	Ab	–	O	O	–	O	–	–	O	–	–	O	–	–	–	–	–	–	–	O	O	–	–	O	O	–
	Mgh	–	–	–	–	–	–	–	–	–	–	–	–	–	–	–	–	O	–	O	–	–	–	–	–	–
	Goe	–	–	–	–	–	–	–	–	–	–	–	–	–	–	–	–	–	–	–	–	–	–	–	–	–
	Gib	O	O	O	O	–	O	O	O	–	O	O	O	O	–	–	O	–	–	–	O	–	O	O	–	O
	Bz	–	–	–	O	O	–	–	–	–	–	–	–	–	–	O	–	–	–	–	–	–	–	–	–	–

TABLE		NORTH MAINLAND (Khambhat - Dahej) SAMPLES NOS.															SOUTH - MAINLAND (Dahej - Hazira) SAMPLES NOS.										
	Minerals	26	27	28	29	30	31	32	33	34	35	36	37	38	39	40	41	42	43	44	45	46	47	48	49	50	51
CLAY MINERALS	Mo	O	O	O	O	O	O	O	O	O	O	O	O	O	O	O	O	O	O	O	O	O	O	O	O	O	O
	Ka	O	–	–	O	O	–	O	–	O	–	O	O	O	O	O	–	O	O	O	–	–	O	O	–	O	–
	Ha	O	–	O	O	O	O	–	–	–	O	O	–	–	O	–	O	O	O	O	–	–	–	O	–	O	–
	Na	–	–	–	–	–	–	–	–	O	–	–	–	–	–	–	–	–	O	–	–	–	O	–	–	–	–
	Dl	–	O	–	–	–	–	–	–	–	–	–	–	O	O	–	–	–	–	–	–	–	O	O	–	–	–
	I	O	O	–	O	O	O	O	O	O	–	O	O	O	O	O	O	O	O	O	O	O	O	O	O	O	O
	Bi	O	–	–	–	–	O	–	–	–	O	O	–	–	–	O	O	O	O	–	–	–	O	–	–	O	–
	Mu	O	–	O	O	O	O	O	O	O	O	O	O	O	O	O	O	O	O	O	O	O	O	–	O	O	O
	Zi	–	–	O	–	–	–	–	–	O	–	O	–	–	O	–	–	–	–	–	–	–	O	–	–	O	–
	Gla	–	–	–	–	–	–	–	–	–	–	–	–	–	–	–	–	–	–	–	–	–	–	–	–	–	–
	Ch	O	O	–	O	O	O	O	O	O	O	O	O	O	O	O	O	O	O	O	O	O	O	O	O	O	O
	Pe	–	–	–	–	O	–	–	–	–	–	–	–	–	–	–	–	–	–	–	–	–	–	–	–	–	–
	Se	O	O	–	–	O	O	–	–	–	–	–	–	O	O	–	O	–	O	O	–	O	O	–	O	–	–
NONCLAY MINERALS	Q	O	O	O	O	O	O	O	O	O	O	O	O	O	O	O	O	O	O	O	O	O	O	O	O	O	O
	Ca	O	O	O	–	O	O	–	–	O	O	–	O	–	O	O	O	O	O	O	O	O	–	O	–	O	O
	Ar	–	–	–	–	–	–	O	O	O	–	–	O	–	–	–	–	–	–	–	–	–	O	–	–	–	–
	Ze	–	O	–	–	O	–	O	O	O	O	O	O	–	–	–	O	–	–	O	–	–	–	O	–	–	–
	Ab	–	–	–	–	O	–	–	–	–	–	–	–	–	O	O	–	–	–	–	–	–	–	–	–	–	–
	Mgh	–	–	–	–	–	–	–	–	O	O	–	–	–	–	–	–	–	–	–	–	–	–	–	–	–	–
	Goe	–	–	–	–	–	–	–	–	–	–	–	–	–	–	–	–	–	–	–	–	–	–	–	–	–	–
	Gib	O	O	–	O	O	–	O	O	O	O	O	–	O	O	O	O	O	–	O	O	O	–	O	O	O	–
	Bz	–	–	–	–	–	O	O	O	O	O	–	–	–	–	–	–	–	O	O	–	–	–	O	O	–	–

O' Present – Absent/under detectable limit

(For location please refer Fig. II)

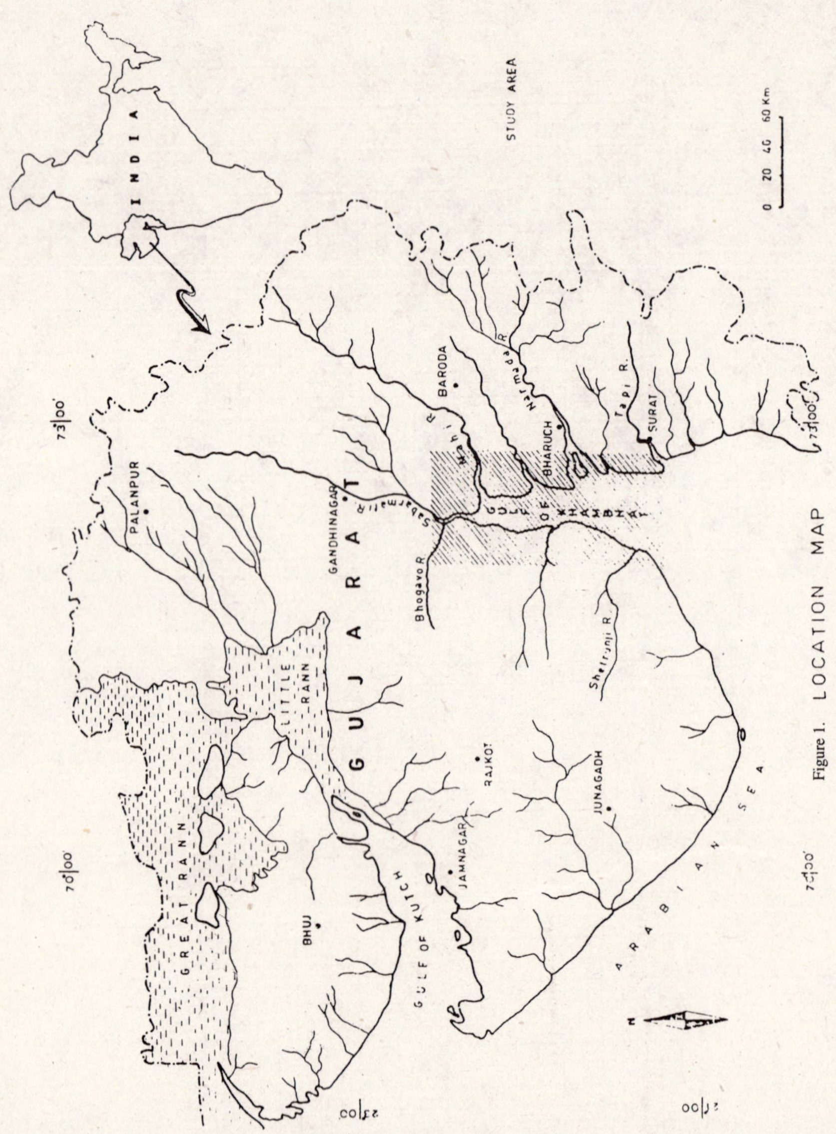

Figure 1. LOCATION MAP

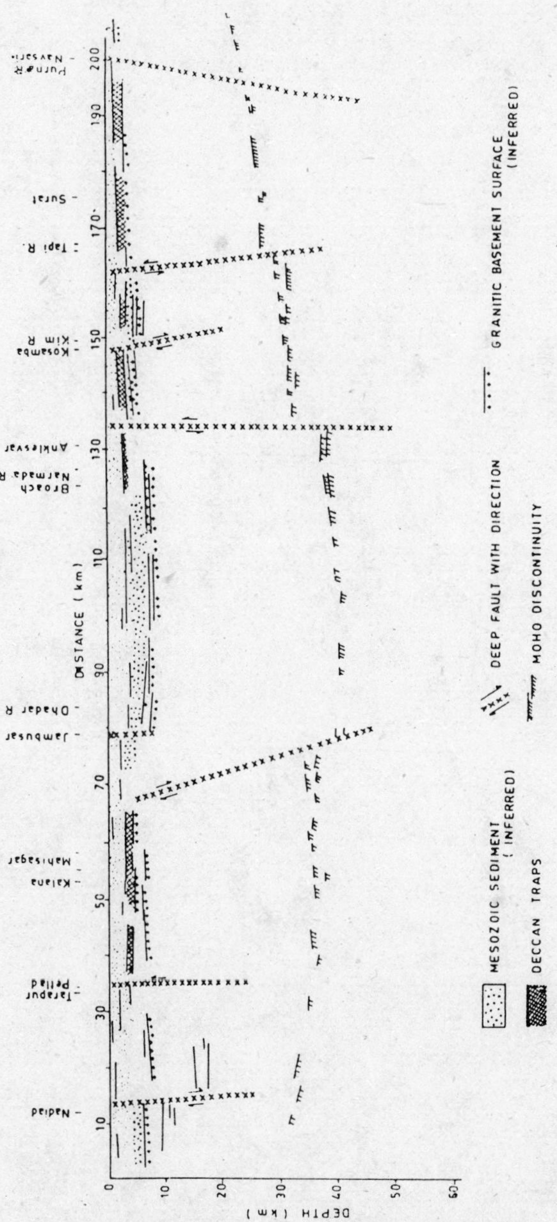

Figure 2. MAJOR BASEMENT LINEAMENTS CUTTING MOHO BOUNDARY
(Based on Kaila et al, 1981)

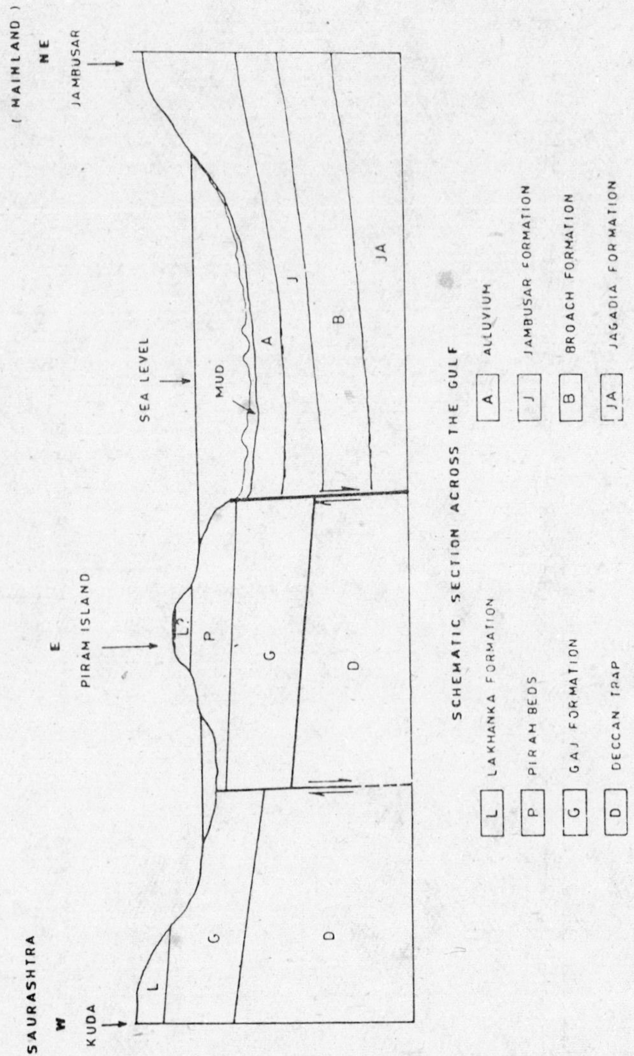
SAURASHTRA
W
KUDA
E
PIRAM ISLAND
SEA LEVEL
MUD
(MAINLAND)
NE
JAMBUSAR
L
G
D
P
G
D
A
J
B
JA
SCHEMATIC SECTION ACROSS THE GULF
L LAKHANKA FORMATION
P PIRAM BEDS
G GAJ FORMATION
D DECCAN TRAP
A ALLUVIUM
J JAMBUSAR FORMATION
B BROACH FORMATION
JA JAGADIA FORMATION

Figure 3.

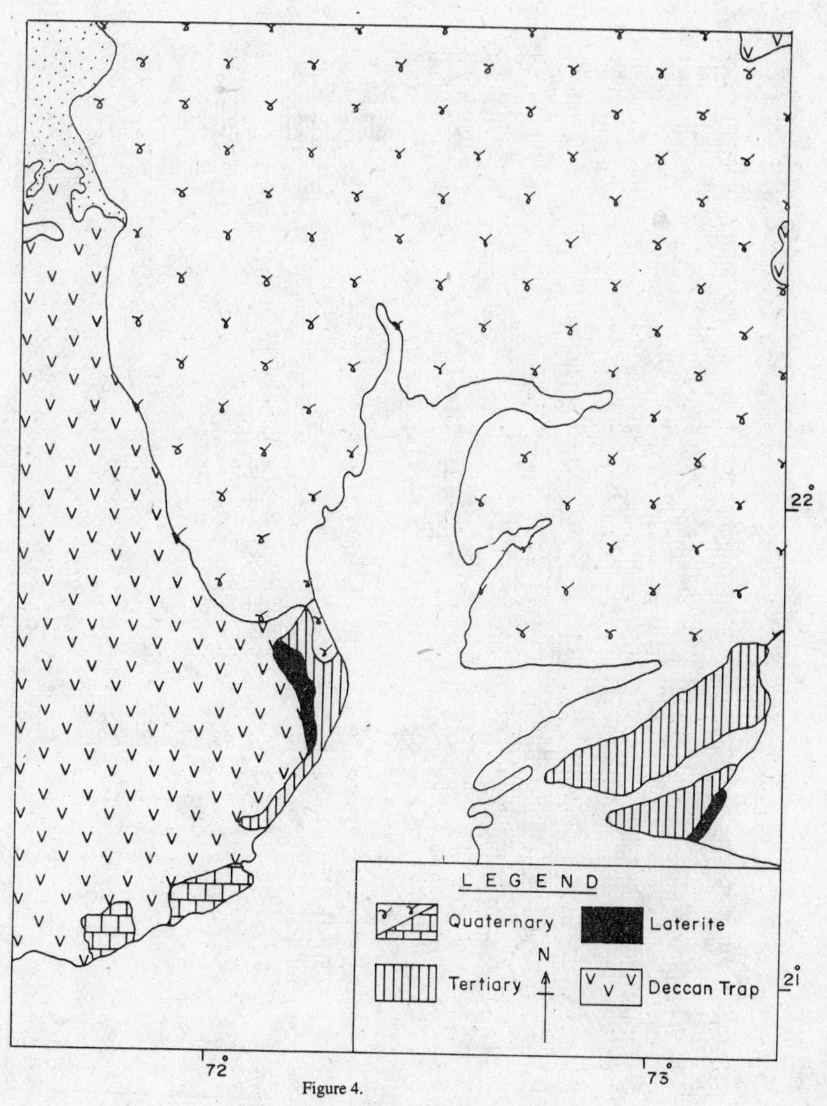
LEGEND
Quaternary
Laterite
N
Tertiary
Deccan Trap
22°
21°
72°
73°

Figure 4.

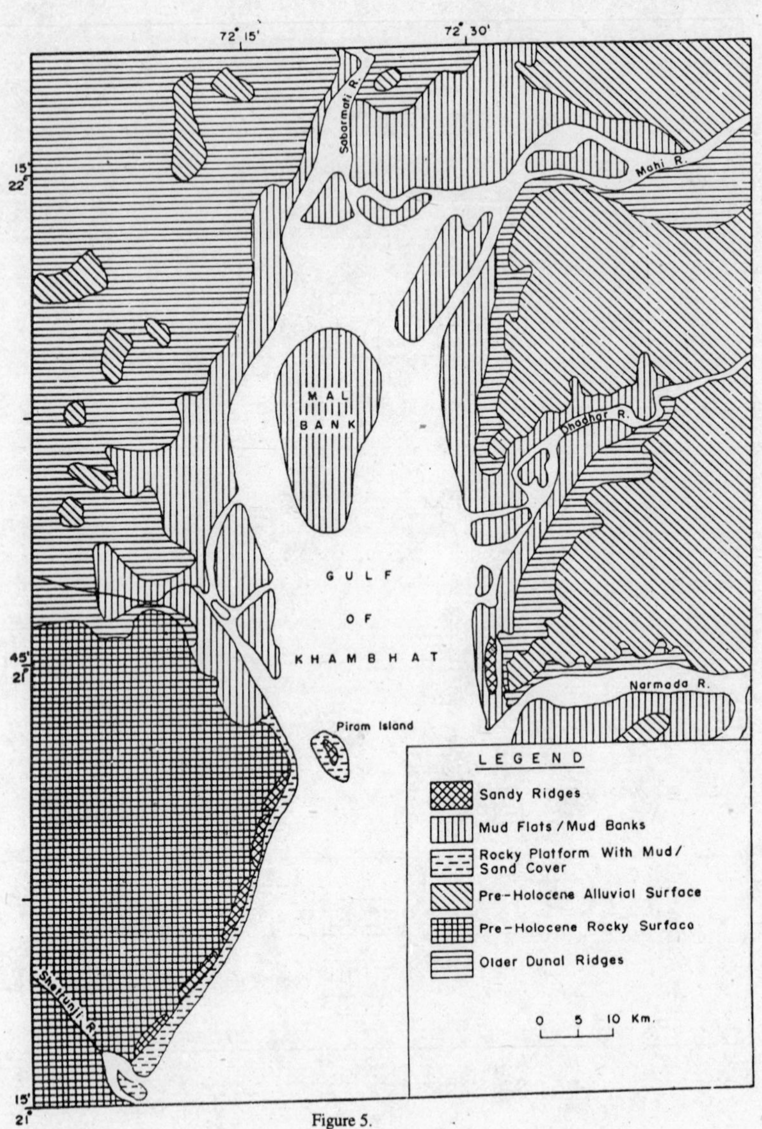
72° 15'
72° 30'
15'
22°
45'
21°
15'
21°
Sabarmati R.
Mahi R.
MAL
BANK
Dhadhar R.
GULF
OF
KHAMBHAT
Narmada R.
Piram Island
Shetrunji R.
LEGEND
Sandy Ridges
Mud Flats / Mud Banks
Rocky Platform With Mud/
Sand Cover
Pre-Holocene Alluvial Surface
Pre-Holocene Rocky Surfaco
Older Dunal Ridges
0 5 10 Km.

Figure 5.

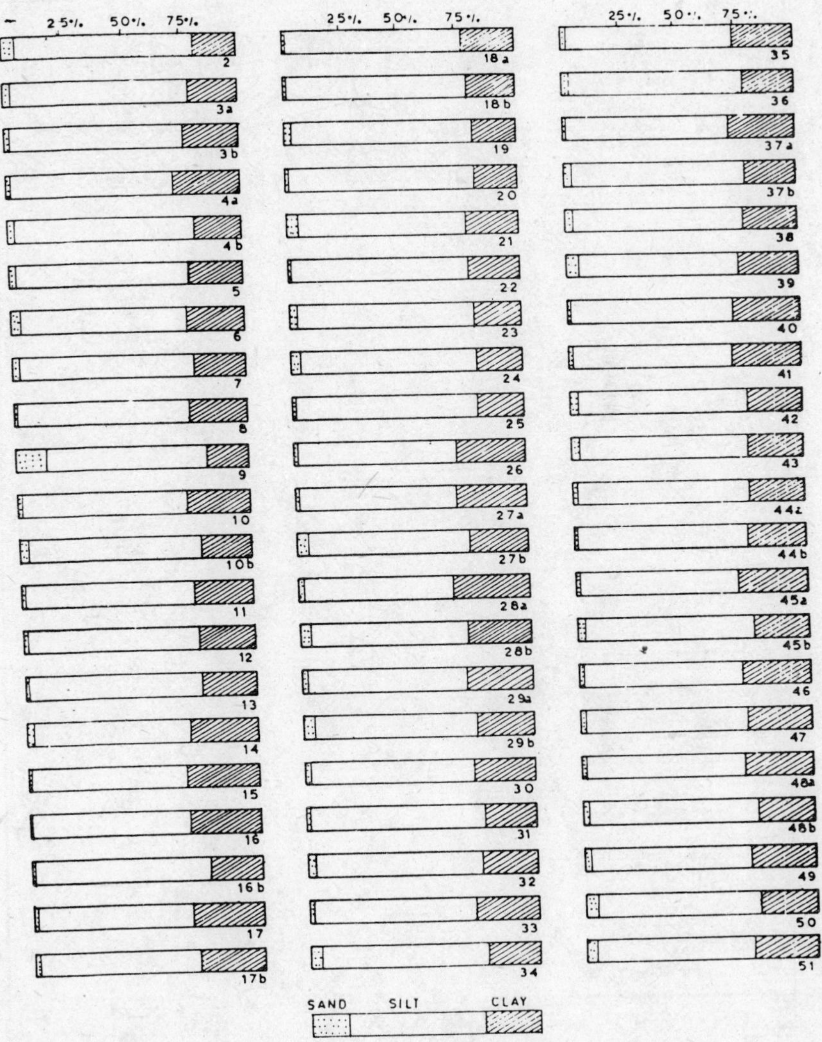

Figure 6. GRAIN SIZE DISTRIBUTION OF TIDAL MUD IN PERCENTAGE

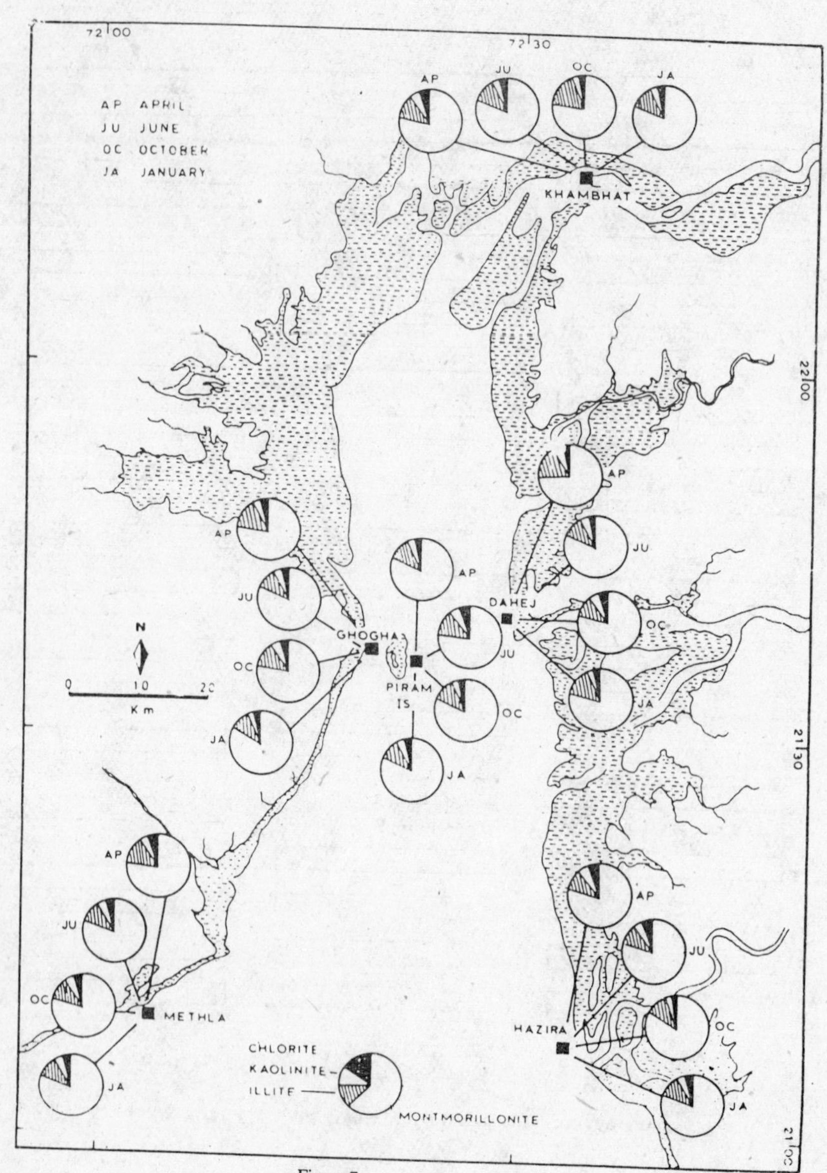
72 00
72 30
22 00
21 30
21 00
AP APRIL
JU JUNE
OC OCTOBER
JA JANUARY
AP
JU
OC
JA
KHAMBHAT
GHOGHA
PIRAM IS
DAHEJ
METHLA
HAZIRA
N
0 10 20
Km
CHLORITE
KAOLINITE
ILLITE
MONTMORILLONITE

Figure 7.

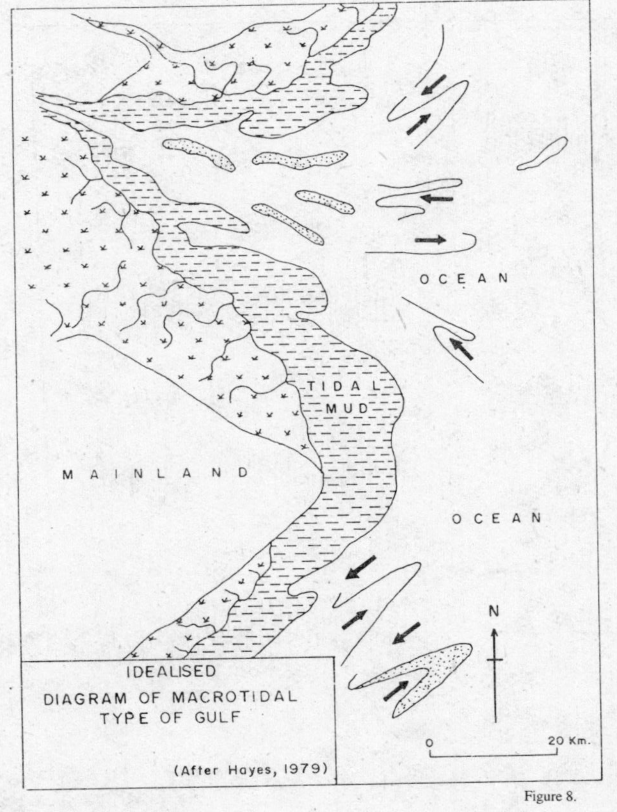

Figure 8.

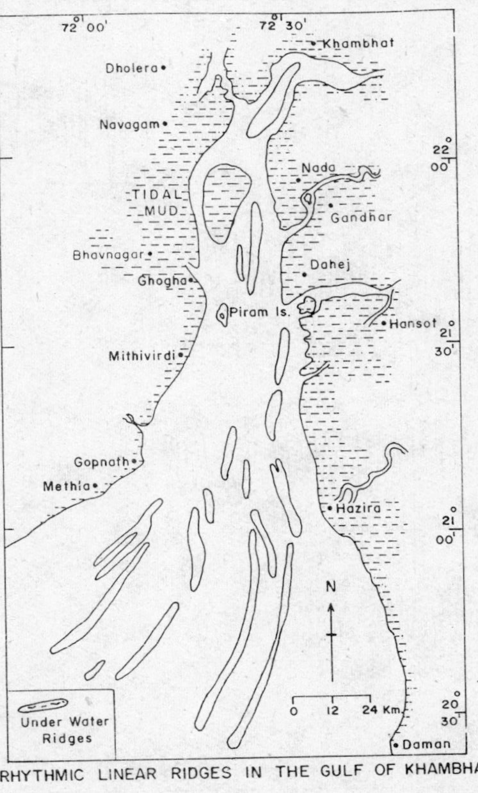

RHYTHMIC LINEAR RIDGES IN THE GULF OF KHAMBHAT

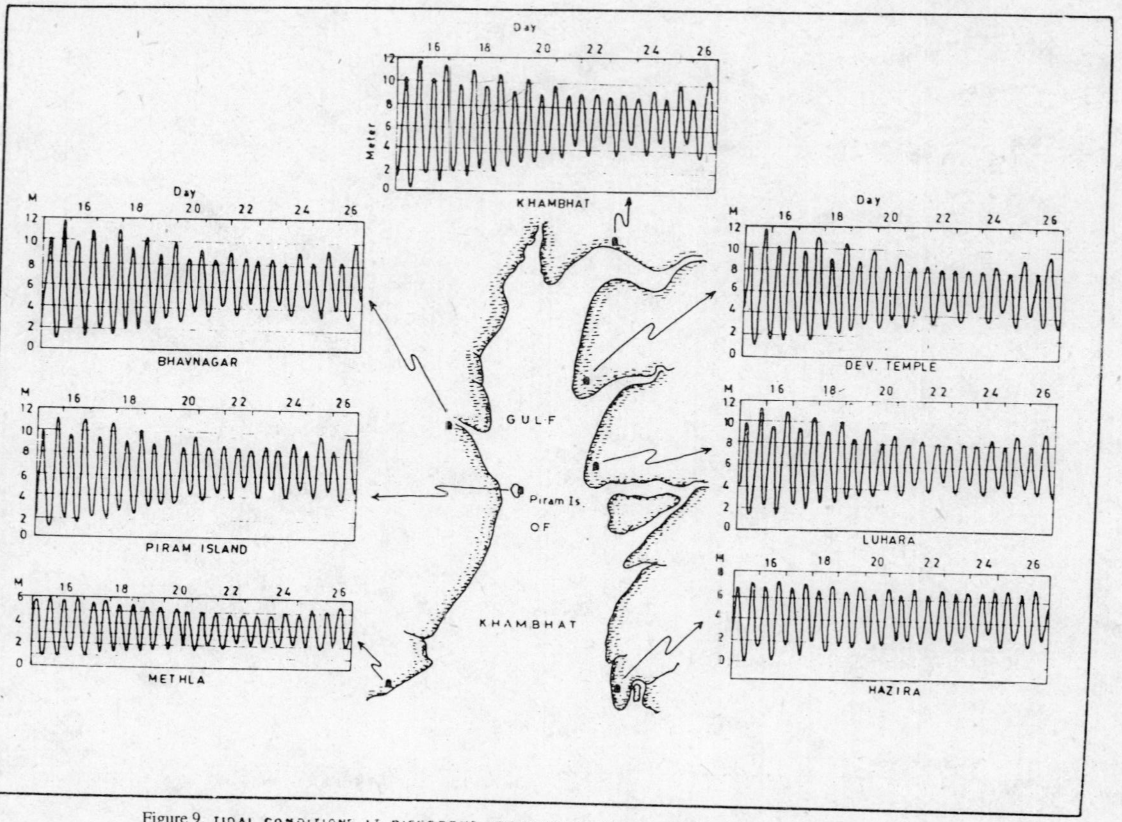

Figure 9. TIDAL CONDITIONS AT DIFFERENT LOCATION IN THE GULF OF KHAMBHAT

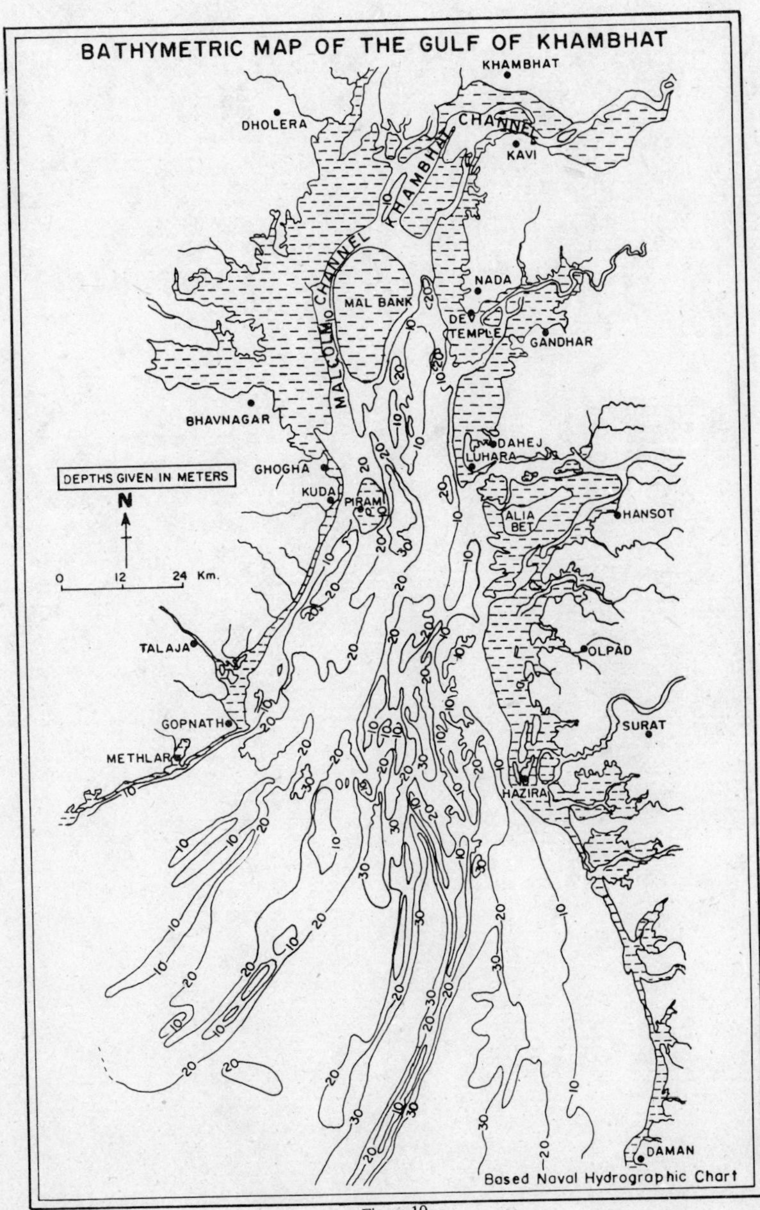
BATHYMETRIC MAP OF THE GULF OF KHAMBHAT
KHAMBHAT
DHOLERA
KHAMBHAT CHANNEL
KAVI
MALCOLM CHANNEL
MAL BANK
NADA
DEV TEMPLE
GANDHAR
BHAVNAGAR
DAHEJ
LUHARA
GHOGHA
DEPTHS GIVEN IN METERS
N
KUDA
PIRAM I.
ALIA BET
HANSOT
0 12 24 Km.
TALAJA
OLPAD
GOPNATH
SURAT
METHLAR
HAZIRA
DAMAN
Based Naval Hydrographic Chart

Figure 10.

APPENDIX II

RECENT ADVANCES IN GEOMORPHIC STUDIES OF PENINSULAR INDIA : A REVIEW*

R. VAIDYANADHAN

Geography Department, Andhra University. Waltair (A. P.)

ABSTRACT

This is a review on selected contributions in the field of geomorphology of Peninsular India published since 1968. Recognition, mapping and correlation of erosion (planar) surfaces were carried out in different regions and it is now generally believed that remnants of surfaces as old as Upper Jurassic are present in parts of Peninsular India. Greater emphasis. of late, is being laid on field studies, which have aided recognition of the effect of neotectonics on the surfaces. Relationship between tectonic trends and landforms have also been studied. The importance of study of laterites is increasingly realised as it is helpful in the recognition of erosion surfaces, besides aiding in the reconstruction of climatic history of the region.

Aerial photographs are being increasingly used in geomorphic studies and parts of the coasts have been mapped using them. Of particular interest are those made of the coasts of Orissa, Andhra Pradesh and Gujarat. Indications are that since Pleistocene, many changes, mostly progradational, have taken place along the Peninsular coast. Among the few studies on fluvial morphology, those of Cauvery in the deltaic portion and the Subarnarekha in Eastern India are noteworthy. Morphometric analyses of a few drainage basins have yielded useful results. The main contribution in the field of climatic geomeorphogy has been by the visiting European geographers. Studies in applied geomorphology cover a very wide field like criteria to recognise evaporites in Rajasthan, residual deposits in eastern and southern India, methods to be adopted to check coastal erosion in parts of West Bengal, identification of ground water aquifers and favourable structures for the accumulation of hydrocarbons.

INTRODUCTION

The varied landforms of India attracted the attention of many early explorers and geologists and their accounts of different parts of the country were essentially descriptive. The Records and Memoirs of the Geological Survey of India, the Proceedings of the Asiatic Society of Bengal and the many travelogues of naturalists contain much information on the landforms of India. The origins ascribed to many of the features were speculative, though later detailed scientific studies proved some of them to be true. It can be said to their credit that much of the description were unbiased as theories were very few then to account for varied types of landforms. However explanation of the origin of landforms on a scientific basis can be said to

---

*reproduced from Indian Journal of Earth Sciences, S. Ray Volume 1977, pp. 13-35 with the consent of the author.

have begun in our country only since the beginning of this century.

The Indian sub-continent is usually divided into (i) the Peninsula, made up mostly of Precambrian formations, Gondwanas (Carboniferous to Cretaceous), Deccan Traps (Cretaceous-Eocene), and Tertiary and Quaternary sediments in parts of its long coast, (ii) the Extra-Peninsula, made up of rocks of ages ranging from Precambrian to Quaternary in different sections of the lofty Himalaya mountains and (iii) the Indo-Gangetic- Brahmaputra Plains in between the two. This review deals only with the study of landforms in Peninsular India.

Among those who have contributed significantly on some aspect or other to our knowledge of the "geomorphology" (quite a few of them have not used this term, though) of parts of Peninsular India since the 1930s are A. M. Heron, D. N. Wadia, J. A. Dunn, W. D. West, J. B. Auden, R. N. P. Arogyaswamy and B. P. Radhakrishna among the geologists and S. P. Chatterjee, H. L. Chibber, S. C. Bose, R. P. Singh, Enayat Ahmed and K. Bagchi among the geographers. Thanks to the initiative of W. D. West, a seminar on Geomorphological Studies in India was held at Sagar in Madhya Pradesh in 1965(the Proceedings were published in 1967), which can be said to have been the first forum where geomorphologist gathered from different parts of India, took stock of our existing knowledge, and suggested some new lines of investigations. Subsequently on the eve of 21st International Geographical Congress held at New Delhi in 1968, Chatterjee (1968) has reviewed the later publications in the field of geomorphology. Hence this review will deal essentially only with the recent advances made in this field in Peninsular India from about 1968 to the present.

A couple of years ago geomorphological studies in India received an impetus with the starting of a Quaternary Geological Division in the Regional offices of the Geological Survey of India. Some geomorphological work has been done also by the Oil and Natural Gas Commission and National Atlas Organisation. During the past two years numerous papers have been presented in a number of seminars. But a majority of them have not been published and hence it has not been possible to include them in the present review, except a few papers that were made available to the author.

Though it is difficult to compartmentalise the contributions as belonging to only a specific sub-field, as a single paper may deal with different aspects of geomorphology of a region, for purposes of the present review it is proposed to group the selected contributions under the following heads : (i) Regional Geomorphology, (ii) Neotectonics, (iii) Laterite, (iv) Coastal geomorphology, (v) Fluvial geomorphology, (vi) Climate geomorphology and (vii) Applied geomorphology. Owing to limitations of space the studies relating to local geomorphology are excluded from the present review.

REGIONAL GEOMORPHOLOGY

One of the usual occupations of the regional geomorphologist has been to take published topographic maps of a region, draw at suitable intervals projected profiles, composite profiles, altimetric frequency graphs, and prepare a variety of relief and slope maps and deduce the possible 'erosion surfaces', in the region. The next step is to try to assign a relative age. Invariably the highest is considered to be the oldest and the lowest, youngest. The next exercise is the study of the major rivers in the region and find out if they were superposed or antecedent or related to structure *etc.*. Some of the workers have taken to actual field studies also to supplement the laboratory studies.

Biswas (1974) has presented the geomorphology of the Kutch region with a detailed study of the geology of the region along with a study of the planation surfaces. The highlands in the central Kutch bear evidences of multiple erosion cycles. Five denudational cycles have been recognised and correlated with major periods of tectonic movements in the region. One interesting feature of the Kutch landscape is the first order topography, *viz,* anticline appearing as hills and synclines as valleys. This is attributed to periodic unidirectional movements and a very late uplift. Sen (1972) has described quite a few geomorphic features from parts of Aravalli Range. Study of altitudes and forms of hills seem to indicate to her that perhaps there has been an arching (cymatogeny?) up of the peneplaned surface in a NW-SE direction, almost perpendicular to the extension of the Aravalli Range. The hilltops however decrease more rapidly in the western part than in the eastern part. This, it is suggested, may be due to differential upwarping along the shear zone which runs almost NE-SW in the Aravalli Range. An interesting suggestion is made that (*op cit* p. 281) the frontal part of the Gondwana supercontinent moving north and carrying Aravalli Range with it, was being swept down under the Tethyan geosyncline by the descending arm of the convective cell thereby causing the warping on an E-W axis. Absence of datable sediments over the recognised peneplaned surfaces exasperates the author (*op cit* p. 283) but perhaps with the type of guidelines provided by Dixey (1970) it should be possible to arrive at the ages of surfaces in Rajasthan as well.

It was West (1962) who reactivated interest in the study of geomorphology in Madhya Pradesh by pointing out that Narmada-Son Line has been an active tectonic element perhaps right from the Precambrian times and has had its influence on the geology and geomorphology of this region in subsequent geological times. Choubey (1968) recognised two erosion surfaces beneath the Deccan basalts, the older capping the higher and resistant ridges and the younger extending over weaker rocks. Both these surfaces existed before the eruption of the basalts. The higher one is around 1950 feet and the lower one on softer rocks around 1450- 1200 feet above M.S.L. in the Katangi area in Madhya Pradesh. Lakshmanan's (1970) studies of the geology and geomorphology of the region on either side of the Kaimur scarp in parts of Satna and Sahdol districts of Madhya Pradesh led him to conclude that the northwest flowing tributaries of the Son river were superposed over the Deccan Traps which might have covered even the Kaimurs (Vindhyans). It was Dixey (1970) who after the perusal of all literature and maps, and traverses in Madhya Pradesh, synthesized the regional geomorphology of Madhya Pradesh, the like of which we do not have of any other part of Peninsular India, except perhaps Kerala-Karnataka- Tamil Nadu. Not only have the surfaces been recognised across various geological formations like Bijawars, Vindhyans, Gondwanas and Deccaan Traps, but also reconstruction of the evolutionary history of the rivers was attempted. According to him (*op cit* p. 219), the major courses of the Narmada and the Son have been strongly influenced by the post-Trap warping of this region. He has opened up a vast area for fruitful investigation, by explaining the means of recognising the surfaces and dating them. He is firmly of the view that the oldest post-Gondwana Bhander surface represents a Jurassic cycle and is post-Jurassic in age *(op cit* p. 215). Choubey (1971) reconstructed the sub-volcanic surface in detail and showed that the current drainage is superimposed on the post-Deccan Trap topography and after the removal of the lavas, the rivers are possibly adjusting to the pre-Deccan Trap drainage system. The current Valley cycle, where the basalts have been removed, is largely coincident with the lower erosion surface, but it is now coming under attack from a newer cycle, working up from Narmada and Ganges rivers. Subramanyan (1975) has made a detailed study of the geomorphology of the area around Sagar and correlated the surfaces recognised by him here with those in the neighbouring regions.

Geomorphology of parts of Bihar have been studied by a number of workers and among them those by Singh (1969) on the Chota Nagpur Plateau, Pathak (1968) on Chhechhari Basin and Mukhopadhyay (1969) on Subarnarekha basin are noteworthy. Singh's studies have been mostly, on conventional lines as listed at the beginning cf this section. He has described the drainage patterns and also the the evolutions of the major rivers. According to Pathak (1968) the three major landforms namely the 'pat', the basin and the scarps, have been evolved in the course of polycyclic evolution of the landscape, interrupted by 'accidents' in the form of diastrophic and volcanic activities. Evans (1970) has described the plateaus of Bihar, recognising a majority of them as erosion surfaces, now gently undulating with a slight general slope towards ESE, SE and SSE. His very tentative classification (*op cit* p. 388) gives the altitudinal range of out of the erosion surfaces recognised in the west as well as the eastern part of Bihar. He has calculated their gradients besides suggesting ages for the surfaces.

Ramana Rao and Vaidyanadhan (1970) recognised three distinct erosion surfaces in the Keonjhar region of Orissa and in the extreme southern part of Bihar, based on conventional methods, study of aerial photographs, recognition of flat areas across varying structural elements and and study of duricrusts. Their altitudinal ranges, nature of duricrusts over them, their gradients and ages are also arrived at, the last on the basis of correlation of the surfaces with the sedimentary record in the Bengal Basin. Attention has been drawn to the fact that the lowest of the surface here is partly erosional and partly depositional (Ramana Rao and Vaidyanadhan 1974a) and hence the nomenclature 'planar surfaces" used instead of "erosion surface".

Erosion surfaces were recognised over the Precambrians of Peninsular India by the early geologists and different origins attributed to them. But it was Radhakrishna's (1952) paper on Mysore Plateau that initiated active interest in the study of erosion surfaces in South India, with particular reference to their origin, evolution and the history of major rivers. Vaidyanadhan's (1967) was perhaps the first attempt to bring together the available data on the surfaces in South India (south of Latitude 18^0 N) and based on his own field observations and the available stratigraphic record of the sediments along the east coast weave a geomorphic history of South India. He also recognised the part played by cymatogeny in modifying the lie of the surfaces and affecting the evolution of the major drainage.

Brunner (1968a) mapped the geomorphic features in parts of the Mysore Plateau and a little southeast of it in Tamil Nadu. The excellent coloured maps accompanying the paper show mostly morphological units, but also present a few faults, recognisable through geomorphic study. The map shows seven slope categories ranging from less than 1^0 to more than 45^0. He is of the view (*op cit* p. 13) that the big step between the Mysore Plateau and the Tamil Nadu Plain does not allow one to suppose denudation to be the sole agent as suggested earlier by Krebs and Budel, but is due to many faults in the region. Gellert (1970) considered the landforms in the Deccan, particularly in Mysore and Tamil Nadu to be of polygenetic and polycyclic character. The step- like character of the surfaces are considered to be the result of uplifts in several places. Based on his personal knowledge of the terrain in Africa, he cautions (*op cit* p. 120) against correlating the surfaces in South India with those in Africa.

Babu (1975a), based on the study of maps, projected profiles and field traverses across Tamil Nadu and Karnataka (south of 13^{0}N Latitude), recognised 'four base levels of erosion. He dated them with the help of detailed subsurface stratigraphic record of the Cauvery Basin. Subsequently Babu (1975b) recognised five erosion surfaces in the Godavari Basin and the adjoining Eastern Ghats. Both these studies confirm the surface recognised earlier by Vaidyanadhan (1967), besides bringing to light a couple of more surface at lower elevations.

For a long time, the stupendous scarp running almost close to the west coast was thought to be a faulted one and representing a fault scarp. Radhakrishna (1966) however is of the opinion that it is the effect of erosion and retreat rather than faulting. The irregular trend of this scarp near the west coast is due to the capturing of the eastward flowing rivers over the plateau by the westward flowing consequents by headwarad erosion (Sharaswati river), reduction of an old plateau to a new surface and late Tertiary upwarps. Radhakrishna (1968) has further gathered evidences to reiterate the original contention of Schlagintweit and Blanford, that charnockite hill massifs in South India were faulted and uplifted. He showed that the petrographic characters of the charnockites are such that they must have originally formed at great depths. They are today where they are in space due to repeated uplifts along major fault zones. Many geomorphologists in the country tend to agree with this view.

The Nilgiri mountains rising to an altitude of over 8,000 feet have attracted the attention of geologists right from the middle of the nineteenth century. The bone of contention has invariably been, if it has come to stay there due to gradual upward or blockfaulted and uplifted. Mounting evidence seems to favour the latter view, though upwarp should have been there to start with. Pardhasaradhi and Vaidyanadhan (1974) recognise two relict levels of erosion over the Nilgiris within a polycylic and polygenetic planar Nilgiri surface. Landforms study suggests a pediment cycle for the lower Ootacamund landforms. The higher, Dodabetta landform, is the remnant of the older planar surface.

Demongeot (1975) studied part of South India between the latitudes of 11^0 and 13^0 and has made a useful contribution to our knowledge of this terrain. He has also recognised almost the same surface, though the ages given to the surfaces are slightly varying. Unlike Babu (1975a) who is firmly of the view that the Nilgiri surface is Pleistocene in age, though it occupies the highest range of altitude in South India, Demongeot (1975, p. 251, 256) feels that the whole morphology is like a Mesozoic one. According to the present author the Nilgiri surface must have been formed in post-Jurassic times, but it comes to occupy the present position in space due to original upwarp, then faulting and uplift in late Tertiary to early Quaternary. An interesting observation of Demongeot (*op cit* p. 241) is that the large basins within the pediplain of Mysore near Nanjangud, Chamrajanagar, Mandya *etc.* are pre-Pliocene.

Kerala in the south wetern part of Peninsular India is another region of great interest. Sections of these have been studied and the origin given to the

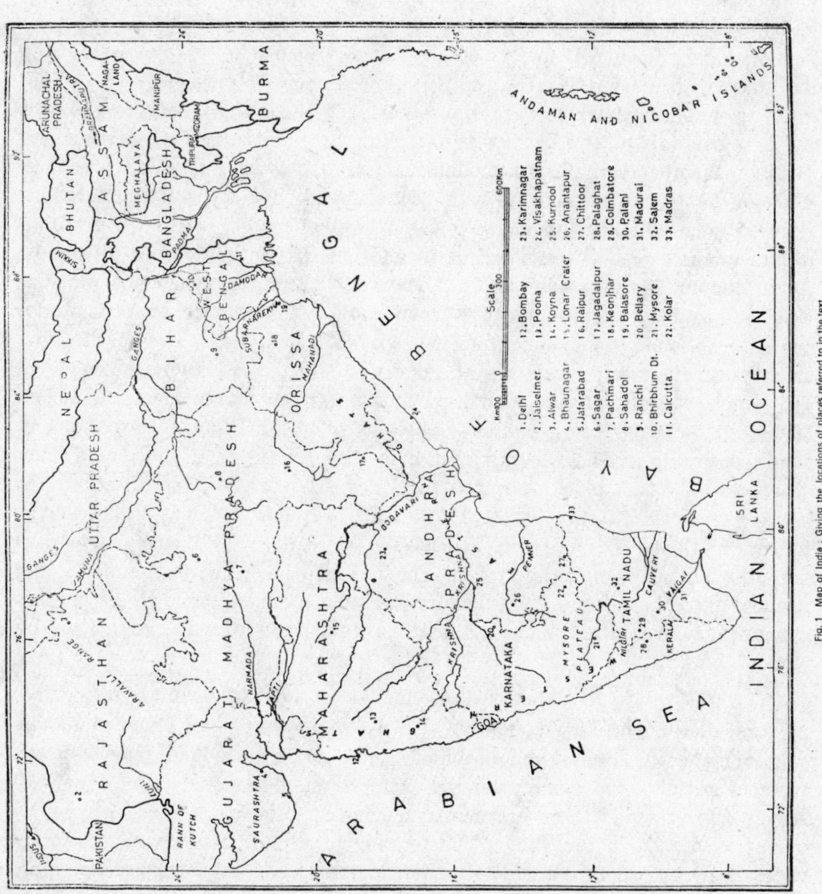

Fig. 1 Map of India : Giving the locations of places referred to in the text

numerous small plateaus and the coast are also different. Demongeot (*op cit* p. 239) is of the view that there are only two surfaces here, the upper one gently sloping from 350 metres at the Palaghat Gap upto 75 metres towards the coast with a lateritic cover, and the lower one, the present coastal plain, moving headward upto Palaghat such that these two merge and are indistinguishable at the Palaghat Gap (*op cit* Fig. 4). However, Tirugnanasambandam (1976) recognises five cycles in this same area and distinguishes the erosional surfaces as occupying distinct altitudinal ranges from 550 metres at the top, to the lowest at around 15 to 70 metres. According to him this is a result of regional uplift with intermittent periods of stillstands. According to Pardhasaradhi (1976), however, the planar surface of Malabar Upland has been monoclinally warped roughly parallel to the coast too, sometimes as stacks and sometimes beneath recent sediments. The lateritisation at successively lower levels may be due to etchplain development (*op cit* p. 76) which permits lateritic formation at successively lower levels. Choubey (1969,80) has recognised similar etchplains over the Bijawar formations in Madhya Pradesh. Backwasting eliminates the top layer of duricrust, then downwasting takes place followed by a second period of lateritisation. On a regional scale Murthy *et al* (1976) recognise five surfaces in Kerala as follows : (i) above 5000 feet, (ii) 2000-3000 feet, (iii) 1100-1300 feet, (iv) 500-700 feet and (v) 200-400 feet. The last two have laterite remnants, but the first three have been identified from a study of maps only. River piracies take place over the ghats by headward erosion of the west flowing consequents.

The following two tables give the relative ages (as suggested by respective authors) of the surfaces recognised in the northern part of Peninsular India (Table1) and in South India (Table 2). The given altitudinal ranges of the surfaces pertain to some specific area within the respective region of survey and is bound to be different when traced elsewhere. The surfaces or cycles as found in the Tables side by side in adjoining columns are not necessarily coeval. It should also be remembered that a surface is more often initiated at the coast and spreads inland and hence has varying ages at different sections, the range transcending sometimes a geological period in the Tertiary. If one adds to this upward and faulting, the complexity is complete! From both the tables it may be noticed that the general view seems to be that in Peninsular India there are definitely erosion surfaces of early Tertiary age, possibly of late Mesozoic also. As Dixey (1970,223) had suggested it should be possible to work on the extension of the surfaces in other parts of the country and their ages as well by going radially from the centre.

TABLE 1

PLANAR SURFACES IN WEST CENTRAL AND EASTERN INDIA

Geological Time	*Kutch* (Biswas 1974)	*Madhya Pradesh* (Dixey 1970)	*Orissa* (Ramana Rao and Vaidyanadhan 1974)	*Bihar* (Evans 1970)
Holocene	3-100'		Alluvial deposits	
Quater- L.Pleis.		Quaternary cycle	Detrital and	
nary M. Pleis.	Kutch Cycle	Marble Rocks	*in situ* laterites	
E. Pleis.	100-400'	cycle (Valleys)		
Late Pliocene				
Middle Pliocene			Jamda cycle	Kodarma
Early Pliocene	Late Tertiary cycle 150-500'		1750-1400'	West : 1200-1700' East : 500-700'
Late Miocene				Lower Ranchi West : 1800'
Middle Miocene			Noamundi cycle 2200-1950'	East : 1400' Upper Ranchi
Early Miocene				West : 2600' East : 2000'
		Jabalpur cycle		
Oligocene	Mid-Tertiary cycle(?) 450-850'	1325-900'	Kiriburu cycle 3000-2750'	Netarhat West : 3200-3900' East : 2000'
Late Eocene				
Middle Eocene	Post-Trap cycle 2100' (?)		Primary cycle more than 3300'	
Early Eocene	Early Teritiary Cycle 1000-1150'			
Palaeocene	(Indian cycle : post - Trap)			
Upper Cretaceous	1200-1500' (pre-Deccam Trap)	Sub Trap cycle		
Upper Jurassic		Bhander cycle 2300-1950' (post-Gondwana)		

TABLE 2

PLANAR SURFACES IN SOUTH INDIA

Geological Time (1975)	Vaidyanadhan, (1967)	Babu (1975a)	Demongeot
Quater nary — Holocene, Pleistocene	Valley, Deltas, Continental deposits	Madurai - Palaghat landscape (150m) Nilgiri-Palani Cardamom and Shevaroy summits (1500-2700m)	Episodes recentes
	Erosion surface 1200' (belts within Eastern Ghats, Tamil Nadu)		
Pliocene	Erosion surface 2000' (parts of Karnataka Andhra Pradesh and Tamil Nadu)	Palani- Tiruppatur landscape (300m)	Cycle du Tamilnad
Miocene	Erosion surface 3000' (Mysore Plateau, Shevroy, Nallamalai, Khondalite	Manantody-Kolar landscape (600m)	Cycle de Mysore
Oligocene Eocene	Hills in Eastern Ghats of Andhra Pradesh)		
Cretaceous	Erosion surface 6000' (Biligirirangan, Bababudan, Agastya malai)	Mysore Plateau Kollimalai lanscape 900m)	Cycle de Moyennes Montagnes (Biligirirangan? Shevaroy)
Jurassic	Erosion surface 8000' (Nilgiri, Anaimalai)	Erosion surface 1640m ?)	Cycle des Nilgiri 2600m

NEOTECTONICS

It is well known that parts of Peninsular India have been uplifted, warped, faulted and modified later by different geomorphic agents to project the present landscape. Areas of cymatogenic upwarps were recognised, besides those where blockfaulting must have taken place. Thanks to the increasing use of aerial photographs in modern geomorphic studies, many lineaments have been recognised and attempts made to identify their significance. Credit should go to the Oil and Natural Gas Commission for laying great emphasis to studies of this kind. The interest of geomorphologist in these studies is to a large extent due to the fact that through such studies he will be able to arrive at the evolution of the landforms through time and also can explain their regional disposition and dispersion. Such studies also give clue to areas of tectonic activity within and at the periphery of the shield areas.

Eremenko (1968) recognised a number of lineaments across South India, some of which have been confirmed by later geophysical surveys. Sastry and Raiverman (1968) made use of the study of fracture pattern in South India passing uninterrupted from the shield to the Cauvery Basin. Raiverman(1969) made use of the study of fracture patterns in South India and arrived at the correct inference that the Cauvery river had abstracted the headwaters of a different river over the Mysore Plateau by the headward erosion of its tributaries along a fracture lane. Biswas *et. al.* (1971) gave a brief account of the Quaternary deposits of India along with a correlation table of the deposits in Extra Peninsula, Peninsula and Indo-Gangetic aluvium drawing particular attention to the Quaternary tectonic movements. A majority of those reported are from the northeastern part of Peninsular India and a few from the northern part of the west coast.

Biswas (1974, 190) has shown that the erosion surfaces recognised in Kutch have been faulted and is placed since post-Palaeocene by 300 feet, since post- Miocene by 250 feet and since post-Pliocene by 100 feet. Differential uplift of highlands took place against the background of regional uplift (cymatogenic) due to which the negative areas, the Rann and Banni plains, are also elevated. Shrivastava (1968) reports that the Saurashtra Peninsula was at a lower elevation during the Pleistocene due to down-faulting enabling deposition around it and subsequently was lifted up. The coastal-deposits have been very little affected by folding or tilting movements, but have been uplifted considerably in the post-Pleistocene times. Further, the pre-existing basement faults are supposed to have been reactivated during the Quaternary period. Babu and Rao (1974) studied the geomorphology of Jaisalmer desert in West Rajasthan and have reported morphostructures which are gentle domal features indicating slow uplift. Fracture zones have also been recognised.

Choubey (1969), following the pre-Deccan Trap surfaces south of Narmada River, noted them to be arched up. The lower surface rises towards Pachmari and the higher surface is dissected and their remnants are the isolated peaks of Dhupgarh (4429'), Mahadeo (4359'), and Chauragarh (4303'). The uparching of Satpura region is post-Deccan Trap in age. Dixey (1970) in his excellent account of the geomorphology of Madhya Pradesh traces each surface across faults and shows how one can arrive at the relative ages of landforms in a tectonically disturbed region. Vaidyanadhan *et al* (1971) by a study of lineaments from aerial photographs showed the recognised lineaments in the Keonjhar region in Orissa and the major physical features.

Das and Rao (1973) opined that the fracture system in the Koyna region of Maharashtra is due to initial monoclinal flexture in the Traps in response to reactivation of the basement fractures during a post-Deccan Trap period. There is no alluvium in the valleys of the streams even after reaching maturity perhaps indicating frequent changes in the gradient due to reactivation. Hot springs are also reported.

As for tectonics in South India, it has been well known that parts of the shield have been faulted, but no systematic documentation was attempted. Geologists used to recognise usually the major and minor faults in the Precambrian rocks and map them very carefully, but were rarely interested to find out their effects on the recent landforms. A careful examination of topographic maps alone of Precambrian terrains will indicate the faulted areas, without recourse to any geological mapping, in some areas (Vaidyanadhan 1968,10,14,16). This is due to landforms responding to faulted structures and possibly there has been reactivation.

Disposition of charnockite massifs at high altitudes led Radhakrishna (1968) to suggest faulting and uplift of those masses and the Palaghat Gap was thought to owe its origin perhaps to a major faulting along its southern edge (Ahmad 1966). Study of Landsat imageries lends support to the above view. Demongeot (1975, 254) has given an example of tilting and faulting of mountains (Shevaroy, 1650m). Another interesting observation of his is that the Mysore surface is subsequent to faulting and uplift of the older surface at Gopalaswamy Beta, south of Mysore. Similarly Srinivasan (1974), based on photogeological study of Attur Block in Tamil Nadu, east of Salem, infers that some of the geomorphic features owe their origin to movement along the faults, both lateral as waell as vertical. If one were to map all the recognised surfaces in Karnataka and Tamil Nadu, below the 13^{0}N latitude, it would be noticed that these would tend to separate out approximately along the latitude of the Palaghat Gap, widening towards east. This may partly be due to the fact that this is a region of tectonic disturbance and partly due to the fact the river Cauvery or an older one and its tributaries have dissected this region since Cretaceous times, supplying sediments to the Cauvery Basin. An inference arrived at from the study of the evolution of the Cauvery river in South India (Vaidyanadhan 1971) based on the study of lineaments, faults and ancient channels and valleys, is that the main trend of the major rivers in South India is between NW-SE and WNW-ESE over the Deccan Traps (if they flow over the Deccan Traps) like the Godavari, Krishna, Bhima *etc*, and also over the coastal alluvium, like the Godavari, Krishna, Palar, Ponnaiyar, Vaigai *etc*. The same rivers or others over the Precambrian rocks flow southwest to northeast (Tungabhadra, Krishna, tributaries of Penner, Palar, Ponnaiyar, Lower, Cauvery, Vaigai *etc*.). It is suggested that the earliest rivers in South India, during the pre-Deccan Trap period, flowed mainly due notheast and perhaps a few flowed east. Major rivers and parts of rivers flowing in other directions may be younger in age.

That the Nilgiri mountains, at least in some sections are bound by faults is well known. Four directions of lineaments have been recognised here (Vaidyanadhan *et al* 1971) and the adjoining low level pleateaus. These do indicate conspicuous changes in landforms along certain sections. Faults and fractures have been traced in the lowlands of the west coast and K. Sriram *(personal communication* 1976) is of the view that the occurrence of fracture traces in Goa in the laterite cappings also along the same directions as those in the basement rocks is indicative of reactivation. Recent studies (Nair 1976) of the northeastern part of Andhra Pradesh leads to the recognition besides lineaments, of certain geomorphic features like ancient beach ridges, abandoned river courses, retreating scarps *etc.* It is true that conventional aerial photographs on scales varying from 1: 25,000 to 1 : 60,000 usually available in our country may show the same features much better. But then the imageries definitely give a regional picture and are perhaps better guides to understanding of the regional tectonics.

LATERITE

Since laterite is usually associated with planar surfaces and occurs within Tertiary and Quaternary sediments, these are of great interest to geomorphologists.

Noel G de Souza (1968) describing the features of laterites of Goa in the west coast of India, disputes the general contention that the high level laterite is 'in situ' and the low level one, transported and dertital. He distinguishes between two levels of laterite on hill tops and the other on the adjoining lowlands and according to him peneplanation did not precede lateritisation in Goa. Laterites formed over isolated ridges. Choubey (1969) reports two distinct ages of lateritisation, the first one over the Deccan Traps and the next one, subsequent to the stripping away of the lateritized Traps and erosion, occur at a lower level usually over the Vindhyas, in Central India. Brunner (1969) based on the study of the physical and chemical properties of the materials in the lateritic profiles in the eastern part of Mysore Plateau, distinguishes three types of weathered products, namely (i) kaolin layer, (ii) an older regolith, rich in the caly and (iii) an younger regolith, poor in clay. The former two are supposed to have been formed during Miocene and Pliocene and the third, during Pliocene/Pleistocene. Detailed description of physical, chemical and textural properties of laterites of Birbhum district of West Bengal is to be found in the paper by Karunakaran and Roy (1970). The earliest laterite here occurs at the base of Miocene and some lateritic horizons occur even within the Miocene sequence. The upper age limit is not clearly known. Tripathi (1970) in his detailed study of the Quaternary deposits of the the Indian subcontinent perhaps errs (p. 167) in continuing with the old idea that the low level laterite is of secondary origin, being derived from the high level laterite through weathering and disintegration.

The low level laterites of Narmada alluvium have yielded a Middle Pleistocene fauna. Choudhury (1973) broadly described the processes of formation of laterites and lateritic soils of parts of West Bengal. Niyogi and Mallick (1973) have made a very detailed study of the laterites of West Bengal. The authors here have presented unequivocal evidences to show that the lateritisation is in the alluvial sequences, that the soil profile is due to pedogenic processes and not due to sedimentational processes and that the low level laterites are not transported. Brunner (1968a, 7) is also of the view that lateritisation can form on the alluvial soils. It is to the credit of Niyogi and Mallick that they could, by detailed mapping and study of laterites and associated sediments establish a stratigraphic record ranging in age from the Middle Pliocene to Holocene in part of West Bengal.

Age of laterites, particularly where they occur over unfossiliferous formations (more often the Precambrains in South India), is very difficult to arrive at. Usually search is made in the adjoining areas to see if similar laterites (not that one can always distinguish between *'in situ'* laterites) occur over datable sediments, and that latieritic layer is projectd over the problem area. One such exercise has led Ramam (1976) to arrive at a Mio-Pliocene age for the Eastern Ghat mountains of Visakhapatnam district of Andhra Pradesh. It may be of interest to note that it is these laterites, extending over tens of kms (detached though) due NE-SW and covering over the Eastern Ghats in parts of Andhra Pradesh and Orissa, that have now been proved to be one of the richest deposits of bauxites in the country. Subramanian and Murthy (1976) have made a detailed study of the bauxite and hematite cappings in the Nilgiris and confirm (*op cit* p. 456) the earlier views of Pardhasaradhi and Vaidyanadhan (1974) regarding the relationship between the planar surfaces and residual deposits in this area. The general observation has been that dominance of diaspore is in Palaeozoic bauxites, boehmite and gibbsite in Mesozoic and Tertiary age to the laterites in the Nilgiris. Nair (1976) observes that the laterite capping the Neogene exposures along the periphery of Kerala Basin appears to plunge towards the deeper part where the laterite and ferruginous sandy clay, probably equivalent to it, are buried under late Neogene or Quaternary black sands. Thus the age of the laterite is arrived at here.

Based on the study of the different types of laterites, their lie and associations, Ramana Rao and Vaidyanadhan 1974b) discuss the significance of these in the recognition of planar surfaces and thus possibly residual deposits. One of the findings is that whereas study of distribution of *'in situ'* laterite helps in the confirmation of the existence of planar surfaces in a particular region, the laterite occurrences by themselves need not automatically be taken as indicative of the existence of planar surfaces.

One of the very few attempts made to study, in depth, the process of weathering of Deccan Trap basalts and arrive at resulting changes from a geochemical point of view is that by Bhattacharya and Sinha (1976). They show that the weathering is dominantly controlled by topography, rainfal and groundwater activity resulting in leaching, oxidation, mineral synthesis and polymerisation. Such studies should help to know better the role of different agencies in the formation of laterites.

COASTAL GEOMORPHOLOGY

Peninsular India has a long coast extending from Kutch in the west to West Bengal in the east, through a distance of about 5,700kms. Numerous articles have been written over the changes in the coastline through the geological past (some speculative), and through the historical past (based on comparatively dependable records). Earlier attempts were confined largely to the study of the different features recognisable in certain sections of our coasts and to pigeon-holing them into those due to transgression or regression. Now detailed mapping with the help of aerial photographs is done and where necessary, field work is also done, besides samples of sediments being collected and analysed. Our knowledge is still confined to a few sections of the west and east coasts.

Shrivastava (1971) considers the Ranns of Kutch to have been recently uplifted as they now have in the upper portions more of fluviatile sediments, but beneath them there are marine deposits. Presently the sediments are formed in tidal channels, lagoon, playa lake and a mudflats within the area of intertidal influence and also in supratidal environment. Gupta (1975) is of the opinion that at least since about 2,000 years ago the Little Rann of Kutch was about 4 metres deep and remained inundated throughout the year. Three time- contemporaneous layers have been identified in the Holocene sediments here. The bottom most is sandy, the next is silty-clay and this is again followed by another silty clay. The lower two gave ages of 9,000-4,200, and 42,00-1,500 years. Glennie and Evans (1976) considered that possibly the Ranns occupy sites of tectonic depressions having been once shallow marine gulfs following the post-glacial rise in sea level. According to them changes in the courses of the Himalayan rivers and diversion away towards the Indus led to the accumulation of sand here.

Saurashtra coast offers the geomorphologists a variety of features both of erosional and depositional origin. Further, the area is known to have been subjcctd to tectonic disturbances, though views may differ as to its nature and magnitude. According to Shrivastava (1968) the position of Miliolitic limestones, the coral reefs and oyster beds, besides ancient beaches, indicate upliftment in Sub-Recent times continuing till historic times. Dixit (1969-71) considers Kharlands, the extensive flat lands close to the coast at two elevations, one around 8-10 metres and the other just above high tide levels, as plains of marine abrasion, mainly due to sea level changes and not related to any tectonic disturbance. To Gupta (1972), the features listed by Shrivastava (1968) indicate a higher level of the sea during the late Quaternary and uplifts of the coast. Radiocarbon dating of samples in these sediments give an age of 5,000-35,000 years B.P. Mathur and Verma (1976) also seem to be of view that the southeastern part of Saurashtra coast has been a stable mass during Quaternary and there have been repeated transgressions. Different periods of dune building is attributed to drier climatic conditions. Bedi and Ramana Rao (*unpub. dissertation*, Ind. Photoint. Inst. 1976), based on a study of aerial photographs followed by field check of the southeastern part of Saurashtra coast from south of Bhavnagar to west of Jafarabad, extending ENE-WSW, with a width ranging from 12 to 19 kms. infer that the coast can be classified as 'Terrigenous coast of primary Morphologic Equilibrium' under which an emergent shoreline of coastal plain type has developed.

According to Arunachalam (1968) the shoreline around Salsette island near Bombay bears evidence of submergence on the eastern side and emergence on the western side. Dixit (1976) has made a study of geomorphic features from Bombay to Goa. On the whole this part of the west coast has undergone submergence and later slight emergence, with a regression of coastline in parts. Evidences in the nature of drowned valleys, rocky islands *etc.* are cited. These features have been reported from this coast by Vaidyanadhan by airphoto study (1973, 4, Figs. 12 and 13) and besides ancient beach ridges further inland are also reported.

Descriptions of features in Kerala coast are many. A series of bays are supposed to have existed which have been subsequently filled, rather quickly. Overlying this are the recent sand deposits. The network of lagoons and estuaries and the rich heavy mineral sands on the barrier beaches today appear to have been built up on these earlier mudflats. The emergence of these, on the basis of radiocarbon dating of molluscan shells, is supposed to be around 4,460 years B. C. corresponding to the Neolithic age (Prabhakar Rao 1968). On the whole no evidence of a western margin fault close to the shore has been found. Besides the present shelf edge, another shelf nearer to the shore has been identified, meaning that the shelf has been prograding outward (Cloos *et al* 1974).

Coming to the east coast, Niyogi (1970a) studied the morphology and evolution of the Balasore shoreline, Orissa, using aerial photographs, followed by field mapping and study of sediments. He recognised a series of ancient beach ridges and off-shore bars. Five previous positions of shorelines were interpreted. The fluviatile landforms close to the shore, in the Subarnarekha and Burhabalang rivers were also mapped. The features, according to Niyogi, reflect a low sea level, about perhaps 25,000 years ago, followed by a rise around 3,000 years B. P. The features also indicate differential uplifts as there are differences in the nature and amount of material deposited in the area immediately below the shoreline. Recent studies of logs of over 1,000 boreholes put near Visakhapatnam, and the geomorphology of the coast has enabled Pridhvi Raju and Vaidyanadhan (1977) to infer that during the commencement of Holocene, the sea level might have been about 7 metres higher than the present sea level before it went down to as low as 25 metres and rising again to the present level. Further south, between Visakhapatnam and Pudimadaka in Andhra Pradesh, Bhaskara Rao and Vaidyanadhan (1975) report, based on airphoto study, that the survival and orientation of landforms along the coast here, like wavecut platforms, wavecut benches, sea cliffs *etc.* are due to lithology, attitude of strata faults and fracture lines. Ancient dune ridges have also been recognised near Pudimadaka. A detailed analysis of the morphology and evolution of the rocky coast in the Visakhapatnam district in Andhra Pradesh indicates that the very projection of a series of hills into the coast in this part of Eastern India is due to a few NNW-SSE to N-S and ENE-WSW faults (recognisable in Landsat imageries) intersecting each other and extending over tens of kms (Bhaskara Rao and Vaidyanadhan, 1977). Indications of translational movements are also pre-present particularly in the ENE-WSW direction.

Babu (1973) inferred two former shorelines of maximum transgression (strand lines) in the Krishna and Godavari deltas, besides a few morphostructures, the latter being geomorphic expressions of buried landscape. A detailed study of Krishna Delta later by Babu (1977c) led him to recognise four groups of ancient beach ridges. Sambasiva Rao and Vaidyanadhan (1977) have made a study of the landforms of the lower parts of the Godavari Delta and the coastal tract and identified series of groups of ancient beach ridges, each group discordantly disposed to earlier one, but subparallel to the present coast. They have recognised three former shorelines in the Godavari Delta also. Changes in the configuration of barriers and spits, particularly within such a short span of about 40 years from the date of the Survey of India toposheets of early 1930s and the aerial photographs of 1970s have been recognised and mapped in detail on the ground. A few lineaments and fractures are traced and it is suspected that slight movements may be taking place along some of the latter. This, however, is yet to be proved. Meijerink (1971) reports a late Pleistocene to early Holocene transgression at the coast adjoining Cauvery Delta. Three marine stages recognised in the cuspate foreland have been correlated with fluviomarine stages in the northern part of the delta.

The first book on Coastal Geomorphology of India was published by Enayat Ahmad (1973). As the author himself puts it, he has tried to gather the data from large scale maps and analysed the same in the light of the latest methodological literature pertaining to the field of geomorphology. This is a welcome addition to our limited literature in this field.

FLUVIAL GEOMORPHOLOGY

Studies in this field can be divided into (i) those pertaining to any river or stream and (ii) drainage basin analyses. The first involves mapping, sampling of sediments, model studies *etc.*, and the second, mathematical analysis of data collected from the topographic sheets and ocasionally from the field or airphotos.

Niyogi (1968) traced seven systems of step-like terraces along the steeply entrenched river Subarnarekha in Bihar and Orissa, on the basis of airphoto study and fluvial characteristics. The altitudes of terraces are successively 75, 60, 50, 34, 20, 12 and 9 metres, the last being the meander belt. Basu and Kar (1968) studied the formation of alluvial terraces in the lower reaches of the Bhagirathi valley, about 80 kms north of Calcutta. The Katwa bend has migrated west in 5 stages. Maps of 1765, 1840, 1916 and 1957 have been studied and the possible future course predicted. Sharma and Sharma's paper (1973) contains a very good airphoto-interpreted geomorphic map of the flood plain of Betwa and Ken rivers in Uttar Pradesh and contains units like flood plain, point bar deposits, natural levee, channel fill, flood plain scroll, meander cut-offs, badlands *etc.* Extent and location of terraces on the upper and lower sections of the Subarnarekha river flowing through Biihar, Orissa and Bengal (Mukhopadhyay 1973) indicate that Tertiary warping has had its effects on these, more on the upper than on the lower reaches.

Laboratory studies indicate that the modern flood or overbank deposits (Muzumdar *et al* 1970) are very much similar to the Older alluvial deposits of the Godavari in their particle size composition and also in their mineral contents, and the latter must have been deposited during monsoon floods of late Pleistocene. The basal portion of such a deposit near Mula dam gave an age (C^{14} date) about 30,000 B.P. and the upper portion at Panthan about 18,000 B. P. Quite a few changes have taken place in the fluvial regime of the the river Krishna and its tributaries in Western Maharashtra (Gupte and Rajaguru 1971). Terraces, meandering courses separated by straight courses, misfits and falls in this region substantiate their inference of possible tectonism and climatic changes in the Quaternary. Babu (1975c) based on airphoto study and subsurface data, delineated the different features in the Krishna Delta like levees, abandoned channels, swamps *etc.*, and traced the migration of the distributaries and the delta lobes.

Meijerink (1971) mapped on air photos, fluvial and coastal features in the Cauvery delta and field check was carried out to confirm the interpreted units. An attempt has been made to arrive at a possible sequence in the lateral shift of the main river system in Holocene. Perhaps Meijerink's paper is the first attempt in this part of the country to tentatively date some of the stages in the fluvial deposits with those dated elsewhere, using different eustatic curves.

Passing on to quantitative studies, there have been many, but most of them were only exercises with limited or negligible geomorphic inferences flowing from them. Singh and Kumar (1969) worked out the geomorphological evolution of the stream order of the Topa and Silpi basins in Ranchi. Subramanyan (1974) presents details of quantitatie analyses of the Karawan and Bankri drainage basins around Sagar, Madhya Pradesh. These data satisfy some of the laws of morphometry by Horton, Strahler, Chorely and Schumm. It is concluded on the basis of these data that the erosional development of the area by the streams has progressed well beyond maturity and that lithology has had a influence in the drainage development. Sen (1971) studied the relation between longitudinal profiles of streams and the rock types over which they flow over around Jawai Bandh, southwestern Rajasthan. The slope of the gully ranges from 2^0 over unconsolidated coarse rubble to 20^0 on massive granite. Padmaja (1975) studied quantitative drainage characteristics in the Dhund Basin, Rajasthan, and Kumar and Pandey (1975) the morphometric properties of a polygenetic drainage basin. A new quasi-statistical method involving the use of Smithson's length-breadth diagrams (adapted by direct analog) has been proposed by Verma and Tandon (1971) for a quick and effective means of appraisal of stream tortuosities. Tortuosity seems to increase with higher order and maturity. Streams of higher order also show a tendency towards a wider scatter. Chakraborty and Ghosh (1974), on the basis of a supposition that there shall be as many profiles of equilibrium as the number of times the cycle of erosion of a given basin has been interrupted, collected data from toposheets of the river Kangsabati flowing from Meghalaya to Bangla Desh, and identified 4 periods of disturbance. Of course, the authors do not justifiably recommend this for deciphering intricate subtelities of landscape evolution.

CLIMATIC GEOMORPHOLOGY

This is one of the branches of geomorphology that needs better attention, situated as we are in different climatic belts within the country with considerable local changes. It is also known that Peninsular India has migrated over thousands of kilometres from a possible polar regions (?) through the equatorial belt, to the present position. Different parts of it must have experienced climatic extremes during this migration and some of the landforms and deposits may contain evidences of this climatic history. During the period under review notable contributions have been from visiting European geographers on this aspect of geomorphology.

Besides analysing the stages of evolution of certain types of dunes in the

Thar desert in Rajasthan, Verstappen (1970) has shown that some of them have been formed under deteriorating climatic conditions and indicate a transition from a former humid climate. Shrivastava (1968) considered the formation of oolites, the precipitation of calcium carbonate presumably from the sea water saturated with it and the lateritisation of the Traps in Saurashtra, as indicating a warm tropical climate during Pleistocene and sub-Recent. Brunner (1968b), based on geomorphic and pedologic studies, found traces of Pleistocene climatic fluctuations in the profiles and sediments of the area southeast of Mysore Plateau in South India. Some of the indications referred to are : truncated lateritic profiles, older and younger lateritic covers, redeposited relicts of older regolith, wind facetted pebbles in stone layers, terrace deposits due to anthropogenous deforestation *etc.* Gellert (1970) considered the inselbergs in Tamil Nadu as those modified from an earlier semi-humid palaeoclimate to the present semi-arid climate. According to Demongeot (1977, 247) the rock penitents, near Karamadai, northeast of Coimbatore in Tamil Nadu, made up of schists prove that a wet tropical weathering preceeded the semiarid phase of the Tamil Nadu planations. Similarly the tors and the perched blocks in many parts of Tamil Nadu are considered inherited from weathering and separation into those forms from an earlier humid period.

Murthy (1975) on the basis of faunal evidence (vertebrates) from the Kurnool caves, suggests a thicker vegetation cover in the region around Kurnool in southwestern Andhra Pradesh, during late Pleistocene. Ramana Rao and Vaidyanadhan (1974c) tracing the evolution of features over the pediments in the Keonjhar region, Orissa, relate certain landforms, deposits and processes as due to successive changes in the climate from post-Pleistocene to the present day.

APPLIED GEOMORPHOLOGY

It needs hardly to be reiterated that any knowledge one can gain on the geomorphology of a region can be utilised for some purpose or other, if not immediately, at least in the near future. Herein a few examples of studies made in Peninsular India are given. Examples are numerous particularly in the field of groundwater and geotechnical investigations.

Ghose and Singh (1968) showed that the distribution of evaporites in the arid zone of Rajasthan is closely related to the distribution of dried up and disorganised drainage system. Accumulations of sodium salts (speciallly chloride) are invariablly found near the confluences, and oxide or iron and carbonate of calcium in the upper reaches of these drainage system. Gypsum occupies the intermediate position in the buried and disorganised channels. Keeping this relationship in view, it is possible to locate these evaporites with the help of aerial photographs which minimises the cost of prospecting for field survey. Vaidyanadhan (1970) has listed a few studies made by him with relevance to prospecting for diamonds in Anantapur district in Andhra Pradesh, excavation of canals on pediments in Bellary district of Karnataka, depth and nature of foundation of the damsite across rivers like Penner and Sileru in Andhra Pradesh and finds of boulders, cobbles and pebbles in ancient channels that can be used for construction and industry. Mallick and Bagchi (1975) traced the sand quarries to be in a belt of old channel of the river Damodar, within the Damodar Deltaic plain in West Bengal. Each quarry appears to be a fossil channel bar or part of it within the buried channel.

Niyogi's (1970b) study of geology and geomorphology of the Digha coast in West Bengal is a very good example of how one should go about in seeking solutions for problems of coastal erosion. The study is a multi-disciplinary one with geomorphic investigations, aided by airphoto study, playing a major role. Das (1975) shows how based on the study of the relative relief of the area, better methods may be adopted to reduce the devastating effects of the floods, the example taken being the Kaliaghai river basin in West Bengal.

Among the many groundwater investigations, Karanth (1969) establishes the aeolian nature of the Quaternary sediments in Alwar valley, Rajasthan, and this enables him to understand the groundwater potentialities of the region. In the intermontane valleys on the leeside of Rajasthan desert, the author suggests that the best site for groundwater is nearer the piedmonts and not in the central tracts. Adyalkar and Radhakrishna (1972) gave a detailed account of the karst features around Jagadalpur in Madhya Pradesh and have tried to relate some of them to potential areas of groundwater development. Nageswara Rao (1975) has delineated the different types of landforms, faults and fractures in the Poini basin in the Chittor district of Andhra Pradesh, and an interesting observation made by him is that the yields of wells located in the area along the right bank of Poini are considerably much higher than those of the wells along the left bank. Photostudy revealed that the intensity of fracturing is quite high on the right bank.

In the Keonjhar region Ramana Rao and Vaidyanadhan (1970, 179,Table 1) established a relation between a mojority of residual deposit occurrences of iron and manganese ores and the erosion surfaces. In other parts of the world such relations are known. Studies by Ramam (1976) of the bauxite deposits of Visakhapatnam district in Andhra Pradesh and those of Subramanian and Murthy (1976) of bauxite and hematite occurrences in the Nilgiris in South India lend support to the relation established.

Last but not the least is the use of geomorphological studies in the investigations of hydrocarbons. Varadarajan (*unpub. manuscript* 1973) on the basis of photostudy demarcated a number of anomalies in the Cauvery Delta besides recognising a few faults and joints. The anomalies are sudden juxtaposition of different units (flood plain, lacustrine surface, coastal belt), differences in sequences and tonal variations. Similarly Roy (1975) studied drainage anomalies in the Upper Assam valley and recognised the departures from normally expected patterns as due to growing domal structures. Some of the anomalies both in the Cauvery Delta and in the Assam valley have been confirmed as sub-surface 'highs' by seismic surveys. Babu and Rao (1974), based on photostudy, inferred morphostructures indicating domal features in the Kishangarh-Ghotaru region in Jaisalmer desert, Rajasthan. The features indicated by them around Bhuttowala, Ghotaru and Bhuner are, according to them, likely to disrupt the strata beneath and form traps for hydrocarbons.

Intergrated surveys of natural resources are being undertaken now and geomorphic studies form a basic part of any such study. Indian Photointerpretation Institute (1975) at Dehra Dhun has completed a survey of Karimnager district of Andhra Pradesh, with a good geomorphic study. This is to some extent similar to the Landsystem surveys undertaken by CSIRO in Australia.

CONCLUSION

Thus it is seen that in almost all the sub-fields of geomorphology, some work has been done during the last decade, though mostly on conventional lines. Mukherjee (1977) has made a forceful plea for the need for pursuing geomorphic studies on a regional scale. He has even suggested how geomorphologists can master enough data from the present topography to aid studies in global tectonics.

The author (Vaidyanadhan 1968) realising the need to bring home to students of earth sciences in our country Indian example of landforms published a booklet giving a few examples of practically every type of landform that can be recognised from topogaphic sheets, from different parts of India. Similarly to show that aerial photograph is a very important tool to be used in the understanding of the landforms, another study was undertaken to illustrate more examples (Vaidyanadhan 1973). Features which may be missed on conventional maps like faults, lineaments, ancient beach ridges, abandoned channels, river terraces, *etc.* have also been recognised and illustrated from different parts on India. Recently Tamasakar and Deshmukh (1974) have brought out a publication illustrating methods of geographical interpretation of a few Indian topographic maps. It is hoped that these publications and more yet to come in the field of geomorphology will enthuse the future student of earth sciences to take a deep interest in this fascinating field of science.

There has been a gradual change in the geomorphic methods of study and concepts (Christofoletti 1975). From mere description of landforms then mapping, the stage has come to study the processes, quantify them and attempts are now being made to explain the landscape as parts of some stochastic process, and their individual differences as related to variations of the energy intensity and to mass distribution. Geomorphologists working in different parts of Peninsular India attribute the causes of formation of any landforms to structure, lithology, climate or some other parameter. But it is now realised (Vaidyanadhan 1975) that no single agent is responsible, but a combination of many contribute to the evolution. It is only a question of the influence on one being more than the other. It is also very likely that future studies may throw up causative forces in the evolution of landforms, we are not yet aware of. The aim of the geomorphologist should be, besides making use of the data from various other discplines like biology, chemistry, geophysics, anthropology, *etc*., to always subscribe to careful field and laboratory work and conservative interpretation, since we are dealing with processes and environments, neither of which have been and are entirely the same through the geological time.

REFERENCES

Adyalker, P. G. and Radhakrishna, T. S. 1972 : Karstic phenomenon in the carbonate rocks of the Indravati Series of Madhya Pradesh with their groundwater possibilities. *Proc. Ind. Nat. Aca. Sci*., 38, Part A, 5 & 6, 133-141.

Ahmad, R. 1966 : Post-Gondwana faulting in Peninsular India and its bearing on the tectonics of the sub-continent. *Annals, Geology Dept., Aligarh Muslim Univ.* II. 40p.

Arunachalam, B. 1968 : Geomorphic features of the Salsette Island, India. *Abstr. 21st Inter. Geog. Congr. New* Delhi. 2-3.

Babu, P. V.L. P. 1973 : Morphostructures of the Krishna and Godavari Delta Region. *Photonirvachak* (Jour. Ind. Soc. Photo. Inter.) 1, 65-67.

1975a : A study of cyclic erosion surfaces and sedimentary unconformities in the Cauvery Basin, South India. *J Goel. Soc*, India., 16(3), 349-353..

1975b : Denudation chronology in and around Godavari Basin. *Sym. on Sediments, Sedimentation and Sedimentary environments, Univ. Delhi*, 351-355.

1975c : Morphological evolution of the Krishna Delta. *Photonirvachak* (Jour. Ind. Soc. Photo-Inter.) 3(1),21-27.

and Rao, P. V. 1974 : A note on the geomorphology of the Jaisalmer desert, West Rajasthan. *Ind. J. Earth Sci*., 1(2), 214-217.

Basu, S. R. and Kar, N. R. 1968 : Meandering, migration and alluvial terraces in Lower Bhagirathi valley, West Bengal. *Abstr. 21st Inter. Geog. Congr. New Delhi*, 4.

Bhaskara Rao, V. U. and Vaidyanadhan, R. 1975 : Photogeomorphic study of constal features between Visakhapatnam and Pudimadaka in Andhra Pradesh. *Photonirvachak* (Jour. Ind. Soc. Photo-Inter.), 3(1), 43-46.

1977 : Morphology and evolution of the rocky coast in Visakhapatnam district, Andhra Pradesh (*under publication*).

Bhattacharya, N and Sinha, A. K. 1976 : Weathering of Deccan Trap basalts. *Ind. J. Earth Sci.* 3 (2), 148-159.

Biswas, A. B. Dassarma, D. C. and Roy, A. K. 1971 : Quaternary deposits of India. *Rec. Geol. Surv. India*, 101(2), 209-228.

Brunner, H. 1968a : Geomorphologische Karte des Mysore plateaus (sudindien). Ein Beitrag zur Methodik der morphologischen Katrierung in den Tropen. Wiss, Veroff, 25, 1-17, Leipzig.

1968b: Pleistocene climatic fluctuations in Mysore Plateau. *Abstr. 21st Inter. Geogr. Congr. New Delhi*, 5-6.

1969 : Verwitterungstypen auf den Granitgneisen (Peninsular Gneiss) des ostilichen Mysore - Plateaus. (Sud-Indian), *Pet. Geogr. Mitt.* 113 (4), 241-248.

Chakraborty, S. C. and Ghosh, A. R. 1974 : Longitudinal profiles in fluvial processes: an analytical study. *Geogr. Rev. India*, 36(1), 31-37.

Chatterjee, S. P. 1968 : Progress of Geography in India (1964-1968). *Indian Science Congress Association, Calcutta*. 82p.

Choubey, V. D. 1969 : Study of the pre-Deccan Trap erosion surface in Central India. *Bull. Geol. Soc. India*, 6 (3), 79-82.

1971 : The geomorphology of the Vindhyan Plateau, Central India. *J. Geo. Soc. India*. 14 (4), 369-377.

Choudhury, R. N. 1973 : Laterites of West Bengal. *Geogr. Rev. Ind.* , 35(1), 61-73.

Christofoletti, A. 1975 : The explanatory finality of geomorphology. *Recent Researches in Geology, Hindusthan Pub. Corpn., Delhi*. 2, 1-5.

Cloos, H. Hari Narain, and Garde, S. C. 1974 : Continental margins of India. *In* geology of the Continental Margins, Eds. Burk, C. A. and Drake, C. L. *Springer-Verlag, Berlin*, 629-639.

Das, P. 1975 : An appraisal of flood incidence of the Kaliaghai River Basin. *Geogr. Rev. India*, 37(1), 73-78.

Das, S. R. and Ray, A. K. 1973 : Photogeological interpretation of structure and tectonics of the Koyna Region of West Coast, Maharashtra. *Rec. Geol. Surv. India*, 105(2), 83-94.

Demongeot, Jean. 1975 : Recherches geomorphologiques en Inde du Sub. *Z. Geomorph. N. F.* 19(3), 229-272.

Dixey, F. 1970 : The geomorphology of Madhya Pradesh. *Studies in Earth Sciences, West Commemoration Volume, Today & Tomorrow publ., New Delhi*, 195-224.

Dixit, K. R. 1969-1971: The morphological features of the Kharlands (saline wastes) of Gujarat. *Ind. J. Geogr.*, 4-6 (1), 21-28.

1976 : Geomorphic features of the West Coast. *Geogr. Rev. Ind.*,38(3), 260-279.

Enayat Ahmad 1973: Coastal Geomorphology of India. *Orient Longmans, Bombay*, 222p.

Eremenko, N. A. 1968: The tectonics and petroleum possibilities of the southern part of India. *Mem. Geol. Soc. India*, 2. 371-379.

Evans, P. 1970: The Tertiary Plateau of Bihar. *Studies in Earth Sciences, West Commemoration Volume, Today & Tomorrow Publ., New Delhi*, 195-224.

Gellert, J. F. 1970 : Beobachtungen und Bemerkungen zur Geomorphologic Sudindiens: *Geog. berichte*, 55(2), 118-132.

Ghose, B. and Singh S. 1968: Geomorphological control on the distribution of evaporites in the arid zone of Rajasthan India. *Abstr. 21st Inter. Geogr. Congress. New Delhi*, 419.

Glennie, K. W. and Evans, G. 1976: A reconnaissance of the Recent sediments of the Rann of Kutch, India. *Sedimentology*, 23(5), 625-647.

Gupta, S. K. 1972: Chronology of the raised beaches and inland coral reefs of the Saurashtra coast. *J Geol.* 80, 357-361.

1975: Silting of the Rann of Kutch during Holocene. *Ind. J. Earth Sci*, 2 (2), 163-175.

Gutpe, R. B. and Rajaguru, S. N. 1971: Late Pleistocene geomorphological history of rivers of Western Maharashtra. Internat. Symp. on Deccan Traps and other flood eruptions. Proc. Pt. 1. *Bull. Volcanology*, 35(3), 686-695.

Indian Photointerpretation Institute 1974: Integrated Resources Survey, A pilot project in part of Karimnagar district Andhra Pradesh.

Karanth, K. R. 1969: Geology and groundwater conditions in a part of Alwar valley, Rajasthan. *Rec. Geol. Surv. India.*, 98 (2), 147-167.

Karunakaran, C. and Roy, S. S. 1970: Laterites of Birbhum district, West Bengal, their origin and age. *Rec. Geol. Surv. India.*, 98(2), 16-40.

Kumar, A. and Pandey, R. N., 1975: Morphometric properties of the polygenetic drainage basin: A case study of Burha Nadi. *Geogr. Rev. India*, 37(4), 329-338.

Lakshmanan, S. 1970: Evolution of the Son drainage. *Proc. Ind. Nat. Sci. Acad.*, 38(1-2) part A, 21-31.

Mallick, S. and Bagchi, K. 1975: Morphology and genesis of buried sand of Magra-Pandua area, West Bengal, India. *Recent Researches in Geology, Hindusthan Pub. Corpn., Delhi.* 2, 212-223.

Mathur, U. and Verma, K. K. 1976: On the Quaternary geology of Saurashtra coast bordering Diu island. *Abstr. Colloquium on Palaeontological studies in Southern Region, Hyderabad* (June 1976), 46.

Meijerink, A. M. J. 1971: Reconnaissance survey of the Quaternary Geology of the Cauvery delta. *J. Geol. Soc. India*, 12(2), 113-124.

Mukherjee, S. K. 1977: Know thy Mother's face: A plea for geomorphological studies in India. *J.Geol. Soc. India*, 18(1), 1-6.

Mukhopadhyay, S. C. 1969: Some aspects of the geomorphology of parts of the Subarnarekha Basin around Mahali Marup, Bihar. *Geogr. Rev. India*, 31 (2), 33-40.

1973: River terraces of the Subarnarekha Basin. *Geogr. Rev. India*, 35 (2), 152-170.

Murthy, M. L. K. 1975: Late Pleistocene fauna of Kurnool caves, South India, *In Archaeological Studies, Ed.* Clason, A.T., North Holland Pub. Co. Amsterdam, Chapter 2.3.

Murthy, Y. G. K., Thampi, P. K. and Nair, N. M. 1976: Geology and geomorphology of Kerala. *Abstr. Seminar on Geology and Geomorphology of Kerala*, Geol. Surv. India, (Kerala Circle), Trivandrum, (1976), 6-9.

Nageswara Rao, P. 1975: Photogeological and geomorphological report on Poini Basin, Chittor district, Andhra Pradesh. *Photogeology Report No. 1, Central Ground Water Board, Southern Region, Hyderabad*, 6p.

Nair, K. M. 1976: Stratigraphic analysis of Kerala Basin. *Abstr. Seminar on Geology and Geomorphology of Kerala, Geol. Surv. India, (Kerala Circle), Trivandrum*, (1976), 9-10.

Niyogi, D. 1968: Morphology of the river Subarnarekha and its terraces. *Abstr. 21st Internat. Geog. Congr., New Delhi*, 13.

1970a: Morphology and evolution of the Balasore shoreline, Orissa, *Studies in Earth Sciences, West Commemoration volume, Today and Tomorrow Publishers, New Delhi*, 289-304.

1970b: Geological background of beach erosion at Digha, West Bengal. *Bull. Geol. Min. Met. Soc. India*, 43, 1-36.

and Mallick, S. 1973: Quaternary Laterite of West Bengal: its geomorphology, stratigraphy and genesis. *Q. J. Geol. Min. Met. Soc. India*, 45(4), 157-174.

Noel, G. de Souza 1968: Geomorphic features of the laterites of Goa. *Abstr. 21st Internat. Geog. Congr., New Delhi*, 21.

Padmaja, G. 1975: Some aspects of quantitative drainage characteristics of the Dhund Basin. *Geogr. Rev. India*, 37(2), 158-164.

Pardhasaradhi, Y. J. 1976: Etchplain development of Malabar Uplands-a suggestion. *J. Geol,. Soc. India*, 17(1), 73-78.

and Vaidyanadhan, R. 1974: Evolution of landforms over the Nilgiri, South India. *J. Geol. Soc. India.*, 15(2), 182-188.

Pathak, M. 1968: Geomorphology of Chhechhari Basin, India, *Abstr. 21st Internat. Geog. Congr., New Delhi*, 14.

Prabhakar Rao, G. 1968: Age of the Warkalli (Varakala) formations and the emergene of the present Kerala coast, India. *Bull. Nat. Inst. Sci.*, 38(1), 449-456.

Pridhviraju, K. N. and Vaidyanadhan, R. 1977: Geomorphology of Visakhapatnam, *Andhra Pradesh. Geol. Soc. India.*, 18 (in press).

Radhakrishna, B. P. 1952: The Mysore Plateau -its physiographic and structual evolution. *Bull. Mysore Geol, Assn.*, No.3.

1966: Geomorphological evolution of the Mysore plateau. *Symp. on Tectonics. Nat. inst. Sci. India. and Ind. Geophysical Union, Hyderabad*, 95-106.

1968: Geomorphological approach to the charnockite problem. *J. Geol,. Soc. India*, 9(1), 67-74.

Raiverman, V. 1969: River capture in the Cauvery. *Sci. & Culture*, 35(1), 29-31.

Ramam, P. K. 1976: The bauxite deposits of Anantagiri, Visakhapatnam district, Andhra Pradesh. *J. Geol. Soc. India*. 17(2), 236-244.

Ramana Rao, K. L. V. and Vaidyanadhan, R. 1970: Recognition, correlation and age of erosion surfaces in the iron Ore Range of Biharand Orissa. *J. Geol,. Soc. India*, 11, 178-181.

1974a: The planar surfaces of the Keonjhar region, Eastern India. *Ind. J. Earth Sci.* 1(1), 84-95.

1974b: Significance of laterites in the recognition of planar surfces in the Keonjhar region, Eastern India. *Geogr. Rev. India*, 36(2), 93-101.

1974c: Evolution of features over the Keonjhar pediplain, Orissa, Eastern India. *Golden Jubilee Volume. Q. J. Geol. Min. Met. Soc. India*. 347-354.

Rao, D. P. 1976: Neotectonic and lineament analysis of Landsat imagery for oil and mineral exploration -an example from Godavari delta-Eastern Ghats area. *Workshops on Coastal. Sedimentaries of India.-South of 18⁰ N Latitude. ONGC (Madras)*, 6p.

Roy, T. K. 1975 : Drainage analysis in the Upper Assam Valley. *Ind. J. Earth Sci*, 2, 39-50.

Sambasiva Rao, M. and Vaidyanadhan, R. 1977: Photogeomorphic study along the coastal tract of Godavari Delta. *Abstr. 64th Session Ind. Sci. Congr. (Geology & Geography Section)*, 23-24.

Sastry, V. V. anda Raiverman, V. 1968: On the basin study programme of the Cretaceous- Tertiary sediments of the Cauvery Basin.*Mem. Geol,. Soc. India*, 2, 143-152.

Sen, Deepawati 1971: Effect of rock type on stream slopes around Jawai Bandh, S. W. Rajasthan. *J. Geol. Soc. India*, 12(2), 189-191.

1972: Geomorphology of the Aravalli Range, Rajasthan, and a reinterpretation of residual erosion surface. *Geogr. Rev. India*, 34(3), 262-285.

Sharma, S. K. and Sharma, S. C. 1973: Hydrological evolution of morphological features of some parts of Hamirpur, Jhansi, Banda and Fatehpur districts, U.P. using photointerpretation technique. *Photonirvachak* (Jour. Ind. Soc. Photo. Inter.), 1(1&2), 43-48.

Shrivastava, P. K. 1968: A note on the Quaternary geology of Saurashtra Peninsula. *Q. J. Geol. Min. Met. Soc. India*, 12(4), 392-395.

Singh, R. P. 1969: Geomorphological evolution of Chotanagpur highlands-India. *Nat. Geog. Soc. India., Research Pub.* No.5.

and Kumar, A. 1969: Geomorphological evolution of stream order of the Topa and Silpi basins in Ranchi. *Nat. Geog. Soc. India*, 15(1) 38-44.

Srinivasan, V. 1974: Geological structures in Attur Valley, Tamil Nadu, based on photointerpretation *J. Geol. Soc. India*, 15(1) 89-93.

Subramanian, K. S. and Murthy, M. V. N. 1976: Bauxite and hematite capping in the Nilgiris, Tamil Nadu Study from geomorphic angle. *J. Geol. Soc. India*, 17(3), 353-358.

Subramanyan, V. 1974: A quantitative analysis of two drainage basins around Sagar, Madhya Pradesh. *Proc. Ind. Nat Sci Acad*, 40(A-1), 76-99.

1975: Evolution of landscape around Sagar, Madhya Pradesh. *Ibid*. 41(A.1), 448-469

Tamaskar, B. G. and Deshmukh V. M. 1974: *Geographical interpretation of Indian toporgraphical maps*, Orient Longman, Bombay, 72p.

Tirugnanasambandam, S. 1976: Geomorphological studies of Bharathapuzha river basin around Ottaapalem and Kuttipuram areas, Kerala. *Abstr. Seminar on Geology and Geomorphology of Kerala. Geol. Surv. India*. (Kerala Circle), Trivandrum, 1976, 28-29.

Tripathi, C. 1970: Quaternary deposits of the Indian subcontinent. *Rec. Geol. Surv. India*, 98(2), 144-174.

Vaidyanadhan, R. 1967: An outline of the geomorphic history of India, south of 180 N Latitude. *Proc. Seminar on Geomorphological Studies in India* 1965, Centre for Adv. Study Univ. Saugar, Sagar, 121-130.

1968: Index to a set of sixty topographic maps illustrating specified phsiographic features from India. *Council Scient. Indust. Res.*, New Delhi 22p.

1970: Studies in Applied Geomorphology. Studies in Earth Sciences, West Commemoration Volume 1. *Today & Tomorrow Publ., New Delhi*, 318-321.

1971: Evolution of the drainage of Cauvery in South India. *J. Geol. Soc. India.* 12(1), 14-23.

1973: Index to a set of seventy aerial stereopairs illustrating physographic features from India. *University Grants Commission, New Delhi,* 39p

1975: Evolution of landforms over Precambrian terrains in India. *Studies in Precambrians, Dept. of Geology, Bangalore* Univ. 283-291.

Vaidyanadhan, R., Ramana Rao, K. L. V. and Pardhasardhi, Y. J. 1971: Lineaments and their importance in landform studies. *J. Geo,. Soc. India,* 12(3), 299-302.

Verma, V. K. and Tandon, S. K. 1971: Stream Tortuosity : a new method of visual estimation. *Z. Geomorph. N. F. Suppl. Bd.* 12, 165-171.

Verstappen, H. Th. 1970: Aeolean geomorphology in the Thar desert and palaeoclimates. *Zitschr. F. Geom. Suppl. Ed.* 10, 104-120.

West, W. D. 1962: The line of Narmada-Son Valley. *Curr. Sci.* 31, 143-144.

SOME SELECTED REFERENCES

Bloom, Arthur L (1978) Geomorphology, Prentice. Hall, Inc., New Jersey, p 510

Butzer, Karl W (1976) Geomorphology from the Earth, Harper and Row Publishers, New York, p. 463

Dayal, P. (1995) A Text Book of Geomorphology, 2nd edition, Shukla Book Depot, Patna, p. 693

Easterbrook, Don, J. (1969) Principles of Geomorphology, McGraw Hill Book Company, New York, p. 462

Ghose, Bimal, Singh, Surendra and Kar, Amal (1977) Desertification around the Thar- A Geomorphological Interpretation, Annals of Arid Zone, Vol. 16, No. 3, pp. 290-301

Gorshkov, G. and Yakushova A (1967) Physical Geology, Mir Publisher, Moscow, p. 596

Rogers, John J.W. and Adams, John A.S. (1996) Fundamentals of Geology, Harper and Row, New York, p. 424

Sharma, V.K. (1986) Geomorphology-Earth Surface Processes and Forms, Tata McGraw Hill Publishing Co. Ltd., New Delhi, p. 244

Thornbury, William D. (1985) Principles of Geomorphology, Wiley Easten India, New Dellhi,. p. 594

Valdiya, K.S. (1987) Environmental Geology : Indian Context, Tata McGraw Hill Publishing Co. Ltd. New Delhi, p. 583

Miller, William J. (1949) An Introduction to Physical Geology. Affiliated East-West Press Pvt. Ltd., New Delhi, p. 482

Verma, V.K. (1973) Map Parameter for Correlation of Marine Terraces in California, Geol Soc. Am. Bull, Vol. 84, pp. 2727-2742

Verma, V.K. (1992) High Altitude Geo-Environment and Resources: Problems and Prognoses, in Dynamics of Mountain Geosystems (ed. RB Singh). pp. 337-352

INDEX